Computing and Simulation for Engineers

Mathematical Engineering, Manufacturing, and Management Sciences

Series Editor:

Mangey Ram

Professor, Assistant Dean (International Affairs), Department of Mathematics, Graphic Era University, Dehradun, India

The aim of this new book series is to publish the research studies and articles that bring up the latest development and research applied to mathematics and its applications in the manufacturing and management sciences areas. Mathematical tool and techniques are the strength of engineering sciences. They form the common foundation of all novel disciplines as engineering evolves and develops. The series will include a comprehensive range of applied mathematics and its application in engineering areas such as optimization techniques, mathematical modelling and simulation, stochastic processes and systems engineering, safety-critical system performance, system safety, system security, high assurance software architecture and design, mathematical modelling in environmental safety sciences, finite element methods, differential equations, reliability engineering, etc.

Applied Mathematical Modeling and Analysis in Renewable Energy
Edited by Manoj Sahni and Ritu Sahni

Swarm Intelligence: Foundation, Principles, and Engineering Applications
Abhishek Sharma, Abhinav Sharma, Jitendra Kumar Pandey, and Mangey Ram

Advances in Sustainable Machining and Manufacturing Processes
Edited by Kishor Kumar Gajrani, Arbind Prasad and Ashwani Kumar

Advanced Materials for Biomechanical Applications
Edited by Ashwani Kumar, Mangey Ram, and Yogesh Kumar Singla

Biodegradable Composites for Packaging Applications
Edited by Arbind Prasad, Ashwini Kumar and Kishor Kumar Gajrani

Computing and Stimulation for Engineers
Edited by Ziya Uddin, Mukesh Kumar Awasthi, Rishi Asthana, and Mangey Ram

For more information about this series, please visit: https://www.routledge.com/Mathematical-Engineering-Manufacturing-and-Management-Sciences/book-series/CRCMEMMS

Computing and Simulation for Engineers

Edited by
Ziya Uddin, Mukesh Kumar Awasthi,
Rishi Asthana, and Mangey Ram

CRC Press
Taylor & Francis Group
Boca Raton London New York

CRC Press is an imprint of the
Taylor & Francis Group, an informa business

First edition published 2022
by CRC Press
6000 Broken Sound Parkway NW, Suite 300, Boca Raton, FL 33487-2742

and by CRC Press
4 Park Square, Milton Park, Abingdon, Oxon, OX14 4RN

CRC Press is an imprint of Taylor & Francis Group, LLC

ISBN: 9781032119427 (hbk)
ISBN: 9781032119434 (pbk)
ISBN: 9781003222255 (ebk)

DOI: 10.1201/9781003222255

Typeset in Times
by codeMantra

Contents

Preface

The mathematical modeling of problems that exist in day-to-day life is a niche area for many engineers and scientists around the globe. Various engineering problems can be modeled mathematically, which can be solved further to understand the engineering phenomena in depth before creating the prototypes. Even to understand biological phenomena or the spread of any kind of pandemic like COVID-19, the mathematical models and advanced computational techniques are very useful. In a broader sense, the bio-mathematical engineering can be used to model various diseases and the impact of involved parameters.

With the advancement in computing technologies, the mathematical modelling and simulations of practical problems have received much attention in recent years. Because of high tech industrial processes, most of the engineering systems today have ever increasing levels of complexity, and due to the complexity of the systems, direct experiments for a new process are very costly in terms of time and money. Therefore, the engineering system and the impact of different complex parameters need to be well understood and analysed before performing the experiments. Considering all these aspects, the mathematical models of the engineering problems are very much helpful.

This work is the culmination of ideas presented by researchers working in the field of Applied Mathematics, computing and simulations. The chapters presented in the book give comprehensive insight about various aspects of mathematical modeling for the issues emerging in different parts of engineering design and processes, including mechanical engineering, computer science engineering, electrical and electronics engineering, civil engineering, and so on. Specific attention is given to numerical treatment of image and signal processing, fluid flows in various geometries, biomechanics, and biological modeling. The book additionally covers the mathematical portrayal of solar cells, analytical and numerical treatment of problems in fracture mechanics, and the stability of electric devices.

The book is intended to provide valuable research reference for academic scholars, graduate students, and applied mathematicians who have strong interest towards the computation and simulation of different applied engineering problems. This book will be a relevant support not only in academic areas, but will also be a tool to the engineering professionals who deal with various aspects of design and innovations in various industries, including nano medicine delivery, cooling of industrial machinery and electronic components, design of solar cells, data security, methods and analysis of pandemic data for policy making, etc. The chapters included here demonstrate the use of computing and simulation techniques to solve various real world problems for the betterment of our society and shaping the future of the mankind.

We wish the readers of a successful study of the material presented, leading to new inspiration, a deepening understanding of the described concepts, and also fruitful applications to the contemporary challenges of science and engineering.

Dr. Ziya Uddin
Dr. Mukesh Kumar Awasthi
Dr. Rishi Asthana
Dr. Mangey Ram

Editors

Dr. Ziya Uddin, Associate Prof. (Mathematics) is Head of Department of Applied Sciences in the School of Engineering & Technology at BML Munjal University, Gurugram, India. He completed his doctorate with a major in Mathematics and minor in Computer Science from G. B. Pant University of Agriculture & Technology, Pantnagar, India, in 2009. He has qualified in various national level competitive exams in the area of Mathematical Sciences. He has been a recipient of a research fellowship under the Council of Scientific and Industrial Research, India during his Ph.D. After his Ph.D., he received the postdoctoral fellowship from the French government and finished his postdoctoral work in 2012 at Université Polytechnique des Hauts-de-France. His research areas include Computational Fluid Dynamics, Soft Computing, Applied Mathematical Modelling, etc. He loves to teach applied mathematics courses to undergraduate, postgraduate, and Ph.D. students. He has served in several academic and administrative positions during his academic career. Dr. Ziya has published and presented his research work in many peer-reviewed journals and conferences of repute. He has also been a member of technical committees of various conferences, chaired many sessions, and delivered various lectures as a guest speaker. Dr. Ziya is also a lifetime member of various scientific and engineering societies. He has also authored a book on Computational Fluid Dynamics. On the personal front, he hails from Uttarakhand and loves traveling and photography.

Dr. Mukesh Kumar Awasthi has done his Ph.D. on the topic "Viscous Correction for the Potential Flow Analysis of Capillary and Kelvin-Helmholtz Instability". He has acquired excellent knowledge in the mathematical modeling of flow problems and he is able to solve these problems analytically as well as numerically. He has a good grasp on subjects such as viscous potential flow, electro-hydrodynamics, magneto-hydrodynamics, heat, and mass transfer. He has excellent communication skills and leadership qualities. He is self-motivated and responds well to feedback and suggestions.

Dr. Awasthi has qualified in the National Eligibility Test (NET) conducted on the all India level in 2008 by the Council of Scientific and Industrial Research (CSIR), and he has gotten the Junior Research Fellowship (JRF) and Senior Research Fellowship (SRF) for his work. He has more than 100 publications to his credit in high impact journals of international repute and has also published 5 books. He has attended many symposia, workshops and conferences in mathematics as well as fluid mechanics. He has gotten four consecutive research awards from the University of Petroleum and Energy Studies, Dehradun, India. He has also received the start-up research fund for his project "Nonlinear Study of Interface in Multilayer Fluid System" from University Grant Commission (UGC), New Delhi.

Dr. Rishi Asthana has done his Ph.D. on the topic "Some Problems on Viscous Potential Flow Analysis of Kelvin-Helmholtz Instability". Dr. Rishi Asthana has qualified in the National Eligibility Test (NET) conducted on the all India level in 2006 by the Council of Scientific and Industrial Research (CSIR). His areas of research interest are hydrodynamic stability, heat and mass transfer, and data analysis. He possesses good knowledge of mathematical modelling of problems related to engineering applications, especially fluid flow. He has publications in refereed journals with good impact factors, such as *Applied Mathematical Modelling*; *International Journal of Engineering Sciences*; *International Journal of Heat and Mass Transfer*; and *Pattern Recognition Letter*, etc. Dr. Asthana has taught courses of Applied Mathematics viz. Numerical Methods, Operations Research, Discrete Mathematics, Probability & Statistics, Linear Algebra and Engineering Mathematics, etc.

Prof. Mangey Ram received the Ph.D. degree with a major in Mathematics and minor in Computer Science from G. B. Pant University of Agriculture and Technology, Pantnagar, India. He has been a faculty member for around thirteen years and has taught several core courses in pure and applied mathematics at undergraduate, postgraduate, and doctorate levels. He is currently the Research Professor at Graphic Era (Deemed to be University), Dehradun, India & Visiting Professor at Peter the Great St. Petersburg Polytechnic University, Saint Petersburg, Russia. Before joining Graphic Era, he was a Deputy Manager (Probationary Officer) with Syndicate Bank for a short period. He is Editor-in-Chief of *International Journal of Mathematical, Engineering and Management Sciences*; *Journal of Reliability and Statistical Studies*; *Journal of Graphic Era University*;

series editor of six book series with *Elsevier, Taylor and Frances Group, Walter De Gruyter Publisher Germany, River Publisher* and the Guest Editor & Associate Editor for various journals. He has published over 250 publications (journal articles/books/book chapters/conference articles) through *IEEE, Taylor & Francis, Springer Nature, Elsevier, Emerald, World Scientific,* and many other national and international journals and conferences. Over 50 of these publications have been books (authored/edited) with international publishers like *Elsevier, Springer Nature, Taylor and Frances Group, Walter De Gruyter Publisher Germany, and River Publisher.* His fields of research are reliability theory and applied mathematics. Dr. Ram is a Senior Member of the IEEE, Senior Life Member of Operational Research Society of India, the Society for Reliability Engineering, Quality and Operations Management in India, and the Indian Society of Industrial and Applied Mathematics. He has been a member of the organizing committee of a number of international and national conferences, seminars, and workshops. He has been recognized by the "*Young Scientist Award*" by the Uttarakhand State Council for Science and Technology, Dehradun, in 2009. He has also been awarded the "*Best Faculty Award*" in 2011; "*Research Excellence Award*" in 2015; and "*Outstanding Researcher Award*" in 2018 for his significant contribution in academics and research at Graphic Era Deemed to be University, Dehradun, India. Recently, he has been received the "*Excellence in Research of the Year-2021 Award*" by the Honorable Chief Minister of Uttarakhand State, India.

1 Solar Cell Modeling

Hamdy Hassan
Egypt-Japan University of Science and Technology (E-JUST)
Assiut University

Tamer F. Megahed
Egypt-Japan University of Science and Technology (E-JUST)
Mansoura University

CONTENTS

NOMENCLATURE

A Area, m^2
C Specific heat, J/kg.K

DOI: 10.1201/9781003222255-1

f_{MM}	Mismatch factor for different types of PV modules
h	Heat transfer coefficient, W/m^2.K
I	Current
k	Thermal conductivity, W/m.K
m	Mass, kg
n	Normal direction
N_{SM}	Number of series solar cell modules in the system
N_{PM}	Number of parallel modules in the solar cell system
N_{PC}	Number of parallel cells in the solar cell module
N_{SC}	Number of series cells in the solar cell module
N_{SBat}	Number of batteries connected in series
N_{PBat}	Number of batteries connected in parallel
P_{cel}	Output power of the solar cell system
P_{inv-ip}	Inverter input power
P_{inv-op}	Inverter output power
P_{iBC-ip}	Charge controller input power
P_{BC-op}	Charge controller output power
$q^{\bullet}$	Heat generation, W/m^3
$\dot{Q}$	Energy rate, W
R_S^M	Total resistance of the solar cell module
R_S^C	Resistance of the solar cell
SOC	Batteries' state of charge
t	Time, s
T	Temperature, K
T_c	Operating cell temperature
T_a	Ambient temperature
T_0^c	Reference operating temperature
V_{oc}^c	Open circuit voltage of photovoltaic cell
$V_{oc,0}^c$	Open circuit voltage at temperature T_0^c
V_{OC}^M	Open circuit voltage of the solar cell module
V_{cel}	Total voltage of the solar cell system
V_{Bat}	Single battery voltage

GREEK SYMBOLS

σ	Stefan–Boltzmann constant, W/m^2.K^4
ε	Emissivity
η	Efficiency
τ	Transmitivity
Δt	Time interval
σ_i	Self-discharge losses of the battery
$\eta_{i(I_{Kolbat}(t))}$	Battery efficiency
η_{inv}	Inverter efficiency

SUBSCRIPTS

a	Ambient
ab	Absorbed

c	Convection
cel	Cell
cd	Conduction
e	Electric
eva	EVA
g	Glass
r	Radiation
s	Sky
sol	Solar
t	Transmitted
ted	Tedlar
v	Velocity
w	Wind

ABBREVIATIONS

AC	Alternating current
DC	Direct current
MPP	Maximum power point
PV	Photovoltaic

1.1 INTRODUCTION

The world is facing a significant increase in energy demand due to the fast growth of world population and their activities. The use of fossil fuels as a source of energy could lead to global warming and serious environmental pollution besides their non-sustainability (Wang et al., 2021; Abd Elbar & Hassan, 2019). These causes have also necessitated the acceleration in research and dependence on cleaner energy solutions such as wind, photovoltaic, biomass, and hydro for home and even industrial usage. Therefore, the necessity of utilizing sustainable and renewable energy resources for power and electricity generation has become an urgent technical issue to avoid the previous serious problems and obtain the benefits of cheaper cost of extraction and less pollution caused by renewable energy generators, especially in sunny areas (Soliman & Hassan, 2018). These days solar energy is considered as one of the greatest vital sustainable and renewable energy resources, which can be utilized by various technologies such as solar cooling, solar heating, and solar power systems and photovoltaics (solar cells) (Soliman et al., 2018). The solar cell or solar panel (photovoltaic panel) can yield electrical energy from the sunlight with the help of photovoltaic effect. Electrical energy can be generated from the incident solar energy by direct transformation to electricity by employing solar panels or by transforming it to thermal energy and then to electricity (Jamil et al., 2016). The thermal modeling and electrical modeling of solar cells have attracted the attention of many researchers to study their performance. In thermal modeling of the solar cell, researchers concentrate on the cell temperature and its relationship with its efficiency and how these models can predict the solar cell performance. An experimental and theoretical analysis of solar cell module temperature at different climate conditions was performed by Kaplani and Kaplanis (2014) with respect to module inclination, direction, and wind speed. The experimental works make use of hourly solar cell temperature data gathered from a double-axis sun-tracking solar cell system for a time period of 1 year. The coefficient value which corresponds to the solar cell module temperature with the solar radiation intensity on the solar cell plane and the atmospheric temperature is evaluated in relation to the solar cell inclination angle, the incidence angle of the wind flowing on the solar cell surface, either back or front, and the wind speed. The coefficient value is assessed both theoretically and experimentally through thermal model on the equation of energy balance. They found that the coefficient value calculated by the simulation model accords with the experimental data for the total range of solar cell wind velocities, inclination angles, and wind directions. A dynamic theoretical model of hybrid PV/thermal (PVT) collector with a tube and sheet thermal absorber was investigated by Guarracino et al. (2016). The model is utilized to assess the annual electrical energy generation along with the national hot water provision from the output of the thermal energy by utilizing actual climate results at high chronological resolution. Their model includes the impact of a non-uniform distribution of temperature on the surface of the PV cell on its output electrical power. The findings reveal that the usage of the real climate data dynamic model at high resolution is of essential importance when assessing the yearly system performance. The outcomes of the dynamic simulation within 1 min inlet data reveal that the thermal system output is extremely dependent on the choice of control parameters (choice of flow rate, differential thermostat controller, pump operation, etc.) in answer to the changing weather conditions. Akhsassi et al. (2018) developed and verified two new solar cell models, and then, they compared the results of ten theoretic models including their models with measured results based on the surrounding solar radiation temperature and with and without the consideration of wind velocity. The tested models were utilized to calculate the solar cell module temperature at the 7.2 kWp stand-alone solar cell power plant mounted in Elkaria, Morocco. The outcomes showed that their model without wind speed produces the highest the correlation coefficient value and the lowest root-mean-square error value. Also, their model with wind velocity yields the best statistical coefficients compared to other models without and with the consideration of wind. Tuncel et al. (2020) presented a dynamic thermal model of the solar cell considering the climate conditions as temperature and wind speed and module parameters. Their model gives a validated performance investigation for a poly-c-Si solar cell module with an estimated heat capacity value installed in Ankara in Turkey, and a sensitivity investigation for the dynamic thermal model heat capacity is achieved. They concluded that the model outcomes are nearly constant under varying values of module heat capacity. Patro and Saini (2020) presented a reliable and robust hybrid technique to forecast the performance of the cell system, which is a grouping of social and analytical learning differential evolution to obtain the unidentified limits of double-diode solar cell model. The performance of four different types of solar cell modules is assessed. They did an experimental work under uncontrolled and controlled ambient conditions to validate the efficiency and accuracy of the proposed method. The outcomes resulted from the proposed method can be helpful to effectively forecast the performance for partial shading conditions and under low solar radiation conditions. Swarupa et al. (2020) presented a MATLAB-Simulink algorithm to calculate the highest power from the cell under supreme power point tracking at various dynamic conditions such as solar energy and temperature. They estimated the output current and voltage of the module. They declared that the simulated data resulted from their algorithm are the same as the data of the reference solar cell modules of different types. Wang et al. (2021) examined the output energy of different PV modules and assessed the precisions of diverse simplistic power forecast of ten PV modules. The radiation of solar, power generation and ambient temperature are considered. The one-diode model established the highest accuracy for poly-Si PV and mono-Si PV modules. Moreover, the accuracies of the tested models are low for the thin-film PV module.

Many photovoltaic battery systems are distinguishable in design; however, the complete dynamic testing of this system incurs higher costs in addition to taking very long time. In some way, irrespective of whether the budget and required time are provided for the system simulation, it is too difficult to examine all the hypothetical problems, which are going to meet with the hybrid system over its life cycle. When the actual performance of the system goes differently than the planned scenarios, the system becomes unpredictable. And in such cases, the optimization of the system becomes difficult. An obtainable model for appreciating the hybrid power system operation at a certain place is an important precondition for raising investment in renewable power systems. This technique is also convenient for comparing the performance of hybrid power systems in certain conditions at specific places (Wu et al., 2011). Therefore, the creation of state-of-the-art software and emulation techniques is vital for the success of this process.

Several researches tested the financial worth and technical possibility of solar cell battery system designs, which were able to obtain a certain level of accuracy and supply energy that matches that demand profile (Cao et al., 2016; Alanne & Cao, 2017). The numbers of cell panels, storage units, demand scenarios and available renewable sources are important factors in measuring the cell energy system.

Deterministic and probabilistic methods are usually used by researchers and scientists to measure and emulate the photovoltaic power system. In the deterministic method, time series data such as solar radiation, wind speed, atmosphere temperature, demand profile, and geographical coordinates and altitude are assumed known. This means that such data should be provided on hourly intervals (at least) before configuring with this method (Mulepati, 2013). It is vital also for such data to be time-coded and recorded in reference to the emulation period step. On the other hand, the probabilistic technique assumes random values for renewable source data and demand scenarios, also with random timings. This method is based on creating hypothetical situations for energy generation and its consumption, and then, such hypothesis is combined to draw a system risk scenario. There were studies that defined two techniques for sizing an autonomous photovoltaic energy system. One technique measures the system performance by calculating the energy-to-demand ratio on an annual average of the monthly operation of the system.

The other technique calculates the storage-to-demand ratio based on the least efficient monthly performance (Park et al., 2012). Yang et al. (2009) introduced a different probabilistic method. They suggested to base their scenarios on a typical meteorological year to come up with a more accurate assumption for the photovoltaic energy system. Tina & Gagliano (2011) sized the solar cell energy system using a probabilistic method founded on a convolution module, including a probability weight factor. When compared, probabilistic and deterministic measuring and emulation techniques have their own pros and cons. Sizing the energy system based on a deterministic technique requires enormous calculations and results in suboptimum models depending on the quality of the determined data. Meanwhile, random values in the probabilistic techniques are simply put and they may base their scenarios on daily or monthly data instead of time-coded data.

This chapter first presents the compositions of the solar cell layers. Then, an overview on the solar cell thermal modeling and electrical modeling is presented, including the main equations governing the thermal and electrical behaviors of the solar cell. Moreover, it presents the optimal sizing of the solar cell, invertor, and battery.

1.2 SOLAR CELL LAYERS

Most solar cell bulk silicon cell modules are composed of a transparent front layer, an encapsulant of the solar cells, a rear layer, and a frame surrounding the outer edge. In most solar cell modules, the front or top layer is glass, while the encapsulant layer is EVA (ethyl vinyl acetate) and the rear layer is Tedlar, as shown in Figure 1.1.

1.2.1 Front Layer

The front surface of a solar cell module must have a high transmission for solar radiation in the wavelength range of 350 to 1200 nm. Moreover, its reflection to incident solar energy must be as low as possible. Additionally, its material should be stable under prolonged UV exposure and impervious to water and should have low thermal resistance and good impact resistance. In most cell modules, the front layer is utilized to provide the cell with the rigidity and mechanical strength; therefore, either the top layer or the rear layer must be mechanically rigid to support the cells and the wiring. Various materials can be chosen for a top layer, including polymers, acrylic and glass. But, tempered, low-iron glass is the mostly recommended as it is strong, of low cost, highly transparent, stable and impervious to gases and water and it has worthy self-cleaning properties.

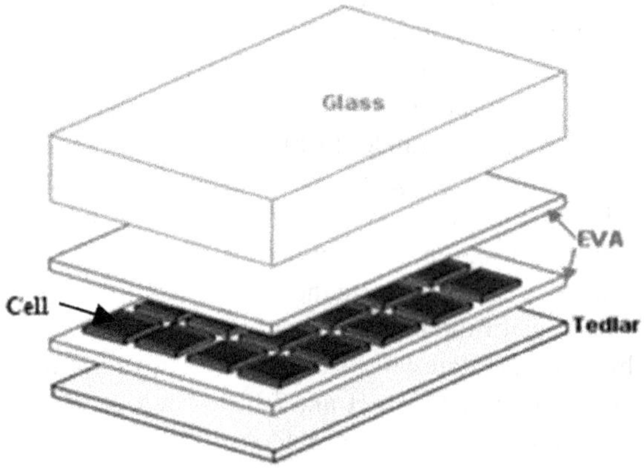

FIGURE 1.1 Solar cell layers.

1.2.2 Silicon Layer (Cell)

The silicon layer is formed of crystalline silicon, either monocrystalline silicon or polycrystalline silicon. Crystalline silicon is considered the foremost semiconducting material utilized in solar cell technology for the fabrication of silicon cells. These cells are combined in solar panels as part of a solar cell system, which is mainly used to generate electric power from the light of sun.

1.2.3 Encapsulant

An encapsulant is utilized to provide adhesion between the cells, the top layer, and the rear one of the cell modules. It should be stable at high UV exposure and elevated temperatures. Additionally, it should also have low thermal resistance and be optically transparent. EVA (ethyl vinyl acetate) is the most commonly recommended encapsulant material. EVA is used in thin sheets, which are inserted between the top layer and the rear one from one side and the solar cells from the other side.

1.2.4 Rear Layer

The important characteristic of the rear layer of the cell module is the thermal resistance, which must be as low as possible, and at the same time, it should prevent the ingress of water vapor and water. In most solar cell modules, a thin polymer sheet, typically Tedlar, is utilized as the rear layer. Some solar cell modules such as bifacial modules are designed to receive sunlight from the front or the rear. In this case, both the rear and the front should be optically transparent.

1.2.5 Frame

Finally, the structural component of the cell module is the module frame. A conventional solar cell module frame is normally fabricated using aluminum. The frame structure must be free of projections that could lead to lodgment of dust, water, or other matter.

1.3 SOLAR CELL MODELING

The solar cell modeling includes the thermal and electrical models. In the thermal model, the energy equation is specified for each solar cell layer described previously considering the incident solar radiation on the front glass, layer properties, and ambient conditions. However, in the electrical model, the main concept of the electrical energy of the solar cell is presented. Also, a method for renewable sources "solar energy, energy storage to supply load is introduced.

1.3.1 Thermal Modeling

In this thermal model, energy balance equations are employed in every layer of the solar cell: glass, EVA, silicon layer (cell) and Tedlar layer. In this model, the thermophysical properties of the material of the cell layers are assumed constant

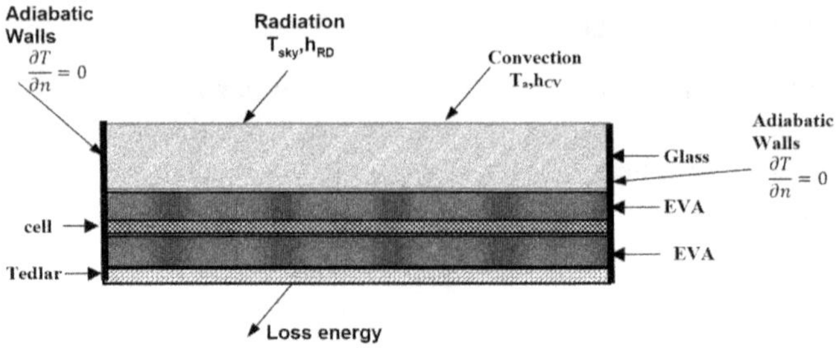

FIGURE 1.2 Solar cell boundary conditions.

and independent of temperature. In this model, the contact thermal resistances between each cell layer are neglected because their values are small. The loss of heat from the sides of cell layers is also neglected because the layer thickness is very small compared to its surface area. Moreover, the radiation loss from the back layer is neglected because the back layer surface temperature is low, as shown in Figure 1.2.

The equation that regulates the transfer of heat within the solar cell layers is the basic heat conduction equation as follows (Soliman et al., 2018; Soliman et al., 2019):

$$\nabla \cdot (k \nabla \mathrm{T}) + q^{\bullet} = 0 \tag{1.1}$$

where T is the layer temperature in K, $q^{\bullet}$ is the entire heat created in the layer in W/m^3, and k is considered the thermal conductivity of every layer.

The solar cell sides are assumed adiabatic (Reddy et al., 2014), which are governed by (Hassan, 2014; Hassan & Harmand, 2013):

$$\frac{\partial T}{\partial n} = 0 \tag{1.2}$$

where n is the normal direction.

1.3.1.1 Glass Cover Energy Balance

The cell glass cover must have greater transmissivity and smaller reflectivity and absorptivity. Consequently, most of the total incident solar radiation on the glass cover is transferred to the cell–Tedlar layers. The remainder of the solar radiation is split into a reflected part and absorbed part. The cell temperature is greater than the temperature of the glass cover. As a result, a conduction transfer of heat happens from the cell surface to the glass. After that, the heat dissipates from the glass cover top surface to the atmosphere via radiation and convection.

The balance of energy of the glass cover shown in Figure 1.3 is written as follows (Soliman & Hassan, 2018; Soliman & Hassan, 2020):

$$\dot{Q}_{\mathrm{sol}} + \dot{Q}_{\mathrm{cd,\,eva-g}} - \dot{Q}_{\mathrm{r,\,g-s}} - \dot{Q}_{\mathrm{c,\,g-a}} - \dot{Q}_{ab,\,g} - \dot{Q}_{ref,\,g} - \dot{Q}_{\mathrm{t,\,g}} = m_g c_g \frac{dT_g}{dt} \tag{1.3}$$

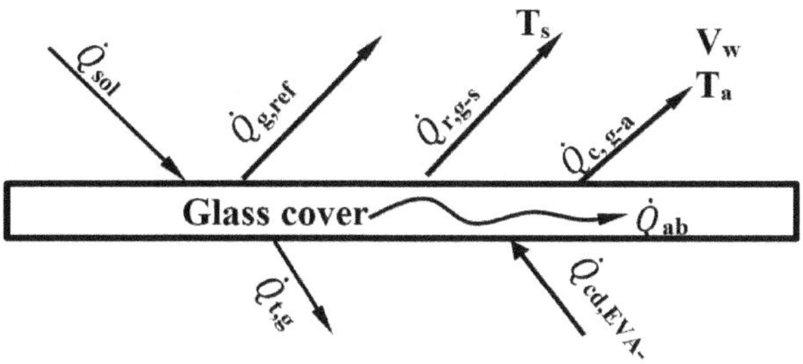

FIGURE 1.3 Glass cover energy balance.

where q_{sol} is the incident solar radiation in W, $\dot{Q}_{\mathrm{cd,\ cel-g}}$ is the conductive heat transfer from cell to glass cover in W, $\dot{Q}_{\mathrm{r,\ g-s}}$ is the radiation heat transfer from glass cover to sky in W, $\dot{Q}_{\mathrm{c,\ g-a}}$ is the convection heat transfer from glass to ambient in W, $\dot{Q}_{ab,g}$ is the absorber solar radiation inside the glass cover, $\dot{Q}_{ref,g}$ is the reflected solar energy due to glass surface in W, $q_{\mathrm{cd,\ eva-g}}$ is the conductive heat transfer from the EVA to the glass, and $\dot{Q}_{\mathrm{t,\ g}}$, is the solar energy transmitted in W.

The heat transfer via radiation from the glass to the sky is calculated by (Incropera et al., 2011):

$$\dot{Q}_{\mathrm{r,\ g-s}} = A_g\left(T_g^4 - T_s^4\right) \tag{1.4}$$

where σ is the Stefan–Boltzmann constant in W/m^2.K^4, ε_g is the glass emissivity, and T_g and T_S are the glass surface and sky temperatures, respectively, in K. The sky temperature is estimated by (Soliman & Hassan, 2018; Nasef et al., 2019):

$$T_S = 0.0552\, T_a^{1.5} \tag{1.5}$$

The convective transfer of heat from the surface of the glass to the surroundings is estimated by (Incropera et al., 2011; Hassan et al., 2021):

$$\dot{Q}_{\mathrm{c,\ g-a}} = Ah_{c,\ g}\left(T_g - T_a\right) \tag{1.6}$$

1.3.1.2 Top EVA Energy Balance

The balance of energy of the top EVA layer as shown in Figure 1.4 is written as follows:

$$\dot{Q}_{\mathrm{t,g}} + \dot{Q}_{\mathrm{cd,\ cel-eva}} - \dot{Q}_{\mathrm{cd,\ eva-g}} - \dot{Q}_{ab,eva} - \dot{Q}_{t,\ eva} - = m_{eva}c_{eva}\frac{dT_{eva}}{dt} \tag{1.7}$$

where $\dot{Q}_{\mathrm{cd,\ eva-g}}$ is the conductive heat transfer from EVA to glass cover in W, $\dot{Q}_{ab,eva}$ is the absorber solar radiation inside the EVA layer, $\dot{Q}_{\mathrm{cd,\ cel-eva}}$ is the conductive transfer of heat from the cell to EVA, and $\dot{Q}_{\mathrm{t,\ eva}}$ is the solar energy transmitted in W.

1.3.1.3 Cell Energy Balance

In the cell layer, the first part of the energy transmitted to this cell is partly transformed to electricity reliant on the cell's efficiency. The second portion is lost through conduction from the surface of the cell to EVA, and another part is lost in heating the cell layer. The final portion is transformed back to the cell, where it is transformed to the bottom EVA and then to the Tedlar layer. The energy balance of the cell as shown in Figure 1.5 is written as follows (Soliman & Hassan, 2018):

$$\dot{Q}_{\mathrm{t,eva}} - \dot{Q}_{\mathrm{cd,\,cel-eva,\,bot}} - \dot{Q}_{\mathrm{cd,\,cel-eva,\,top}} - \dot{Q}_{ab,\,cel} - \dot{Q}_{t,\,cel} = m_{cel}c_{cel}\frac{dT_{cel}}{dt} \tag{1.8}$$

where $\dot{Q}_{\mathrm{cd,\,cel-eva,\,bot}}$ is the conductive heat transfer from cell to the bottom EVA, $\dot{Q}_{\mathrm{cd,\,cel-eva,\,top}}$ is the conductive heat transfer from cell to the top EVA, $\dot{Q}_{ab,\,cel}$ is the absorbed energy inside the cell, and $\dot{Q}_{t,\,cel}$ is the transmitted energy from the cell.

1.3.1.4 Bottom EVA–Tedlar Energy Balance

The energy balance through bottom EVA–Tedlar as shown in Figure 1.6 is written as:

$$\dot{Q}_{\mathrm{t,cel}} - \dot{Q}_{\mathrm{cd,\,cel-eva,\,bot}} - \dot{Q}_{\mathrm{c,\,ted-f}} - \dot{Q}_{ab,ted} - \dot{Q}_{ab,\,eva} = m_{eva-ted}c_{eva-ted}\frac{dT_{eva-ted}}{dt} \tag{1.9}$$

where $\dot{Q}_{\mathrm{c,\,ted-f}}$ is the convective heat transfer from the Tedlar layer surface to the cooling fluid, $\dot{Q}_{ab,\,ted}$ is the absorbed energy inside the Tedlar, and $\dot{Q}_{ab,\,eva}$ is the absorbed energy inside the EVA layer.

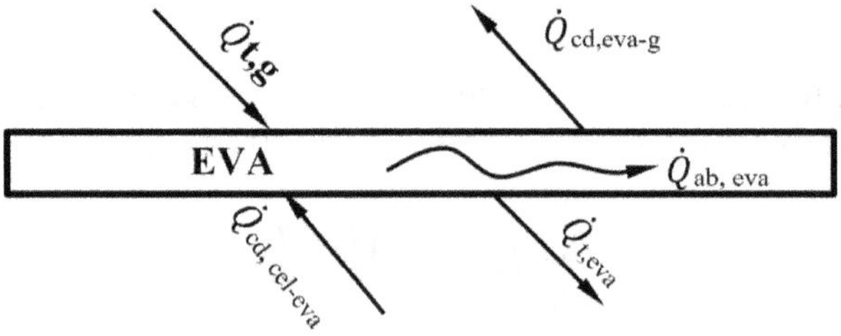

FIGURE 1.4 EVA energy balance.

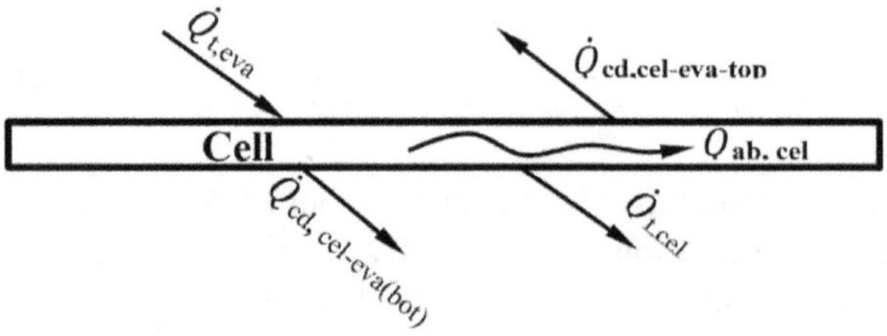

FIGURE 1.5 Cell energy balance.

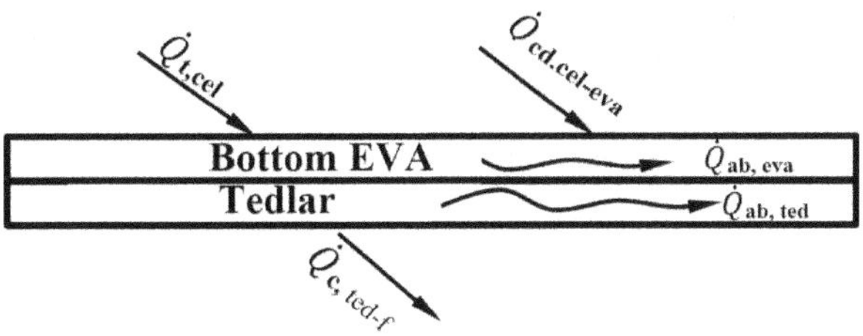

FIGURE 1.6 Bottom EVA–Tedlar energy balance.

1.3.1.5 Solar Cell Power

The solar cell output electrical power is calculated by (Emam et al., 2017):

$$P_e = \eta_{cel}\, \tau_g\, \dot{Q}_{sol} \tag{1.10}$$

where τ_g is the glass transmitivity and η_{cel} is the cell electrical efficiency, which is estimated by (Xu & Kleinstreuer, 2014; Soliman & Hassan, 2019):

$$\eta_{cel} = \eta_{ref}\left(1 - \beta_{ref}\left(T_{cel} - T_{ref}\right)\right) \tag{1.11}$$

where η_{ref} and β_{ref} are supplied by the company in the data sheet at a reference temperature of T_{ref}= 25 °C and T_{cel} is the average cell temperature.

1.4 ELECTRICAL MODELING

Combining energy generation from both sources (solar and energy storage) in one energy system creates a sustainable model that can compete with depletable sources and can sometimes generate surplus energy. This is because the two sources alternate in their intensity and sequel each other on a day and night basis or from season to season, thus making this integrated power system more streamlined and less disruptive. That said, the sizing, optimization and control of such integrated systems have become more complicated, given the inconsistent and non-similar nature of its components (solar and storage). A complex sizing formula must be derived, for example, to create a generation model that would be equivalent to the consumption profile of a certain area.

1.4.1 Solar Cell/Battery System

The main components of photovoltaic energy systems are demonstrated in Figure 1.7. Photovoltaic panels generate energy to meet the demand, and when the energy generated is higher or lower than the demand, storage units save the excess energy or supply energy to meet the load. Controllers for charging the storage units maintain the battery voltage in a specific range, which helps avoid discharge or overcharge cases. To prevent overcharging, the charging controller isolates the batteries from the DC bus if the bus voltage exceeds its maximum value, which means that the storage system is fully charged, and the load energy is lower than the energy supplied by the

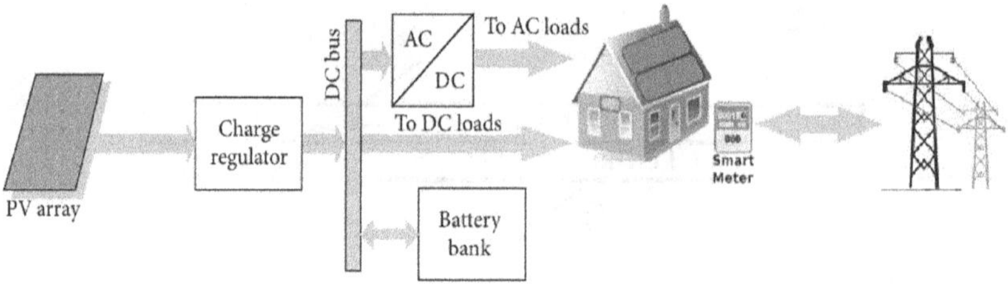

FIGURE 1.7 Energy system components.

renewable energy sources. The storage will be connected again if the DC bus voltage decreases under the allowable minimum value, which means that the load requires energy more than the energy generated by the sources. The load is disconnected if its current exceeds the generated current and at the same time the bus voltage decreases under the minimum value; this action is performed to protect the storage batteries from excessive discharge. The load is connected again if the bus voltage exceeds the minimum value. The function of the inverter is to convert DC energy to AC energy to be suitable for AC loads.

1.4.2 Solar Cell Module Performance Model

Several solar cells connected in parallel and series to get the desired voltage and current output levels make a photovoltaic module. The solar cell is made up of a P–N junction, which generates electrical energy from the solar radiation [9]. The single-diode calculation formula is used to emulate silicon photovoltaic cells made of a photocurrent resource, internal resistances, and a nonlinear diode, as shown in Figure 1.8. In the datasheets provided by the manufacturers, usually they give a basic electrical characteristics module. Short-circuit current as well as open-circuit voltage may be obtained, in addition to the maximum power point (MPP) voltage and current values found in the data form (Ram et al., 2017). The manufacturer's data form contains photovoltaic cell temperature coefficients called current and voltage. The model voltage and current values depend on irradiance and temperature. The coefficient of the cell temperature, which is a reference to voltage, is large and negative. From another point of view, the current temperature coefficient is small and positive. In some photovoltaic cell models, we can see the variations of current with temperature.

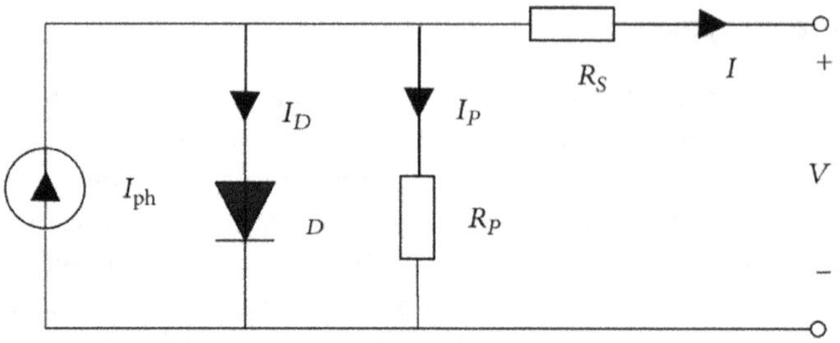

FIGURE 1.8 Single-diode calculation method of a photovoltaic cell.

It can be defined as the variations of voltage with temperature for a photovoltaic module that consists of cells connected in series (Fudholi et al., 2014).

The operating cell temperature that is not like the ambient temperature determines the voltage of the open circuit. The operating cell temperature can be calculated using Equation (1.12). The ambient temperature is given as (Castaner and Luis, 2002):

$$T_{cel} = T_a + 0.03 \cdot G_a \tag{1.12}$$

where T_c is the operating cell temperature; T_a is the ambient temperature; and G_a is irradiation.

In this case, the ambient temperature in kelvin and the operating cell temperature are the same; also, the cell open-circuit voltage in W/m^2 can be calculated using Equation (1.13) (Megahed & Radwan, 2020).

$$V_{oc}^{c} = V_{oc,0}^{c} + (-2.3mV / C)\left(T_{cel} - T_0^{c}\right) \tag{1.13}$$

where V_{oc}^{c} is the open-circuit voltage of the photovoltaic cell; $V_{oc,0}^{c}$ is the open-circuit voltage at temperature T_0^c; and T_0^c is the reference operating temperature.

In the photovoltaic cell, the SC current is directly proportional to the solar radiation, whereas the open-circuit voltage is a logarithmic function of the current. Based on Equation (1.14), the SC current of a photovoltaic cell can be calculated from the irradiance and is given as (Engin, 2013):

$$I_{SC}^{C} = \frac{I_{SC,0}^{C} G_a}{G_{a,0}} \tag{1.14}$$

where I_{SC}^{C} is the short-circuit current of the photovoltaic cell.

The photovoltaic panel SC current is proportional to the single module current multiplied by the number of modules connected in parallel (Engin, 2013):

$$I_{SC}^{M} = N_{PC}.I_{SC}^{C} \tag{1.15}$$

where I_{SC}^{M} is the short-circuit current of the module and N_{PC} is the number of parallel cells in the module.

By using Equation (1.16), the voltage of the open circuit can be calculated.

$$V_{OC}^{M} = N_{SC} \cdot V_{OC}^{C} \tag{1.16}$$

where V_{OC}^{M} is the open-circuit voltage of the module and N_{SC} is the number of series cells.

The module's equivalent series resistance can be calculated using the following equation.

$$R_S^M = \frac{N_{SC}}{N_{PC}}.R_S^C \tag{1.17}$$

where R_S^M is the total resistance of the module and R_S^C is the resistance of the cell.

The photovoltaic module's current under virtual operating condition is illustrated in the following equation.

$$I^M = I_{SC}^M \left[1 - \exp\left(\frac{V^M - V_{OC}^M + R_S^M . I^M}{N_{SM} V_t^C} \right) \right] \tag{1.18}$$

where I^M is the module current.

The photovoltaic modules are connected in series to reach the necessary bus voltage. The bus voltage shown in Equation (1.19) is directly proportional to the number of series modules and the single module voltage.

$$V_{cel} = V_{OC}^M . N_{SM} \tag{1.19}$$

where V_{cel} is the total voltage of the photovoltaic system and N_{SM} is the number of series modules in the system.

The output current of a photovoltaic array at time t is related to the number of parallel strings as (Sawle et al., 2018):

$$I_{cel}(t) = I^M(t) . N_{PM} . f_{MM} \tag{1.20}$$

where I_{PV} is the total current of the photovoltaic system; N_{PM} is the number of parallel modules in the photovoltaic system; and f_{MM} is the mismatch factor for different types of modules.

The output energy of the photovoltaic array at time *t* would be

$$P_{cel}(t) = I^M(t) . N_{PM} . f_{MM} \tag{1.21}$$

where P_{PV} is the output power of the photovoltaic system.

1.4.3 Battery Performance Model

To match the batteries' voltage with the load voltage, the batteries are connected in a series string (Sawle et al., 2018). To calculate the number of batteries in a single string, the bus voltage is divided by the voltage of a single battery as in Equation (1.22). The batteries should be symmetrical to avoid the damage of the batteries in the string.

$$N_{SBat} = \frac{V_{cel}}{V_{Bat}} \tag{1.22}$$

where V_{Bat} is the single battery voltage and N_{SBat} is the number of series batteries.

There are many types of battery banks in compound system. The calculation of the battery state of charge at a time point is based on adding the charge or discharge current (positive or negative, respectively) to the storage charging state at the previous time point promptly. This calculation should consider the situations of self-discharging and battery charging losses (Goel & Sharma, 2017):

$$SOC(t+1)=\sum_{i=0}^{BatBan}\left[SOC_i(t)\cdot\sigma_i+I_{Bat}(t)\cdot\Delta t\cdot\eta_{i(I_{Kolbat}(t))}\right]\Delta N_{PBat} \quad (1.23)$$

where SOC is the batteries' state of charge; σ_i is the self-discharge losses of the battery; I_{Bat} is the battery current; Δt is the time interval; N_{PBat} is the number of batteries connected in parallel; and $\eta_{i(I_{Kolbat}(t))}$ is the battery efficiency.

1.4.4 Inverter, Charger and Demand Performance Model

The advantages of the inverter can be described in terms of its input–output relation (Sawle et al., 2018). Transformation losses can cause loss of some of the power transferred to the inverter, and such loss is called inverter efficiency loss, η_{inv}:

$$P_{inv-ip}\cdot\eta_{inv}=P_{inv-op} \quad (1.24)$$

where P_{inv-ip} is the inverter input power; η_{inv} is the inverter efficiency; and P_{inv-op} is the inverter output power.

It can be noticed that the charge regulators are in the form of a switch that connects or disconnects the generation unit to or from the storage unit or demand source based on the battery state of charging, temperature, or discharge. The battery charger power output is equal to the power input multiplied by the efficiency losses while converting the energy. Efficiency losses are based nonlinearly on the DC output energy and thus depend nonlinearly on the battery charger DC output current:

$$P_{iBC-ip}\cdot\eta_{BC}=P_{BC-op} \quad (1.25)$$

where P_{iBC-ip} is the charge controller input power and P_{BC-op} is the charge controller output power.

The losses of efficiency can be calculated by comparing efficiency losses to output power lines provided by the manufacturers. In many situations, there would be two types of demand, DC appliances working on 48V, 24V or 12V, and 220V AC appliances. The evaluated energy consumption must be provided in time spans of years, days or hours. If both a DC and an AC generator exist, some of the DC generators energy can be routed through the inverter to the AC loads:

$$I_{GDC}\geq I_{dcl}\ \text{or}\ I_{GDC}=I_{dcL}+I_{inv-ip} \quad (1.26)$$

where I_{GDC} is the DC generator current; I_{dcl} is the DC load current; and I_{inv-ip} is the inverter input current.

1.5 CONCLUSIONS

A solar thermal modeling is presented, including thermal and electrical modeling. The thermal model considers the energy balance within each layer of the solar cell (glass cover, EVA, cell and Tedlar) in the transient form. Moreover, the energy equation that is used to solve the solar cell component temperature as a function of time is presented. Also, the mathematical equation used to calculate the solar cell output power is given. In addition, this chapter presented solar cell design. Detailed sizing steps are proposed for the solar cell grid-connected AC system with battery backup to supply a certain electric load.

REFERENCES

Abd Elbar, A. R., & Hassan, H. (2019). Experimental investigation on the impact of thermal energy storage on the solar still performance coupled with PV module via new integration. *Solar Energy*, *184*(April), 584–593. https://doi.org/10.1016/j.solener.2019.04.042.

Akhsassi, M., El Fathi, A., Erraissi, N., Aarich, N., Bennouna, A., Raoufi, M., & Outzourhit, A. (2018). Experimental investigation and modeling of the thermal behavior of a solar PV module. *Solar Energy Materials and Solar Cells*, *180*(March 2017), 271–279. https://doi.org/10.1016/j.solmat.2017.06.052.

Alanne, K., & Cao, S. (2017). Zero-energy hydrogen economy (ZEH2E) for buildings and communities including personal mobility. *Renewable and Sustainable Energy Reviews*, *71*(December 2016), 697–711. https://doi.org/10.1016/j.rser.2016.12.098.

Cao, X., Dai, X., & Liu, J. (2016). Building energy-consumption status worldwide and the state-of-the-art technologies for zero-energy buildings during the past decade. *Energy and Buildings*, *128*, 198–213. https://doi.org/10.1016/j.enbuild.2016.06.089.

Castaner, L., and Santiago, S. (2002). *Modelling Photovoltaic Systems using PSpice*. John Wiley and Sons.

Emam, M., Ookawara, S., & Ahmed, M. (2017). Performance study and analysis of an inclined concentrated photovoltaic-phase change material system. *Solar Energy*, *150*, 229–245. https://doi.org/10.1016/j.solener.2017.04.050

Engin, M. (2013). Sizing and simulation of PV-wind hybrid power system. *International Journal of Photoenergy*, *2013*(March 2013). https://doi.org/10.1155/2013/217526.

Fudholi, A., Sopian, K., Yazdi, M. H., Ruslan, M. H., Ibrahim, A., & Kazem, H. A. (2014). Performance analysis of photovoltaic thermal (PVT) water collectors. *Energy Conversion and Management*, 78, 641–651. https://doi.org/10.1016/j.enconman.2013.11.017

Goel, S., & Sharma, R. (2017). Performance evaluation of stand alone, grid connected and hybrid renewable energy systems for rural application: A comparative review. *Renewable and Sustainable Energy Reviews*, *78*(October 2016), 1378–1389. https://doi.org/10.1016/j.rser.2017.05.200.

Guarracino, I., Mellor, A., Ekins-daukes, N. J., & Markides, C. N. (2016). Dynamic coupled thermal-and-electrical modelling of sheet-and-tube hybrid photovoltaic / thermal (PVT) collectors. *Applied Thermal Engineering*, *101*, 778–795. https://doi.org/10.1016/j.applthermaleng.2016.02.056.

Hassan, H. (2014). Heat transfer of Cu-water nanofluid in an enclosure with a heat sink and discrete heat source. *European Journal of Mechanics, B/Fluids*, 45, 72–83. https://doi.org/10.1016/j.euromechflu.2013.12.003.

Hassan, H., & Harmand, S. (2013). 3D transient model of vapour chamber: Effect of nanofluids on its performance. *Applied Thermal Engineering*, *51*(1–2), 1191–1201. https://doi.org/10.1016/j.applthermaleng.2012.10.047.

Hassan, H., Yousef, M. S., Mohamed, S. A., & Abo-Elfadl, S. (2021). Enhancement of the daily performance of solar still by exhaust gases under hot and cold climate conditions. *Environmental Science and Pollution Research*. https://doi.org/10.1007/s11356-021-15261–y

Incropera, F. P., Dewitt, D. P., Bergman, T. L., & Lavine, A. S. (2011). *Fundamentals of Heat and Mass Transfer*. Wiley.

Jamil, M., Adamu, I., Azwadi, N., Sidik, C., Noor, M., Muhammad, W., Mamat, R., & Naja, G. (2016). The use of nano fluids for enhancing the thermal performance of stationary solar collectors : A review. *63*, 226–236. https://doi.org/10.1016/j.rser.2016.05.063.

Kaplani, E., & Kaplanis, S. (2014). Thermal modelling and experimental assessment of the dependence of PV module temperature on wind velocity and direction, module orientation and inclination. *Solar Energy, 107*, 443–460. https://doi.org/10.1016/j.solener.2014.05.037

Kumar Patro, S., & Saini, R. P. (2020). Mathematical modeling framework of a PV model using novel differential evolution algorithm. *Solar Energy, 211*(September), 210–226. https://doi.org/10.1016/j.solener.2020.09.065.

Lakshmi Swarupa, M., Vijay Kumar, E., & Sreelatha, K. (2020). Modeling and simulation of solar PV modules based inverter in MATLAB-SIMULINK for domestic cooking. *Materials Today: Proceedings, 38*, 3414–3423. https://doi.org/10.1016/j.matpr.2020.10.835.

Megahed, T. F., & Radwan, A. (2020). Performance investigation of zero-building-integrated photovoltaic roof system: A case study in Egypt. *Alexandria Engineering Journal, 59*(6), 5053–5067. https://doi.org/10.1016/j.aej.2020.09.031.

Mulepati, S. (2013). *A Case Study of Zero Energy Home Built for Solar Decathlon Competition 2013*. Howard R. Hughes College of Engineering, December.

Nasef, H. A., Nada, S. A., & Hassan, H. (2019). Integrative passive and active cooling system using PCM and nanofluid for thermal regulation of concentrated photovoltaic solar cells. *Energy Conversion and Management, 199*(September), 112065. https://doi.org/10.1016/j.enconman.2019.112065.

Park, K., Lim, J., & Kang, C. (2012). Design of optimal sizing for new and renewable hybrid generation system for building traffic information system in energy-isolated areas. *International Conference on Computer and Computing Technologies in Agriculture*, 165–172.

Ram, J. P., Babu, T. S., & Rajasekar, N. (2017). A comprehensive review on solar PV maximum power point tracking techniques. *Renewable and Sustainable Energy Reviews, 67*, 826–847. https://doi.org/10.1016/j.rser.2016.09.076.

Reddy, K. S., Lokeswaran, S., & Mallick, T. K. (2014). Numerical investigation of micro-channel based active module cooling for solar CPV system. *54*, 400–416. https://doi.org/10.1016/j.egypro.2014.07.283.

Sawle, Y., Gupta, S. C., & Bohre, A. K. (2018). Review of hybrid renewable energy systems with comparative analysis of off-grid hybrid system. *Renewable and Sustainable Energy Reviews, 81*(June 2017), 2217–2235. https://doi.org/10.1016/j.rser.2017.06.033.

Soliman, A.M.A., Hassan, H., Ahmed, M., & Ookawara, S. (2018). A 3d model of the effect of using heat spreader on the performance of photovoltaic panel (PV). *Mathematics and Computers in Simulation*. https://doi.org/10.1016/j.matcom.2018.05.011.

Soliman, A.M.A., & Hassan, H. (2019). Effect of heat spreader size, microchannel con figuration and nanoparticles on the performance of PV-heat spreader-microchannels system. *Solar Energy, 182*(February), 286–297. https://doi.org/10.1016/j.solener.2019.02.059.

Soliman, Aly M.A., & Hassan, H. (2020). An experimental work on the performance of solar cell cooled by flat heat pipe. *Journal of Thermal Analysis and Calorimetry*. https://doi.org/10.1007/s10973-020-10102-5

Soliman, Aly M. A., & Hassan, H. (2018). 3D study on the performance of cooling technique composed of heat spreader and microchannels for cooling the solar cells. *Energy Conversion and Management, 170*(March), 1–18. https://doi.org/10.1016/j.enconman.2018.05.075.

Soliman, Aly M.A., Hassan, H., & Ookawara, S. (2019). An experimental study of the performance of the solar cell with heat sink cooling system. *Energy Procedia, 162*, 127–135. https://doi.org/10.1016/j.egypro.2019.04.014

Tina, G. M., & Gagliano, S. (2011). Probabilistic modelling of hybrid solar/wind power system with solar tracking system. *Renewable Energy*, *36*(6), 1719–1727. https://doi.org/10.1016/j.renene.2010.12.001.

Tuncel, B., Ozden, T., Balog, R. S., & Akinoglu, B. G. (2020). Dynamic thermal modelling of PV performance and effect of heat capacity on the module temperature. *Case Studies in Thermal Engineering*, *22*(August), 100754. https://doi.org/10.1016/j.csite.2020.100754.

Wang, M., Peng, J., Luo, Y., Shen, Z., & Yang, H. (2021). Comparison of different simplistic prediction models for forecasting PV power output: Assessment with experimental measurements. *Energy*, *224*, 120162. https://doi.org/10.1016/j.energy.2021.120162

Wu, S. Y., Zhang, Q. L., Xiao, L., & Guo, F. H. (2011). A heat pipe photovoltaic/thermal (PV/T) hybrid system and its performance evaluation. *Energy and Buildings*, *43*(12), 3558–3567. https://doi.org/10.1016/j.enbuild.2011.09.017.

Xu, Z., & Kleinstreuer, C. (2014). Concentration photovoltaic – thermal energy co-generation system using nanofluids for cooling and heating. *Energy Conversion and Management*, *87*, 504–512. https://doi.org/10.1016/j.enconman.2014.07.047.

Yang, H., Wei, Z., & Chengzhi, L. (2009). Optimal design and techno-economic analysis of a hybrid solar-wind power generation system. *Applied Energy*, *86*(2), 163–169. https://doi.org/10.1016/j.apenergy.2008.03.008.

2 An Explicit Modal Discontinuous Galerkin Approach to Compressible Multicomponent Flows

Application to Shock–Bubble Interaction

Satyvir Singh
Nanyang Technological University

CONTENTS

NOMENCLATURE

$\mathbf{U}$	Vector of conservative variables
$\mathbf{U}_h$	Approximate solution to $\mathbf{U}$
$\mathbf{F}, \mathbf{G}$	Vectors of convective fluxes
$\hat{\mathbf{F}}, \hat{\mathbf{G}}$	Numerical fluxes for convective fluxes $\mathbf{F}, \mathbf{G}$

DOI: 10.1201/9781003222255-2

$\mathbf{u}$, v	Velocity components in x- and y-directions
p	Pressure
E	Total energy density
C_p, C_v	Specific heat coefficients at constant pressure and volume
N_x, N_y	Number of mesh points in x- and y-directions
I_i, J_j	One-dimensional ith and jth element in x- and y-directions, respectively
$\mathbf{x}$, y	Spatial directions
M_s	Shock Mach number
r	Bubble radius
u_r	Reference velocity
x_c, y_c	Bubble center
T_0, P_0	Initial temperature and pressure
Δt	Local time step
M	Orthogonal mass matrix
$\mathbf{L}$	Residual function
a_s	Sound speed
(x, y)	Physical space
N_k	Number of degrees of freedom
$\mathbb{S}^k(I_i)$	Space of polynomial functions of degree at most k on I_i element
T_{ij}	Two-dimensional local element
$\mathbb{F}^k(T_{ij})$	Space of polynomials of degree at most k on T_{ij}
CFL	Courant–Friedrichs–Lewy number
ρ	Density
ϕ	Mass fraction
γ_{mix}	Specific heat ratio of a mixture
Ω	Full computational domain
Ω_x, Ω_y	Computational domain in x- and y-directions
Ω_e	Physical element
Ω_e^{st}	Standard local element
$[\Im]$	Family of partitions in the physical domain
ω	Vorticity
$\lambda^{\max}$	Maximum wave velocity of convective fluxes
φ_i	ith basis function
(ξ, η)	Computational space
$\psi_i(\xi), \psi_j(\eta)$	Principal functions in the computational space

2.1 INTRODUCTION

The compressible multicomponent flows have garnered a lot of attention in a variety of natural and industrial applications. These applications cover the shock-interface interactions (Brouillette, 2002), evolution of heart valves and pumps in cavitating flows (Brennen, 2015), reducing injury because of blast trauma (Laksari et al., 2015), shock-burst wave lithotripsy treatments (Pishchalnikov, 2003), underwater explosions (Etter, 2013), hydraulic machinery (Streeter, 1983), separation of jets and liquid droplets (Meng and Colonius, 2018), airplane surface erosion from supersonic light

(Joseph et al., 1999), shock wave attenuation of nuclear blasts (Chauvin et al., 2016), needle-free medication delivery (Tagawa et al., 2013), and many others.

The shock–bubble interaction has long been an interesting subject for describing the turbulence generation and mixing process in compressible multicomponent flows. When an incident shock wave collides with a bubble interface at various material properties, the flow field is dominated by the coupling between them, resulting in a complicated flow pattern. Then, a disturbance is created that moves along with the shock waves, resulting in a substantial deformation of the bubble shape and subsequent turbulent mixing along the gas interface due to a mismatch of acoustic impedances inside and outside the bubble. In this phenomenon, the gas bubble is compressed and accelerated at first and then stretched by creating vortices at the interface of two fluids with different densities, resulting in the Richtmyer–Meshkov (RM) instability (Richtmyer, 1960; Meshkov, 1969). The shock–bubble interaction is one of the closest models to the RM instability. Numerous studies have been carried out over few decades to illustrate the physical features of shock–bubble interaction experimentally (Haas and Sturtevant, 1987; Jacobs, 1992; Jacobs, 1993; Ranjan et al., 2007; Layes et al., 2009), theoretically as well as numerically (Quirk and Karni, 1996; Zabusky and Zeng, 1998; Bagabir and Drikakis, 2001; Niederhaus et al., 2008; Singh, 2020; Singh and Battiato, 2021; Singh et al., 2021; Singh, 2021a, b).

The numerical simulations of compressible multicomponent flows require a computational method that retains numerical stability while maintaining discrete conservation, suppressing oscillations near discontinuities. For this purpose, we require an efficient, accurate, and higher-order numerical algorithm that is away from discontinuities. There has been a growing interest in modeling the higher-order numerical algorithm for the compressible multicomponent flows during the last few decades (Coquel et al., 1997; Shyue, 1998; Abgrall and Karni, 2001; Coralic and Colonius, 2014). Generally, single-fluid algorithms frequently fail in multicomponent flows due to nonphysical oscillations at the interface separating the various components. Shankar et al. (2011) presented the numerical simulations for the compressible multicomponent viscous flows with an ideal non-reactive gas mixture based on a sixth-order compact differencing algorithm.

The discontinuous Galerkin (DG) approach to solving hyperbolic conservation laws has been increasingly popular in recent decades. Initially, Reed and Hill (1973) proposed the DG method to solve the unsteady neutron transport equation. Afterward, several DG algorithms were proposed and developed numerically for solving nonlinear hyperbolic system by Cockburn and Shu (1989, 1998, 2001). The DG approaches integrate the advantages of finite element and finite volume methods and have effectively been applied to a broad spectrum of scientific problems, including computational fluid dynamics, plasma physics, and quantum physics (Le et al., 2014; Raj et al., 2017; Singh and Myong, 2017; Singh, 2018; Chourushi et al., 2020; Singh and Battiato, 2020; 2021; Singh et al., 2021; Singh, 2021a, b, c). DG methods have a number of essential features that make them appealing for use in applications. These features include their ability to readily address complicated geometry and boundary conditions, their flexibility for easy hp-adaptivity, their ability to deal with nonconforming elements having hanging nodes, and efficient parallel implementation along with time-stepping algorithms.

In this chapter, an explicit modal DG scheme with third-order accuracy is developed for solving two-dimensional compressible multicomponent inviscid flows. The scheme employs a spatial DG formulation that converts the given conservation laws to a system of semi-discrete ODEs. After that, the resulting semi-discrete ODEs are solved numerically with an explicit strong stability-preserving Runge–Kutta (SSP-RK33) method. For validation purpose, the obtained numerical results are compared with the available experimental results for shock–bubble interaction problem. Furthermore, some numerical results based on flow field visualization and vorticity generation of shock–bubble interaction are presented for both light and heavy bubbles.

2.2 COMPUTATIONAL MODELING FOR MULTICOMPONENT COMPRESSIBLE FLOWS

2.2.1 Governing Equations

In this work, the multicomponent flows are simulated with the compressible Euler equations for a gas mixture and the effects of viscosity, surface tension, gravity, and chemical reaction are assumed to be negligible here (Shankar et al., 2011). In a two-dimensional space, the multicomponent compressible Euler equations are given in conservative form as

$$\frac{\partial \mathbf{U}}{\partial t}+\frac{\partial}{\partial x}\mathbf{F}(\mathbf{U})+\frac{\partial}{\partial y}\mathbf{G}(\mathbf{U})=0, \tag{2.1}$$

where

$$\mathbf{U}=\begin{bmatrix}\rho\\ \rho u\\ \rho v\\ \rho E\\ \rho\phi\end{bmatrix},\quad \mathbf{F}(\mathbf{U})=\begin{bmatrix}\rho u\\ \rho u^2+p\\ \rho uv\\ (\rho E+p)u\\ \rho u\phi\end{bmatrix},\quad \mathbf{G}(\mathbf{U})=\begin{bmatrix}\rho v\\ \rho uv\\ \rho v^2+p\\ (\rho E+p)v\\ \rho v\phi\end{bmatrix}.$$

Here, ρ is the mass density and u and v are the velocity components in x- and y-directions. E is the total energy density, ϕ is the mass fraction, and p is the static pressure determined by the ideal gas law as

$$p=(\gamma_{mix}-1)\left(\rho E-\frac{1}{2}\left(u^2+v^2\right)\right), \tag{2.2}$$

where γ_{mix} represents the specific heat ratio of a gas mixture, which is evaluated from the following mathematical expression (Marquina and Mulet, 2003) as

$$\gamma_{mix}=\frac{C_{p1}\phi+C_{p2}(1-\phi)}{C_{v1}\phi+C_{v2}(1-\phi)} \tag{2.3}$$

In this expression, subscripts 1 and 2 are meant for bubble and ambient gas, respectively. C_p and C_v represent the specific heat coefficients at constant pressure and volume, respectively.

2.2.2 Explicit Modal Discontinuous Galerkin Approach

In this work, a two-dimensional system of multicomponent compressible Euler equation (2.1) is computed by an explicit modal discontinuous Galerkin approach based on rectangular elements. For the discretization of equation (2.1), a rectangular element based on the tensor product of grids in x- and y-directions is utilized. Let $[\mathfrak{I}]$ be a family of partitions of the physical domain $\Omega = \Omega_x \times \Omega_y$, which is divided into $N_x \times N_y$ uniform rectangular elements as

$$\begin{aligned} &x_L = x_{1/2} < x_{3/2} < \cdots\cdots < x_{i-1/2} < x_{i+1/2} < \cdots\cdots < x_{N+1/2} = x_R, \\ &y_B = y_{1/2} < y_{3/2} < \cdots\cdots < y_{j-1/2} < y_{j+1/2} < \cdots\cdots < y_{N+1/2} = y_U. \end{aligned} \tag{2.4}$$

A two-dimensional Cartesian mesh $\mathfrak{I}$ is defined as

$$\begin{aligned} &\mathfrak{I} := \{T_{ij} = I_i \times J_j, 1 \le i \le N_x, 1 \le j \le N_y\}, \\ &I_i := [x_{i-1/2}, x_{i+1/2}], \quad \forall i = 1 \cdots\cdots N_x; \\ &J_j := [y_{j-1/2}, y_{j+1/2}], \quad \forall j = 1 \cdots\cdots N_y, \end{aligned} \tag{2.5}$$

Here, $(x_i, y_j) = \left(\frac{1}{2}(x_{i-1/2} + x_{i+1/2}), \frac{1}{2}(y_{j-1/2} + y_{j+1/2})\right)$ are the center points of the elements. For the Ω_x – domain, the piecewise polynomial space of the functions $\xi : \Omega_x \to \Re$ is defined as

$$Z_h^k = \{\xi : \xi|_{\Omega_x} \in \mathbb{S}^k(I_i), \; i = 1 \cdots\cdots N_x\}, \tag{2.6}$$

Here, $\mathbb{S}^k(I_i)$ is the space of polynomial functions of degree at most k on I_i. For the domain $\Omega = \Omega_x \times \Omega_y$, it is defined $\varphi : \Omega \to \Re$ as

$$V_h^k = \{\varphi : \varphi|_{\Omega} \in \mathbb{F}^k(T_{ij}), \; 1 \le i \le N_x, 1 \le j \le N_y\}, \tag{2.7}$$

where $\mathbb{F}^k(T_{ij}) = \mathbb{S}^k(I_i) \otimes \mathbb{S}^k(J_j)$ denotes the space of polynomials of degree at most k on T_{ij}. The number of degrees of freedom of $\mathbb{F}^k(T_{ij})$ can be calculated by $N_k = (k+1)(k+2)/2$. After that, the exact solution of $\mathbf{U}$ is approximated by the DG polynomial approximation of $\mathbf{U}_h \in V_h^k$ as

$$\mathbf{U}_h(x,y,t) = \sum_{i=0}^{N_k} \hat{U}_h^i(t) \varphi_i(x,y), \quad \forall (x,y) \in T_{ij}, \tag{2.8}$$

where $\hat{U}_h^i$ denotes the unknown coefficients for $\mathbf{U}$ to be enhanced with time and $\varphi_i(x,y)$ is the basis function. In this study, the two-dimensional scaled Legendre basis functions are constructed as a tensor product of the so-called principal functions defined as (Singh, 2018)

$$\varphi_k(\xi,\eta)=\psi_i(\xi)\otimes\psi_j(\eta), \tag{2.9}$$

with

$$\psi_i(\xi)=\frac{2^i(i!)^2}{(2i)!}P^{0,0}(\xi),\ -1\leq\xi\leq 1,$$

$$\psi_j(\eta)=\frac{2^i(j!)^2}{(2j)!}P^{0,0}(\eta),\ -1\leq\eta\leq 1,$$

where $P^{0,0}(\xi)$ is the Legendre polynomial function. A standard rectangular element Ω_e^{st} is defined using a local Cartesian coordinate system $(\xi,\eta)\in[-1,1]$, which is illustrated in Figure 2.1. The standard element can be mapped from the computational space (ξ,η) to an arbitrary rectangular element in the physical space (x,y) under the linear transformation $T:\Omega_e^{st}\rightarrow\Omega_e$ defined by

$$\begin{aligned} x&=\frac{1}{2}\left[(1-\xi)x_1+(1+\xi)x_2\right],\\ y&=\frac{1}{2}\left[(1-\eta)y_1+(1+\eta)y_2\right], \end{aligned} \tag{2.10}$$

where $(x_i,y_i:i=1,4)$ are the physical coordinates of the vertices of Ω_e.

Then, the DG discretization of the conservation laws in (2.1) is obtained by replacing the exact solutions with the corresponding approximation defined in equation (2.8)

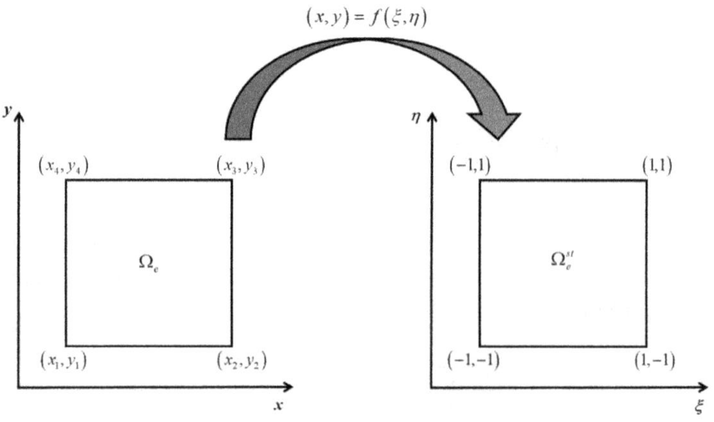

FIGURE 2.1 Elemental transformation from physical element to reference element in a 2D computational domain.

and multiplying by a test function $\varphi_h(x,y)$ and then integrating by parts over the element $T_{ij} \in \Im$. Taking $\varphi_h \in V_h^k(\Im)$ and $\mathbf{U}_h \in V_h^k(\Im)$, we obtain

$$\frac{\partial}{\partial t}\int_{\Omega_e} \mathbf{U}_h\varphi_h\, dV - \int_{\Omega_e} \nabla\varphi_h \cdot \mathbf{F}(\mathbf{U}_h) dV + \int_{\partial\Omega_e} \varphi_h\, \mathbf{F}(\mathbf{U}_h)\cdot \mathbf{n}\, d\Gamma$$
$$- \int_{\Omega_e} \nabla\varphi_h \cdot \mathbf{G}(\mathbf{U}_h) dV + \int_{\partial\Omega_e} \varphi_h\, \mathbf{G}(\mathbf{U}_h)\cdot \mathbf{n}\, d\Gamma = 0, \tag{2.11}$$

where $\mathbf{n}$ denotes the outward unit normal vector and V and Γ are the volume and boundary of the element Ω_e, respectively. The interface fluxes are not uniquely defined because of the discontinuity in the solution $\mathbf{U}_h$ at the elemental interfaces. The functions $\mathbf{F}\cdot\mathbf{n}$ and $\mathbf{G}\cdot\mathbf{n}$ appearing in equation (2.11) are substituted by a single-valued function defined at the elemental interfaces; the so-called numerical fluxes are denoted by $\hat{\mathbf{F}}$ and $\hat{\mathbf{G}}$, respectively. For the convective fluxes, the local Lax–Friedrichs flux is used as

$$\mathbf{F}(\mathbf{U}_h)\cdot\mathbf{n} = \hat{\mathbf{F}}\left(\mathbf{U}_h^-, \mathbf{U}_h^+\right) = \frac{1}{2}\left[\mathbf{F}\left(\mathbf{U}_h^-\right) + \mathbf{F}\left(\mathbf{U}_h^+\right)\right] - \frac{1}{2}C\left(\mathbf{U}_h^- + \mathbf{U}_h^+\right),$$
$$\mathbf{G}(\mathbf{U}_h)\cdot\mathbf{n} = \hat{\mathbf{G}}\left(\mathbf{U}_h^-, \mathbf{U}_h^+\right) = \frac{1}{2}\left[\mathbf{G}\left(\mathbf{U}_h^-\right) + \mathbf{G}\left(\mathbf{U}_h^+\right)\right] - \frac{1}{2}C\left(\mathbf{U}_h^- + \mathbf{U}_h^+\right), \tag{2.12}$$
$$C = \max\left(\left|\mathbf{U}_h^-\right| + a_s^-, \left|\mathbf{U}_h^+\right| + a_s^+\right),$$

where $a_s^\pm = \sqrt{\gamma R T^\pm}$ illustrates the sound speed at the elemental interface, while the superscripts $(-)$ and $(+)$ are the left and right states of the interface. As a result, the DG weak formulation is obtained as

$$\frac{\partial}{\partial t}\int_{\Omega_e} \mathbf{U}_h\varphi_h\, dV - \int_{\Omega_e} \nabla\varphi_h \cdot \mathbf{F}(\mathbf{U}_h) dV + \int_{\partial\Omega_e} \varphi_h\, \hat{\mathbf{F}}\, d\Gamma$$
$$- \int_{\Omega_e} \nabla\varphi_h \cdot \mathbf{G}(\mathbf{U}_h) dV + \int_{\partial\Omega_e} \varphi_h\, \hat{\mathbf{G}}\, d\Gamma = 0. \tag{2.13}$$

In expression (2.13), the emerging volume and surface integrals are approximated by the Gaussian–Legendre quadrature rule within the elements to ensure the high-order accuracy (Cockburn and Shu, 2001). Furthermore, the nonlinear total variation bounded limiter (Cockburn and Shu, 1998) is utilized to remove the artificial oscillations arising in the numerical solution.

Finally, the DG spatial discretization (2.13) can be expressed in semi-discrete ODE form as

$$\mathbf{M}\frac{d\mathbf{U}_h}{dt} = \mathbf{L}(\mathbf{U}_h), \tag{2.14}$$

where $\mathbf{M}$ and $\mathbf{L}(\mathbf{U}_h)$ are the orthogonal mass matrix and the residual function, respectively. Here, an explicit form of strong stability-preserving Runge–Kutta method with third-order accuracy (Shu and Osher, 1988) is adopted as the time-marching scheme.

$$\begin{aligned} \mathbf{U}_h^{(1)} &= \mathbf{U}_h^n + \Delta t\,\mathbf{M}^{-1}\mathbf{L}(\mathbf{U}_h), \\ \mathbf{U}_h^{(2)} &= \frac{3}{4}\mathbf{U}_h^n + \frac{1}{4}\mathbf{U}_h^{(1)} + \frac{1}{4}\Delta t\,\mathbf{M}^{-1}\mathbf{L}\left(\mathbf{U}_h^{(1)}\right), \\ \mathbf{U}_h^{n+1} &= \frac{1}{3}\mathbf{U}_h^n + \frac{2}{3}\mathbf{U}_h^{(2)} + \frac{2}{3}\Delta t\,\mathbf{M}^{-1}\mathbf{L}\left(\mathbf{U}_h^{(2)}\right). \end{aligned} \tag{2.15}$$

The local time step Δt is evaluated by the relationship

$$\Delta t = \frac{CFL}{(2k+1)} \frac{\min(\Delta x, \Delta y)}{\left|\lambda^{\max}\right|}, \tag{2.16}$$

where CFL is the Courant–Friedrichs–Lewy number and $\lambda^{\max}$ is the maximum wave velocity of convective fluxes, respectively.

2.3 PROBLEM SETUP FOR SHOCK–BUBBLE INTERACTION

A schematic diagram of the computational model is shown in Figure 2.2 to describe the flow physics originating in the shock–bubble interaction. In this model, an incident shock wave with the strength $M_s = 1.21$ transports from left to right direction and interacts with a stationary cylindrical bubble of radius $r = 25$ m, which is surrounded by the ambient gas. The computational domain is considered as a rectangular with $[0,220]\,\text{m} \times [0,90]\,\text{m}$. The primary location of the incident shock wave is fixed at $x = 30\,\text{m}$, while the location of the bubble center is established at $(x_c, y_c) = (65,45)\,\text{m}$. The primary temperature and pressure are examined as T_0= 273 K and P_0= 101,325 Pa, respectively, around the bubble. In the hydrodynamic instability community, helium (He) and sulfur hexafluoride (SF_6) gases are often used as light and heavy gases. As a result, this research also considers He and SF_6 as light and heavy gases, respectively, while nitrogen (N_2) gas is used for ambient state. Table 2.1 summarizes the relevant physical properties of these used gases. As for the

TABLE 2.1
Physical Gas Properties for the Numerical Simulations

Gas	Specific Heat Ratio (γ)	Density (ρ)	Specific Heat Coefficient at Constant Pressure (C_p)	Specific Heat Coefficient at Constant Volume (C_v)
Helium (He)	1.67	0.16	520	320
Sulfur hexafluoride (SF_6)	1.10	5.97	97	106.7
Nitrogen (N_2)	1.40	1.25	1040	743

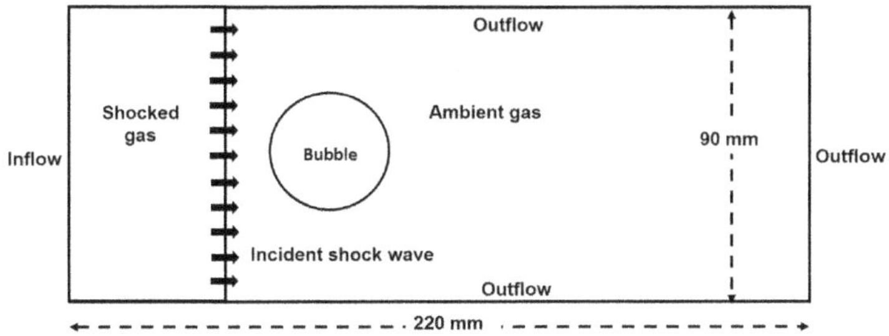

FIGURE 2.2 Schematic representation of the computational model.

computational specifications, the boundary conditions were set to inlet at the left (see Figure 2.2), of the computational domain, respectively, while the upper, bottom, and right parts of the domain are treated as outflow condition.

The standard Rankine–Hugoniot conditions are used for initializing the flow field in the computational simulations, which are given as

$$M_2^2 = \frac{1 + \frac{(\gamma - 1)}{2} M_s^2}{\gamma M_s^2 - \frac{(\gamma - 1)}{2}}, \quad \frac{p_2}{p_1} = \frac{1 + \gamma M_s^2}{1 + \gamma M_2^2}; \quad \frac{\rho_2}{\rho_1} = \frac{\gamma - 1 + (\gamma + 1)\frac{p_2}{p_1}}{\gamma + 1 + (\gamma - 1)\frac{p_2}{p_1}}. \tag{2.17}$$

In the above expressions, subscripts 1 and 2 illustrate the left- and right-hand sides of the incident shock wave, respectively. In order to construct the non-dimensionalization governing equation, the bubble radius is considered as the reference length $(L = r)$, while the reference velocity (u_r) is chosen to represent the sound velocity prior to the shock wave. The non-dimensional equation of state can be written as $\rho_r = p_r / T_r$, where $p_r = 101{,}325$ Pa and $T_r = 273$ K are considered for the reference pressure and temperature, respectively.

2.4 NUMERICAL RESULTS AND DISCUSSION

2.4.1 Grid Refinement and Validation Study

In this section, a grid refinement study is presented to show that the grid used is adequate for accurately capturing the complicated structure of the flow field and the evolution process of the interface. For this purpose, two different test problems are considered on a shock-accelerated light and heavy bubbles surrounded by nitrogen gas with $M_s = 1.21$. The domain is discretized into six different uniform meshes, which are labeled from "Mesh 1" to "Mesh 6" corresponding to the mesh points 100×50, 200×100, 400×200, 800×400, $1{,}000 \times 500$, and $1{,}200 \times 600$, respectively. These six different grids are utilized to compare the effects of different mesh resolutions on the computed results. Figure 2.3 shows the profiles for the density distribution profiles, which are extracted along the centerline of the computed light and heavy bubbles. The results show that the dissipations of the density are lower with high resolution. Therefore, in the present subsequent computations, the "Mesh 6" with $1{,}200 \times 600$ cells are endorsed to ensure numerical accuracy.

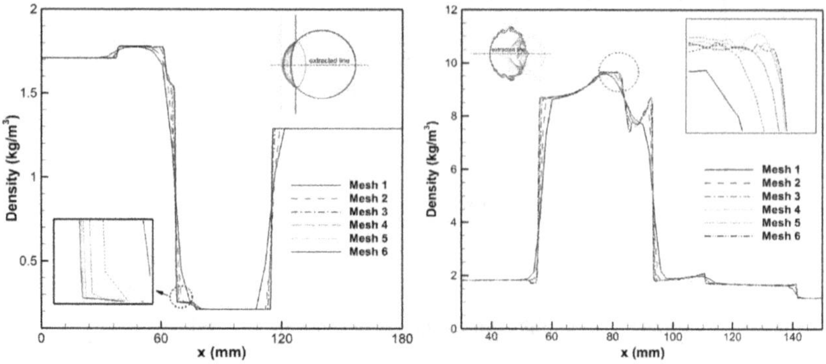

FIGURE 2.3 Density distribution profiles for different mesh sizes of the computed (a) light gas and (b) heavy gas bubbles.

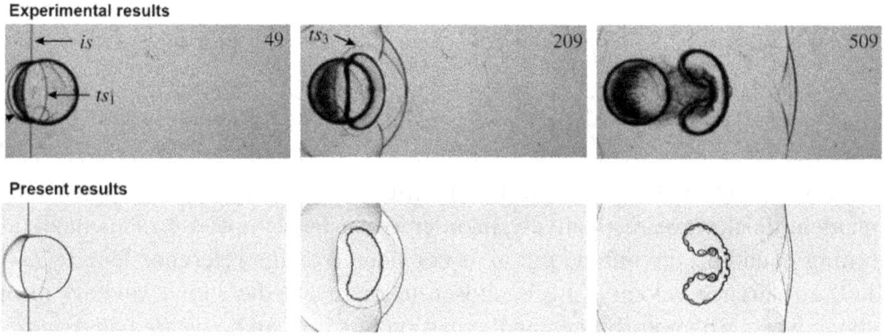

FIGURE 2.4 Comparison of numerical schlieren images between the experimental results [reproduced with permission from Ding et al. "On the interaction of a planar shock with a three-dimensional light gas cylinder," J. Fluid Mech. 828, 289 (2017). Copyright 2017 Cambridge University Press] (Ding et al., 2017), and the present numerical results for a two-dimensional N_2 cylinder surrounded by SF_6 with $M_s = 1.29$ at different time instants.

To demonstrate the reliability of the present numerical method and model, computational results of the interaction between the incident shock wave and the cylindrical light as well as heavy gas bubbles are contrasted with the experimental results of Ding et al. (2017), and Hass and Sturtevant (1987). Figure 2.4 illustrates the comparisons of the schlieren images between the experimental result of Ding et al. (2017) and the numerical simulation results during the interaction between the incident shock wave of $M_s = 1.29$ and a two-dimensional N_2 cylinder surrounded by SF_6. In Figure 2.5, the computational results are also compared with the experimental results of Hass and Sturtevant (1987) at $M_s = 1.22$, where the gas cylindrical bubble is filled with Refrigerant-22 (R_{22}) surrounded by air. As seen from Figures 2.4 and 2.5, the obtained results are very close to the experimental results, indicating that the present computational method is able to capture the flow fields in the shock–bubble interaction accurately. In addition, Figure 2.6 illustrates the spacetime diagram for the

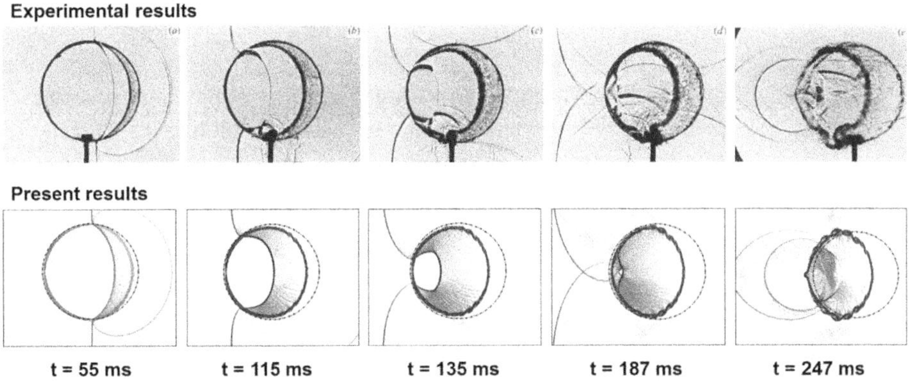

FIGURE 2.5 Comparison of numerical schlieren images between the experimental results [reproduced with permission from Haas and Sturtevant "Interaction of weak shock waves with cylindrical and spherical gas inhomogeneities," J. Fluid Mech. 181, 41 (1987). Copyright 1987 Cambridge University Press] (Hass and Sturtevant, 1987), and the present numerical results for a shock-accelerated R_{22} cylindrical bubble surrounded by air at different time instants.

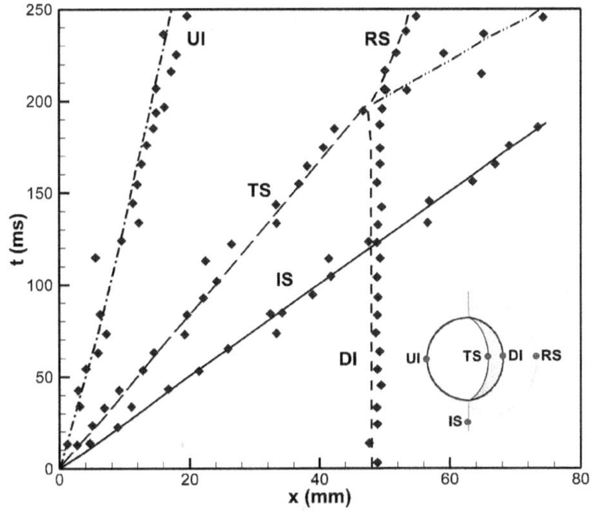

FIGURE 2.6 Comparison of computed characteristic interface points (IS, UI, DI, TS, and RS) between the experimental results (Hass and Sturtevant, 1987) and the present numerical results for a shock-accelerated R_{22} cylindrical bubble surrounded by air at different time instants. The diamond symbol represents the experimental data, whereas the solid line represents the current computational results.

various characteristic interface points such as incident shock (IS), reflected shock (RS), transmitted shock (TS), upstream interface (UI), and downstream interface (DI). These computational and the experimental data (Hass and Sturtevant, 1987) are found to be in a close agreement.

2.4.2 Visualization of Flow Fields

At successive time instants, Figure 2.7 depicts a sequence of density contours for a cylindrical light gas bubble surrounded by nitrogen gas driven by a planar IS wave. The initial state of the bubble interface may be seen before interacting with the IS wave $(t = 0)$. The bubble begins to compress as the IS wave approaches the interface. A transmitted shock wave (TS1) propagates downstream inside the bubble, while a rarefaction shock wave (RS1) propagates upstream $(t = 1 - 1.5)$. Because of the low acoustic impedance, the propagation speed of the IS wave inside the bubble is slower than that of the surrounding gas. As a result, the created TS1 inside the bubble follows the IS wave faster. The IS and TS1 waves generate a triple shock in the gas as the interaction progresses, revealing uneven refraction. At $t = 2$, a Mach reflection configuration with a Mach stem (MS), a triple point (TP), and a slip surface (ss) is thus generated. Later on, near the downstream contact, a secondary transmitted shock wave (TS2) is seen $(t = 2.5)$. A second rarefaction shock (RS2) is generated in the bubble upstream when the TS1 wave passes through the center section of the bubble. This is caused by the distorted interface caused by the differential in acoustic impedance between the interior and exterior gases $(t = 3.5)$. As time passes, another rarefaction shock (RS3) wave within the bubble rises and collides with the upstream interface, resulting in a third transmitted shock wave (TS3) at $t = 4.5$. At the same time, the flow fields grow more convoluted at the downstream interface, and a fourth transmitted shock (TS4) wave emerges $(t = 5)$. The bubble gradually deforms during the encounter. The initial density inhomogeneity is accelerated, and the upstream bubble contact is flat at $t = 3 - 5$. The developing bubble interface then begins to take on the shape of a mushroom $t = 5$, and after that, a re-entrant gas jet head is created near the bubble's center at $t = 6$. The jet eventually catches up to the downstream

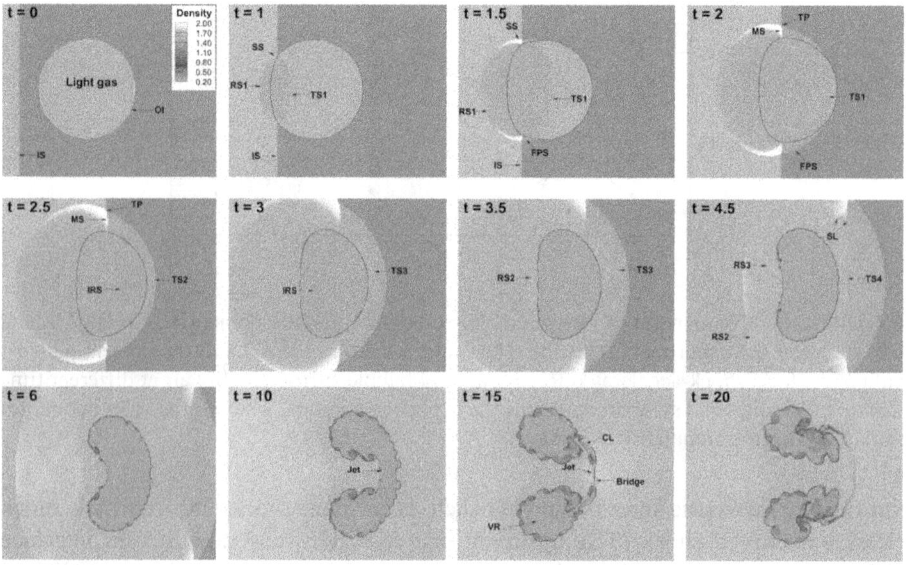

FIGURE 2.7 Time evolution of the flow fields in the shock–helium bubble interaction: density contours at different time instants.

bubble interface, and a pair of vortex rings (VR) joined by a bridge forms, growing almost symmetrically at $t = 10 - 20$. Eventually, the flow field is completely controlled by the VR. The VR eventually has complete control over the flow field.

Figure 2.8 shows the density contours of a cylindrical SF_6 bubble surrounded by nitrogen gas that has been accelerated by a planar incident shock wave at various time intervals. Due to the higher acoustic impedance, the incoming shock wave propagates slower inside the cylindrical heavy bubble than in the surrounding medium. Because of the difference in acoustic impedance, a curved transmitted shock (TS1) wave is observed inside the cylindrical bubble when the IS wave travels through it. The initial state of the cylindrical bubble interface can be seen clearly before interaction with the IS wave $(t = 0)$. When the IS wave hits the bubble interface, the bubble compresses inward, causing the TS1 wave to propagate downstream inside the bubble and a reflected shock (RS) wave to propagate upstream at the same time $(t = 2)$. During this time, an unshocked zone (UZ) develops within the bubble. The TS1 wave is somewhat converging at first and can be seen clearly inside the compressed volume, which produces a crescent shape $(t = 3 - 5)$. The curvature of the TS1 wave increases when the IS wave passes through the top point of the bubble interface. Outside the bubble, a curved diffracted shock (DS) wave is attached to the top and bottom ends of the two straight sections of the IS wave. The curved TS1 constantly pushes the outside bubble's IS wave back. Further outward from the bubble, the ends of the DS wave that were coupled with the IS wave become severely deformed. When the TS1 wave crosses the bubble, the disturbed interface behind it is thicker than the undisturbed interface in front. This is due to the roll-up of small-scale vortices caused by the baroclinic vorticity phenomenon, i.e., the misalignment of the local pressure and density gradients.

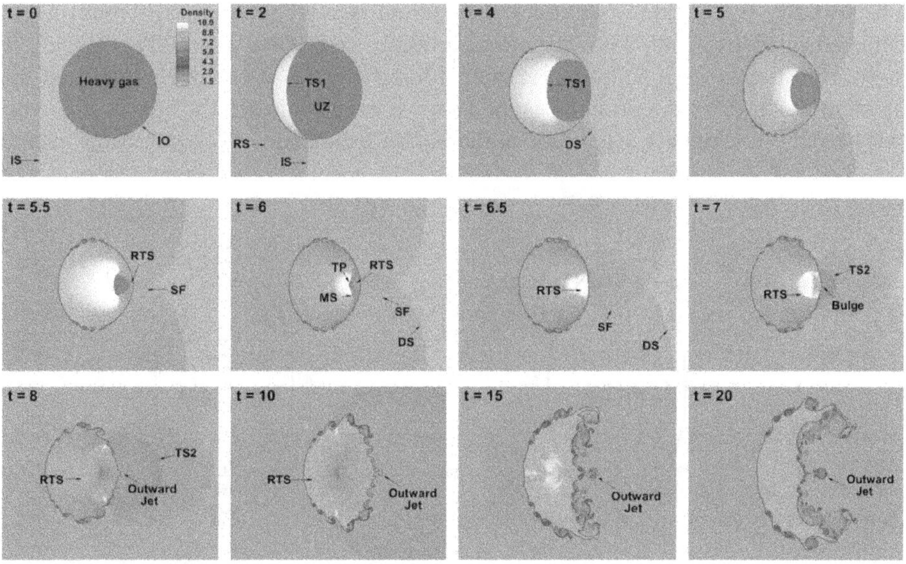

FIGURE 2.8 Time evolution of the flow fields in the shock–SF_6 bubble interaction: density contours at different time instants.

2.4.3 Dynamics of Vorticity Production

The vorticity created by the misalignment of the density and pressure gradients in the process of shock–bubble interaction is critical to understanding the flow mechanism phenomena. For inviscid flows, the vorticity transport equation can be expressed as follows:

$$\frac{D\omega}{Dt} = (\omega \cdot \nabla)\boldsymbol{u} - \omega(\nabla \cdot \boldsymbol{u}) + \frac{1}{\rho^2}(\nabla\rho \times \nabla p), \tag{2.18}$$

where $\omega (= \nabla \times \boldsymbol{u})$ and ρ, **u**, and p represent the vorticity, density, velocity, and pressure, respectively. In equation (2.18), the term $(\omega \cdot \nabla)\boldsymbol{u}$ illustrates the vortex stretching of vorticity because of velocity gradients, which is important for 3D problems. The next term $\omega(\nabla \cdot \boldsymbol{u})$ represents the fluid convection and thermal expansion due to flow compressibility. Finally, the baroclinic vorticity $(\nabla\rho \times \nabla p) / \rho^2$ is depicted in the last term, which accounts for variations in vorticity caused by the intersection of density and pressure surfaces. Furthermore, due to the high misalignment of density and pressure gradients, this term is most noticeable at the top and bottom ends of a vertical bubble diameter.

Figure 2.9 shows the schematic representation of the vorticity production at the interfaces of light and heavy gas bubbles after initial IS wave transits. When the IS wave comes into contact with the upper and lower interfaces, where the pressure and density gradients are exactly aligned, a little amount of vorticity is formed. Mach reflection occurs as the IS wave propagates along the upper and lower interfaces, with the Mach stem connecting the IS wave to the interfaces. As the IS wave moves upward through the interfaces, the MS supplies the pressure gradient for vorticity generation, and the baroclinic vorticity term is therefore triggered gradually. The vorticity production of cylindrical light and heavy bubbles at different time is depicted in Figure 2.10. The vorticity is initially zero everywhere. The baroclinic vorticity is deposited locally on the bubble interface, where the bubble gas and the ambient gas have a discontinuity, as the IS wave travels through the bubble. The vorticity reaches its highest magnitude at the bubble's top and bottom interfaces, where

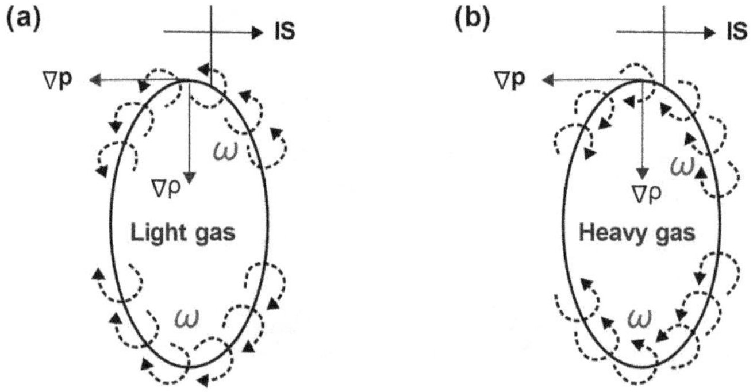

FIGURE 2.9 Schematic diagram of the vorticity generation on the interface of a cylindrical bubble after initial shock wave transit: (a) light gas and (b) heavy gas bubbles.

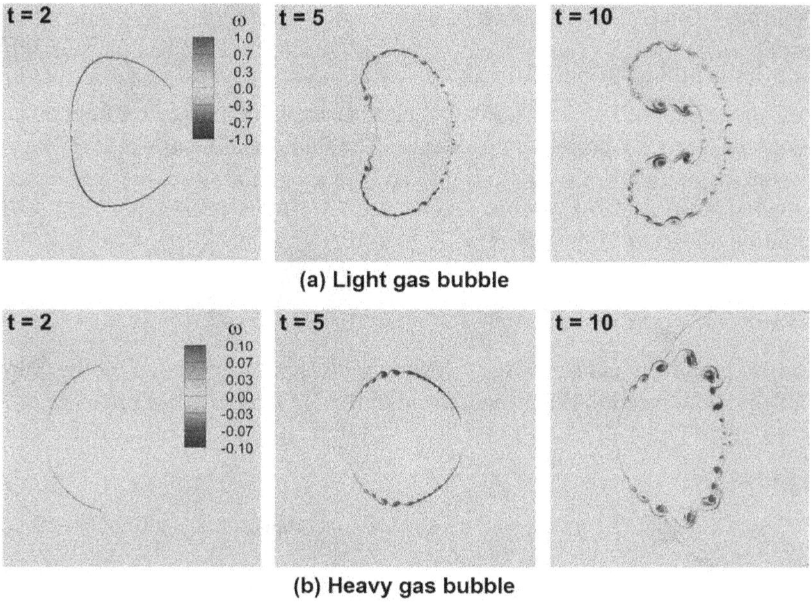

FIGURE 2.10 Contours of vorticity distribution in the shock–bubble interaction at $t = 10$: (a) light gas and (b) heavy gas bubbles.

the density and pressure gradients are orthogonal. In the portions of the interface where the density and pressure gradients are collinear, the vorticity is zero. The upper half of the light (heavy) bubble interface generates a considerable amount of positive (negative) vorticity, while the lower half of the light (heavy) bubble interface generates a significant amount of negative (positive) vorticity. This is due to the fact that the IS wave propagates along the bubble interface from left to right. Thus, the density gradient extends radially outward along the bubble interface, whereas the pressure gradient operates across the upstream IS wave. The transmitted shock wave generates a minor amount of vorticity in the filaments linking the rolled up vortices after passing through the bubble. On the upper half-plane (lower half-plane) of the generated inward (outward) jet head, some negative (positive) vorticity is localized. Following the interaction, there are significant gaps in the vorticity distribution for the light and heavy gas bubbles.

2.5 CONCLUDING REMARKS

An explicit modal discontinuous Galerkin method based on rectangular meshes is developed to solve the compressible multicomponent gas flows. A two-dimensional system of compressible equations for a gas mixture is considered for the multicomponent gas flows. The scaled Legendre polynomials with third-order accuracy are used for spatial discretization, while an explicit strong stability-preserving Runge–Kutta scheme with third-order accuracy is adopted for the temporal discretization. Numerical experiments are carried out for the shock–bubble interaction to validate the present numerical method. The obtained numerical results are compared with the available experimental results. A close agreement is observed between the numerical and experimental results, which show that the present method has the capability to capture sharp discontinuities. Finally, some numerical results of the shock–bubble interaction with both light and heavy bubbles are explained based on flow field visualization and vorticity production in detail. The numerical results reveal that in the interaction of the shock wave with light and heavy bubbles, flow morphologies with complicated wave patterns, vortex development, vorticity production, and interface deformation are completely different.

ACKNOWLEDGMENTS

The author gratefully acknowledges the financial support provided by the Nanyang Technological University, Singapore, through the NAP-SUG grant program.

REFERENCES

Abgrall, R., & Karni, S. (2001). Computations of compressible multifluids. *Journal of Computational Physics, 169*, 594–623.

Bagabir, A., & Drikakis, D. (2001). Mach number effects on shock-bubble interaction. *Shock Waves, 11*, 209–218

Brennen, C.E. (2015). Cavitation in medicine. *Interface Focus, 5*(5), 20150022.

Brouillette, M. (2002). The Richtmyer-Meshkov instability. *Annual Review of Fluid Mechanics, 34*, 445–468.

Chauvin, A., Daniel, E., Chinnayya, A. Massoni, J., & Jourdan, G. (2016). Shock waves in sprays: numerical study of secondary atomization and experimental comparison. *Shock waves, 26*(4), 403–415.

Chourushi, T., Rahimi, A., Singh, S., & Myong, R.S. (2020). Computational simulations of near-continuum gas flow using Navier–Stokes-Fourier equations with slip and jump conditions based on the modal discontinuous Galerkin method. *Advances in Aerodynamics, 2*, 8.

Cockburn, B., & Shu, C.-W. (1989). TVB Runge-Kutta local projection discontinuous Galerkin finite element method for conservation laws II, General framework. *Mathematics of computation, 52*, 411–435.

Cockburn, B., & Shu, C.-W. (1989). TVB Runge-Kutta local projection discontinuous Galerkin finite element method for conservation laws III: one-dimensional systems. *Journal of Computational Physics, 84*, 90–113

Cockburn, B., & Shu, C.-W. (1998). The Runge-Kutta discontinuous Galerkin method for conservation laws V: Multidimensional systems. *Journal of Computational Physics, 141*(2), 199–224.

Cockburn, B., & Shu, C.-W. (2001). Runge–Kutta discontinuous Galerkin methods for convection-dominated problems. *Journal of Scientific Computing, 16*(3), 173–261.

Coralic, V., & Colonius, T. (2014). Finite-volume WENO scheme for viscous compressible multicomponent flows. *Journal of Computational Physics, 274*, 95–121.

Coquel, F., Almine, K.E., Godlewski, E., Perthame, B., & Rascle, P. (1997). A numerical method using upwind schemes for the resolution of two-phase flows. *Journal of Computational Physics, 136*, 272–288.

Ding, J., Si, T., Chen, M., Zhai, Z., Lu, X., & Luo, X. (2017). On the interaction of a planar shock with a three-dimensional light gas cylinder. *Journal of Fluid Mechanics, 828*, 289.

Etter, P.C. (2013). *Underwater Acoustic Modeling and Simulation*, CRC Press.

Haas J.F., & Sturtevant, B. (1987). Interaction of weak shock waves with cylindrical and spherical gas inhomogeneities. *Journal of Fluid Mechanics, 181*, 41.

Jacobs, J.W. (1992). Shock-induced mixing of a light-gas cylinder. *Journal of Fluid Mechanics, 234*, 629–649.

Jacobs, J.W. (1993). Shock-induced mixing of a light-gas cylinder. *Physics of Fluids, 5*, 2239–2247.

Joseph, D.D., Belanger, J., & Beavers, G.S. (1999). Breakup of a liquid drop suddenly exposed to a high-speed airstream. *International Journal of Multiphase Flow, 25*(6–7), 1263–1303.

Laksari, K., Assari, S., Seibold, B., Sadeghipour, K., & Darvish, K.B. (2015). Computational simulation of the mechanical response of brain tissue under blast loading. *Biomechanics and Modeling in Mechanobiology, 14*(3), 459–472.

Layes, G., Jourdan, G., & Houas, L. (2009). Experimental study on a plane shock wave accelerating a gas bubble. *Physics of Fluids, 21*, 074102.

Le, N.T.P., Xiao, H., & Myong, R.S. (2014). A triangular discontinuous Galerkin method for non-Newtonian implicit constitutive models of rarefied and microscale gases. *Journal of Computational Physics, 273*, 160–184.

Meng, J.C., & Colonius, T. (2018). Numerical simulation of the aerobreakup of a water droplet. *Journal of Fluid Mechanics, 835*, 1108–1135.

Meshkov, E.E. (1969). Instability of the interface of two gases accelerated by a shock wave. *Fluid Dynamics, 4*(5), 101–104.

Niederhaus, J.J., Greenough, J. A., Oakley, J.G., Ranjan, D., Anderson, M.H., & Bonazza, R. (2008). A computational parameter study for the three-dimensional shock-bubble interaction. *Journal of Fluid Mechanics, 594*, 85.

Pishchalnikov, Y. A., Sapozhnikov, O.A., Bailey, M.R., Williams, J.C., Cleveland, R.O., Colonius, T., Crum, L.A. Evan, A.P., & McAteer, J. A. (2003). Cavitation bubble cluster activity in the breakage of kidney stones by lithotripter shock waves. *Journal of Endourology, 17*(7), 435–446.

Quirk J.J., & Karni, S. (1996). On the dynamics of a shock–bubble interaction. *Journal of Fluid Mechanics, 318*, 129–163.

Raj, L.P., Singh, S., Karchani, A., & Myong, R.S. (2017). A super-parallel mixed explicit discontinuous Galerkin method for the second-order Boltzmann-based constitutive models of rarefied and microscale gases. *Computers & Fluids, 157*, 146–163.

Ranjan, D., Niederhaus, J.H.J., Motl, B., Anderson, M.H., Oakley, J., & Bonazza, R. (2007). Experimental investigation of primary and secondary features in high-Mach-number shock-bubble interaction. *Physical Review Letters, 98*, 024502.

Reed, W.H., & Hill, T.R. (1973). Triangular mesh methods for the neutron transport equation. *Los Alamos Scientific Laboratory Report LA-UR, 13*(7), 73–79

Richtmyer, R.D. (1960). Taylor instability in shock acceleration of compressible fluids. *Communications on Pure and Applied Mathematics, 13*, 297.

Shankar, S.K., Kawai, S., & Lele, S.K. (2011). Two-dimensional viscous flow simulation of a shock accelerated heavy gas cylinder. *Physics of Fluids, 23*, 024102.

Shu, C.-W., & Osher, S. (1988). Efficient implementation of essentially non-oscillatory shock-capturing schemes. *Journal of Computational Physics, 77*(2), 439–471.

Shyue, K.M. (1998). An efficient shock-capturing algorithm for compressible multicomponent problems. *Journal of Computational Physics, 142*, 208–242.

Singh, S. (2020). Role of Atwood number on flow morphology of a planar shock-accelerated square bubble: A numerical study. *Physics of Fluids, 32*, 126112.

Singh, S., & Battiato, M. (2021). Behavior of a shock-accelerated heavy cylindrical bubble under nonequilibrium conditions of diatomic and polyatomic gases. *Physical Review Fluids, 6*, 044001.

Singh, S., Battiato, M. & Myong, R.S. (2021). Impact of bulk viscosity on flow morphology of shock-accelerated cylindrical light bubble in diatomic and polyatomic gases. *Physics of Fluids, 33*, 066103.

Singh, S. (2021a). Numerical investigation of thermal non-equilibrium effects of diatomic and polyatomic gases on the shock-accelerated square light bubble using a mixed-type modal discontinuous Galerkin method. *International Journal of Heat and Mass Transfer, 169*, 121708.

Singh, S. (2021b). Contribution of Mach number on the evolution of Richtmyer-Meshkov instability induced by a shock-accelerated square light bubble. *Physical Review Fluids, 6*, 104001.

Singh, S., & Myong, R.S. (2017). A computational study of bulk viscosity effects on shock-vortex interaction using discontinuous Galerkin method. *Journal of Computational Fluids Engineering, 22*, 066103. 86–95.

Singh, S. (2018). Development of a 3D discontinuous Galerkin method for the second-order Boltzmann-Curtiss based hydrodynamic models of diatomic and polyatomic gases. (Doctoral dissertation, Gyeongsang National University South Korea. Department of Mechanical and Aerospace Engineering).

Singh, S., & Battiato, M. (2020). Strongly out-of-equilibrium simulations for electron Boltzmann transport equation using explicit modal discontinuous Galerkin method. *International Journal of Applied and Computational Mathematics, 6*, 133.

Singh, S., & Battiato, M. (2020). Effect of strong electric fields on material responses: The Bloch oscillation resonance in high field conductivities. *Materials, 13*, 1070.

Singh, S., & Battiato, M. (2021). An explicit modal discontinuous Galerkin method for Boltzmann transport equation under electronic nonequilibrium conditions. *Computers & Fluids, 224*, 104972.

Singh, S. (2021c). Mixed-type discontinuous Galerkin approach for solving the generalized FitzHugh-Nagumo reaction-diffusion model. *International Journal of Applied and Computational Mathematics*, 7, 207.

Streeter, V.L. (1983). Transient cavitating pipe flow. *Journal of Hydraulic Engineering, 109*(11), 1407–1423.

Tagawa, Y., Oudalov, N., Ghalbzouri, A.E., Sun, C., & Lohse, D. (2013). Needle-free injection into skin and soft matter with highly focused microjets. *Lab on a Chip, 13*(7), 1357–1363.

Zabusky, N.J., & Zeng, S.M. (1998). Shock cavity implosion morphologies and vortical projectile generation in axisymmetric shock-spherical fast/slow bubble interactions. *Journal of Fluid Mechanics, 362*, 327–346.

3 A New High-Order Compact Finite Difference Scheme for One-Dimensional Helmholtz Equation using Dirichlet Boundary Conditions

Friday Ighaghai Oyakhire and Omenyi Louis
Alex Ekwueme Federal University

Jogendra Kumar
DIT University

CONTENTS

3.1 INTRODUCTION

The aim of this article is to derive, implement, and test novel and improved sixth-order, eighth-order, and tenth-order compact difference schemes for solving the Helmholtz equation in one dimension using Dirichlet boundary conditions. Nowadays, there are so many powerful computing machines available, which improve the numerical techniques such as finite difference methods. Most physical problems which arise in many branches of continuum physics, e.g., wave mechanics, fluid dynamics,

DOI: 10.1201/9781003222255-3

diffusion, heat flow, elastic vibration (Wang et al., 2017; Harari & Turkel, 1995), bio-engineering, solid physics, chemistry (Singer & Turkel, 1998; Hirsh, 1975), and population dynamics (Lele, 1992) are being solved with finite difference schemes. Further, many other physical phenomena such as electromagnetic waves and acoustics may be described by Helmholtz equations. Finite difference schemes (Spotz & Carey, 1995; Singer & Turkel, 1998), boundary element methods (Oyakhire et al., 2019; Li et al., 1995), and finite element methods (Córdova et al., 2014; Liu et al., 2009; Spotz, 1995) are being employed to solve the Helmholtz equations.

The mentioned study addresses the one-dimensional case, which is solved using finite difference method of the second order. The solution to Helmholtz equations is highly oscillatory for high wave number on fourth order in space.

On fourth order in space, the main difficulty in solving Helmholtz equations is that the solution is highly oscillatory for high wave number. The phrase error (pollution) of the calculated low-order discretization solution is large, unless thin meshes are used per wavelength (Jiang & Ge, 2020). As of late, high-request exact techniques, for example, ghastly strategies and high-request limited distinction schemes, have effectively been evolved in addressing the Helmholtz condition when k is steady (Li et al., 2015; Babuska & Sauter, 1997; Li & Tang, 2001; Spotz & Carey, 2001). Conservative schemes have been produced for an assortment of elliptic conditions (Al-Shibani et al., 2013; Tian & Ge, 2003; Sanyasiraju & Manjula, 2005), and they have likewise been effectively used to settle the Navier–Stokes equations in Zhang (2006). A conservative fourth-request limited distinction strategy was proposed to settle the Helmholtz condition with steady wave number in ongoing work (Shen & Wang, 2005; Erturk & Gökçöl, 2006).

A compact fourth-order finite difference scheme involves the least number of grid points. Thus, compact schemes result in matrices that have a smaller brand – compared with the non-compact schemes that involve more grid points. One of the advantages of compact schemes is that they can produce a highly accurate numerical solution without requiring extra boundary conditions. There are fewer works considering numerical methods for the Helmholtz equation, i.e., one spatial dimension in higher order. Finite difference approaches that are used are often standard, low-order methods (second order (Amini et al., 2021), fourth order, and sixth order in space) and provide little numerical analysis or convergence results.

Recently, intensive research has been performed to develop competent and exact numerical schemes to solve the Helmholtz equation (Dastour & Liao, 2021; Cocquet et al., 2021). An important research in the field has been done (Fang et al., 2007; Nabavi et al., 2007; Xiu & Shen, 2007; Sutmann, 2007), where compact finite difference schemes of sixth order were developed for the 2D Helmholtz equation. The higher-order correction terms have been calculated using Helmholtz equation. A domain decomposition solver in layered media for acoustic scattering by elastic objects was obtained (Ito et al., 2008); further, high-order finite difference schemes were derived in case of non-uniform grid points (Hermanns & Hernandez, 2008; Fu, 2008; Okoro & Owoloko, 2010). For Helmholtz equations with large wave numbers, a compact fourth-order finite difference scheme was devised based on non-uniform grid point distributions (Singer & Turkel, 2006; Wong & Li, 2011; Erlangga & Turkel, 2012). Dissection of compact finite differences of high order in one direction for 2D Poisson equation, and some exact finite difference and iteration schemes for high-order distribution for exterior for solving Helmholtz equation were studied (Oyakhire, 2019; Elconsul & Lagha, 2013; Córdova et al., 2016). Recently, studies of compact finite difference schemes of higher order for 1D Helmholtz equation have been presented (Duressa et al., 2016; Turkel et al., 2013; Oyakhire, 2017). In Li (2018), Pade approximations are used to solve 1D, 2D, and 3D Helmholtz equation with different compact finite difference schemes.

Consider the non-homogeneous 1D Helmholtz equation

$$\left.\begin{aligned} u'' + k^2 u = f(x), \quad x \in \Omega \\ u(a) = \alpha, \quad u(b) = \beta \end{aligned}\right\} \tag{3.1}$$

where $U\left(x_{i-1}, t^m\right)$ is a constant and $\left(x_j t^m\right)$ is studied in numerous studies.

This is an attempt to obtain accurate compact finite difference schemes of higher order for the Helmholtz equation using Dirichlet boundary conditions on the interval Ω. The derived schemes and test problems were compared to (Singer & Turkel, 2016; Córdova et al., 2016; Li, 2018) sixth order on one-way dissection compact finite difference schemes to solve 2D Poisson equation, eighth order on one-dimensional Helmholtz equation, and sixth order for 1D, 2D, and 3D Helmholtz equations using Pade approximations.

This paper is structured with different sections: Section 3.2 explains the derivation of sixth-order, eighth-order, and tenth-order compact finite difference schemes and their implementations on Helmholtz equations. Section 3.3 presents the convergence analysis. Section 3.4 contains numerical experiments and results, and Section 3.5 describes the conclusions.

3.2 METHOD

Under this section, we try to design sixth-order, eighth-order, and tenth-order compact finite difference methods to solve equation (3.1) with Dirichlet boundary conditions at one interval end. N uniform segments are utilized in the interval $U_{xx}\left(x_{j\pm1}, t^m\right) = U_{xx}\left(x_j, t^m\right) \pm \Delta x U_{xxx}\left(x_j, t^m\right) + \frac{\Delta x^2}{2} U_{xxxx}\left(x_j, t^m\right) \pm \frac{\Delta x^3}{3!} U_{xxxxx}\left(x_j, t^m\right) + 0\left(\Delta x^4\right)$. with a uniform grid, i.e., the grid spacing is uniform, i.e., Δx^0, Δx, Δx^2, and the mesh points are $a = x_1 < x_2 < \cdots < x_{n-1} = b$, where $x_{i+1} = x_i + i\,h, i = 1,2,3,\ldots,N-1,N$. Let u_i, i.e., $u(x_i)$, be the solution to non-homogeneous one-dimensional Helmholtz equation.

At $\frac{1}{10}(u_{xx})_{j-1}^m + (u_{xx})_j^m + \frac{1}{10}(u_{xx})_{j+1}^m = \frac{6}{5\Delta x^2}\left(u_{j-1}^m - 2_j^m + u_{j+1}^m\right)$, and $u_n(x_i)$ denotes its $(u_{xx})_j = \frac{u_{j+1} - 2u_j + u_{j-1}}{\Delta x^2} + 0\left(\Delta x^2\right)$ derivative at $(u_{xx})\left(x_1, t^m\right)$. Also, f_i and f_i^n are used to denote x_M and its $\left(x_1, t^m\right)$ derivative at $\frac{b^*}{\Delta x} U_x\left(0, t^m\right) + a^* U_{xx}\left(x_1, t^m\right) + c^* U_{xx}\left(x_2, t^m\right) = \frac{1}{\Delta x^2}\left(e^* U\left(x_1, t^m\right) + f^* U\left(x_2, t^m\right)\right)$, that is a^*, b^*, c^*, e^* and $f_i^n = f^n(x_i)$.

To obtain an approximate compact scheme, write Taylor series expansion for the discretized field u_i.

$$u_{i+1} = u_i + \frac{hu_i^{(1)}}{1!} + \frac{h^2u_i^{(2)}}{2!} + \frac{h^3u_i^{(3)}}{3!} + \frac{h^4u_i^{(4)}}{4!} + \frac{h^5u_i^{(5)}}{5!} + \frac{h^6u_i^{(6)}}{6!}$$

$$+ \frac{h^7u_i^{(7)}}{7!} + \frac{h^8u_i^{(8)}}{8!} + \frac{h^9u_i^{(9)}}{9!} + \frac{h^{10}u_i^{(10)}}{10!} + \frac{h^{11}u_i^{(11)}}{11!} + \frac{h^{12}u_i^{(12)}}{12!} + \cdots \quad (3.2)$$

$$u_{i-1} = u_i - \frac{hu_i^{(1)}}{1!} + \frac{h^2u_i^{(2)}}{2!} - \frac{h^3u_i^{(3)}}{3!} + \frac{h^4u_i^{(4)}}{4!} - \frac{h^5u_i^{(5)}}{5!} + \frac{h^6u_i^{(6)}}{6!}$$

$$- \frac{h^7u_i^{(7)}}{7!} + \frac{h^8u_i^{(8)}}{8!} - \frac{h^9u_i^{(9)}}{9!} + \frac{h^{10}u_i^{(10)}}{10!} - \frac{h^{11}u_i^{(11)}}{11!} + \frac{h^{12}u_i^{(12)}}{12!} - \cdots \quad (3.3)$$

From equations (3.2) and (3.3), we obtain the $(\delta_x^{'}u_i)$ of the first derivative of $U(x_1,t^m) = U(0,t^m) + \Delta x U_x(0,t^m) + \frac{\Delta x^2}{2}U_{xx}(0,t^m) + \frac{\Delta x^3}{6}U_{xxx}(0,t^m) + 0(\Delta x^4)$. and the standard second-order central difference $(\delta_x^2 u_i)$ of the second derivative of $U_{xx}(x_1,t^m) = U_{xx}(0,t^m) + \Delta x U_{xxx}(0,t^m) + 0(\Delta x^2)$ as

$$\delta_x^{'}u_i = \frac{u_{i+1} - u_{i-1}}{2h} = u_i' + O(h^2) \quad (3.4)$$

$$\delta_x^2 u_i = \frac{u_{i+1} - 2u_i + u_{i-1}}{h^2} = u_i^{(2)} + O(h^2) \quad (3.5)$$

Also using equations (3.2) and (3.3), we obtain

$$\delta_x^{'}u_i = \frac{u_{i+1} - u_{i-1}}{2h} = u_i^{(2)} + \frac{h^2u_i^{(3)}}{6} + \frac{h^4u_i^{(5)}}{120} + \frac{h^6u_i^{(7)}}{5040} + \frac{h^8u_i^{(9)}}{3628800} + 0(h^{10}) \quad (3.6)$$

$$\delta_x^2 u_i = \frac{u_{i+1} - 2u_i + u_{i-1}}{h^2} = u_i^{(2)} + \frac{h^2u_i^{(4)}}{12} + \frac{h^4u_i^{(6)}}{360} + \frac{h^6u_i^{(8)}}{20160} + \frac{h^8u_i^{(10)}}{1814400} + O(h^{10}) \quad (3.7)$$

An estimate of sixth-order accurate finite difference of equation (3.5) is taken from the first three terms of equation (3.7) as

$$\delta_x^2 u_i = u_i^{(2)} + \frac{h^2u_i^{(4)}}{12} + \frac{h^4u_i^{(6)}}{360} + O(h^6) \quad (3.8)$$

Both $O(h^2)$ and $O(h^6)$ terms are retained in equation (3.8) to approximate them to construct a sixth-order accurate scheme. Operating δ_x^2 to $u_i^{(4)}$, we get

$$u_i^{(6)} = \delta_x^2 u_i^{(4)} + O(h^2) \quad (3.9)$$

Substitution of equation (3.9) into equation (3.7) yields

$$\delta_x^2 u_i = u_i^{(2)} + \frac{h^2 u_i^{(4)}}{12} + \frac{h^4}{360}\left(\delta_x^2 u_i^{(4)} + O(h^2)\right) + O(h^6) \tag{3.10}$$

To obtain a compact $O(h^6)$ approximation, from equation (3.1), we may have

$$u_i^{(4)} = -k^2 u_i^{(2)} + f_i^{(2)} \tag{3.11}$$

where $f_i = f(x_i)$ and $f_i^{(2)} = f^{(2)}(x_i)$. Inserting equation (3.11) into equation (3.10), we obtain

$$\delta_x^2 u_i = u_i^{(2)} + \left(\frac{h^2}{12} + \frac{h^4}{360}\delta_x^2\right)\left(-k^2 u_i^{(2)} + f_i^{(2)}\right) + O(h^6) \tag{3.12}$$

With a sixth-order accuracy, a compact (implicit) approximation for $u_i^{(2)}$ will be obtained as

$$u_i^{(2)} = \frac{\delta_x^2 u_i - \frac{h^2}{12}\left(1 + \frac{h^2}{30}\delta_x^2\right) f_i^{(2)}}{\left(1 - \frac{k^2 h^2}{12}\left(1 + \frac{h^2}{30}\delta_x^2\right)\right)} + O(h^6) \tag{3.13}$$

Using this estimate to consider the discrete solution to equation (3.10), we get

$$k^2\left(1 - \frac{k^2 h^2}{12}\right) U_i + \left(1 - \frac{k^4 h^4}{360}\right)\delta_x^2 U_i = \left(1 - \frac{k^2 h^2}{12}\right) f_i - \frac{k^4 h^4}{360}\delta_x^2 f_i + \frac{h^2}{12} f_i^{(2)} + \frac{h^4}{360}\delta_x^2 f_i^{(2)} \tag{3.14}$$

where U_i is the discrete approximation of u_i and satisfies the discrete formulation of equation (3.14). We have

$$u_i = U_i + 0(h^6), \qquad \delta_i^2 f^{(2)} = \left(f_{i+1}^{(2)} - 2 f_i^{(2)} + f_{i-1}^{(2)}\right) / h^2 \tag{3.15}$$

The scheme given in equation (3.14) can be expressed as

$$d_{10} U_i + d_{11}\left(U_{i-1} + U_{i+1}\right) = b_{10}\left(f_{i-1} + f_{i+1}\right) + b_{12} f_i^{(2)} + b_{13}\left(f_{i+1}^{(2)} + f_{i-1}^{(2)}\right) \tag{3.16}$$

where

$$d_{10} = -2 + h^2 k^2\left(1 - \frac{28 k^2 h^2}{360}\right), \; d_{11} = 1 - \frac{k^4 h^4}{360},$$
$$b_{10} = h^2\left(1 - \frac{28 k^2 h^2}{360}\right), \; b_{11} = -\frac{k^2 h^4}{360}, b_{12} = \frac{28 h^4}{360}, \; b_{13} = \frac{h^4}{360} \tag{3.17}$$

The right-hand side of equation (3.16) may be found for all nodes as $f^{(2)}$ and f are known at each mesh point. Thus, the system given by equation (3.16) may be written for each node; as a result, a linear system of equation may be acquired. A fourth-order approximation of $f^{(2)}$is required if f is unknown analytically, which may be achieved using

$$f_i^{(2)} = \left(-f_{i-1} + 16f_{i-1/2} - 30f_i + 16f_{i+1/2} - f_{i+1}\right)/12h^2 u_i^{(6)} = -k^2 u_i^{(4)} + f_i^{(4)} \quad (3.18)$$

For eighth-order accurate approximation, equation (3.7) will be recast as

$$\delta_x^2 u_i = \frac{u_{i+1} - 2u_i + u_{i-1}}{h^2} = u_i^{(2)} + \frac{h^2 u_i^{(4)}}{12} + \frac{h^4 u_i^{(6)}}{360} + \frac{h^6 u_i^{(8)}}{20160} + O(h^8) \quad (3.19)$$

Writing equation (3.7) in the discretized form as equation (3.1) and equation (3.9), we have

$$u_i^{(6)} = -k^2 u_i^{(4)} + f_i^{(4)} \quad (3.20)$$

Substituting equations (3.9), (3.11), and (3.20) into equation (3.19) and performing some algebraic manipulations, we get

$$\delta_x^2 u_i = \left(1 - \frac{k^2h^2}{12} + \frac{k^4h^4}{360} + \frac{k^4h^6}{20160}\delta_x^2\right)u_i^{(2)} + \frac{h^2}{12}\left(1 - \frac{k^2h^2}{12} - \frac{k^2h^4}{1680}\delta_x^2\right)f_i^{(2)}$$
$$-\frac{h^4}{360}\left(1 + \frac{h^2}{56}\delta_x^2\right)f_i^{(4)} \quad (3.21)$$

An implicit approximation for eighth-order accuracy is given as

$$u_i^{(2)} = \frac{\delta_x^2 u_i - \frac{h^2}{12}\left(1 - \frac{h^2k^2}{12} - \frac{h^2k^4}{1680}\delta_x^2\right)f_i^{(2)} + \frac{h^4}{360}\left(1 + \frac{h^2}{56}\delta_x^2\right)f_i^{(4)}}{\left(1 - \frac{h^2k^2}{12} + \frac{h^4k^4}{360} + \frac{h^6k^4}{20160}\delta_x^2\right)} \quad (3.22)$$

Let U_i denote the eighth-order approximation of u_i, i.e., $u_i = U_i + O(h^2)$.

Substituting $u_i^{(2)}$ from equation (3.22) into equation (3.1) with $\delta_x^2 f_i^{(2)} = \frac{f_{i+1}^{(2)} - 2f_i^{(2)} + f_{i-1}^{(2)}}{h^2}$ and $\delta_x^2 f_i^{(4)} = \frac{f_{i+1}^{(4)} - 2f_i^{(4)} + f_{i-1}^{(4)}}{h^2}$, we obtain a three-point compact finite difference scheme of eighth order for 1D Helmholtz equation.

$$a_{81}U_{i-1} + a_{80}U_i + a_{81}U_{i-1} = b_{81}f_{i-1} + b_{80}f_i + b_{81}f_{i+1}$$
$$+c_{81}f_{i-1}^{(2)} + c_{80}f_i^{(2)} + c_{81}f_{i+1}^{(2)} + d_{81}f_{i+1}^{(4)} + d_{80}f_i^{(4)} + d_{81}f_{i-1}^{(4)} \quad (3.23)$$

where

$$a_{80} = -2 + k^2h^2 - \frac{k^4h^4}{12} + \frac{3h^6k^4}{1120},\ a_{81} = 1 + \frac{k^6h^6}{20160},\ b_{80} = h^2\left(1 - \frac{h^2k^2}{12} + \frac{3h^4k^4}{1120}\right),$$
$$b_{81} = \frac{k^4h^4}{20160},\ c_{80} = \frac{h^4}{12}\left(1 - \frac{9k^2h^2}{280}\right),\ c_{81} = \frac{k^2h^6}{20160},\ d_{80} = \frac{3h^2}{1120},\ d_{81} = -\frac{h^6}{20160} \quad (3.24)$$

To obtain the tenth-order compact finite difference scheme, equation (3.7) will be considered using the Dirichlet boundary conditions, and using equation (3.1), we apply δ_x^2 to $u_i^{(8)}$.

$$u_i^{(10)} = \delta_x^2 u_i^{(8)} + O(h^2) \quad (3.25)$$

Substituting equation (3.25) into equation (3.7), we have

$$\delta_x^2 u_i = u_i^{(2)} + \frac{h^2u_i^{(4)}}{12} + \frac{h^4u_i^{(6)}}{360} + \frac{h^6u_i^{(8)}}{20160} + \frac{h^8}{181440}\left(\delta_x^2 u_i^{(8)} + O(h^2) + O(h^{10})\right) \quad (3.26)$$

Writing equation (3.1) in the discretized form as it appears in equations (3.7), (3.20), and (3.25), we have

$$u_i^{(8)} = -k^2u_i^{(6)} + f_i^{(6)} \quad (3.27)$$

Substituting equations (3.20), (3.25), and (3.27) into equation (3.26), we obtain

$$\begin{aligned}\delta_x^2 u_i &= \left(1 - \frac{k^2h^2}{12} + \frac{k^4h^4}{360} - \frac{k^6h^6}{20160} - \frac{h^8k^6}{1814400}\delta_x^2\right)u_i^{(2)} \\ &+ \left(\frac{h^2}{12} - \frac{k^2h^4}{360} + \frac{k^4h^6}{20160} + \frac{h^8k^4}{1814400}\delta_x^2\right)f_i^{(2)} \\ &+ \left(\frac{h^4}{360} - \frac{k^2h^6}{20160} - \frac{h^8k^2}{1814400}\delta_x^2\right)f_i^{(4)} \\ &+ \left(\frac{h^6}{20160} + \frac{h^8}{1814400}\delta_x^2\right)f_i^{(6)} + O(h^{10})\end{aligned} \quad (3.28)$$

With tenth-order accuracy, an implicit approximation for $u_i^{(2)}$ is given as

$$u_i^{(2)} = \frac{\delta_x^2 u_i - \frac{h^2}{12}\left(1 - \frac{h^2k^2}{30} + \frac{h^4k^4}{1680} + \frac{h^6k^4\delta_x^2}{151200}\right)f_i^{(2)} - \frac{h^4}{360}\left(1 - \frac{h^2k^2}{56} - \frac{h^4k^2\delta_x^2}{5040}\right)f_i^{(4)} - \frac{h^6}{20160}\left(1 + \frac{h^2\delta_x^2}{90}\right)f_i^{(6)}}{1 - \frac{h^2k^2}{12} + \frac{h^4k^4}{360} - \frac{h^6k^6}{20160} - \frac{h^8k^6\delta_x^2}{1814400}} - 0(h^{10}) \quad (3.29)$$

Let u_i denote the tenth-order approximation of u_i, that is $u_i = u_i + O(h^{10})$. Substituting $u_i^{(2)}$ from equation (3.29) into equation (3.26),

$$\delta_x^2 u_i - \frac{h^2}{12}\left(1 - \frac{h^2k^2}{30} + \frac{h^4k^4}{1680} + \frac{h^6k^4\delta_x^2}{151200}\right) f_i^{(2)} - \frac{h^4}{360}\left(1 - \frac{h^2k^2}{56} - \frac{h^4k^2\delta_x^2}{5040}\right) f_i^{(4)}$$

$$- \frac{h^6}{20160}\left(1 + \frac{h^2\delta_x^2}{90}\right) f_i^{(6)}$$

$$= k^2 u_i + \frac{h^2k^2}{12} - \frac{h^4k^6}{360} u_i + \frac{h^6k^8}{20160} u_i + \frac{h^8k^8}{1814400} \delta_x^2 u_i$$

$$+ f_i - \frac{h^2k^2}{12} f_i + \frac{h^4k^6}{360} f_i - \frac{h^6k^8}{20160} f_i + \frac{h^8k^8}{1814400} \delta_x^2 f_i \tag{3.30}$$

Rearranging equation (3.30) and using (3.26) together with $\delta_x^2 f_i^{(2)} = \dfrac{f_{i+1}^{(2)} - 2f_i^{(2)} + f_{i-1}^{(2)}}{h^2}$, $\delta_x^2 f_i^{(4)} = \dfrac{f_{i+1}^{(4)} - 2f_i^{(4)} + f_{i-1}^{(4)}}{h^2}$, and $\delta_x^2 f_i^{(6)} = \dfrac{f_{i+1}^{(6)} - 2f_i^{(6)} + f_{i-1}^{(6)}}{h^2}$, we obtain

$$\frac{u_{i+1} - 2u_i + u_{i-1}}{h^2} + k^2\left(1 - \frac{h^2k^2}{12} + \frac{h^4k^4}{360} - \frac{h^6k^6}{20160}\right) - \frac{h^8k^6}{1814400}\left(\frac{u_{i+1} - 2u_i + u_{i-1}}{h^2}\right)$$

$$= \left(1 - \frac{h^2k^2}{12} + \frac{h^4k^4}{360} - \frac{h^6k^6}{20160}\right) f_i - \frac{h^6k^6}{184400}\left(\frac{f_{i+1} - 2f_i + f_{i-1}}{h^2}\right)$$

$$+ \frac{h^2}{12}\left(1 - \frac{h^2k^2}{30} + \frac{h^4k^4}{1680}\right) + \frac{h^2}{12}\left\{\frac{h^6k^4}{151200}\left(\frac{f_{i+1}^{(2)} - 2f_i^{(2)} + f_{i-1}^{(2)}}{h^2}\right)\right\} \tag{3.31}$$

$$+ \frac{h^4}{360}\left(1 - \frac{h^2k^2}{56}\right) f_i^{(4)} - \frac{h^4}{360}\left\{\frac{h^4k^2}{5040}\left(\frac{f_{i+1}^{(4)} - 2f_i^{(4)} + f_{i+1}^{(4)}}{h^2}\right)\right\}$$

$$- \frac{h^6}{20160}\left\{1 + \frac{h^2}{90}\left(\frac{f_i^{(6)} - 2f_i^{(6)} + f_{i-1}^{(6)}}{h^2}\right)\right\}$$

For one-dimensional Helmholtz equation, the three-point tenth-order compact finite difference scheme is obtained from (3.31) after some algebraic manipulation

$$a_{101}\, u_{i+1} + a_{100}\, u_i + a_{101}\, u_{i-1} = b_{101}\, f_{i+1} + b_{100}\, f_i + b_{101}\, f_{i-1} + c_{101}\, f_{i+1}^{(2)}$$

$$+ c_{100}\, f_i^{(2)} + c_{101}\, f_{i-1}^{(2)} + d_{101}\, f_{i+1}^{(4)} + d_{100}\, f_i^{(4)} \tag{3.32}$$

$$+ d_{101}\, f_{i-1}^{(4)} + e_{101}\, f_{i+1}^{(6)} + e_{100}\, f_i^{(6)} + e_{101}\, f_{i-1}^{(6)}$$

where

$$\left.\begin{aligned}
&a_{101}=1+\frac{h^8k^6}{1814400},\ a_{100}=-2+k^2h^2\left(1-\frac{h^2k^2}{12}+\frac{h^4k^4}{360}-\frac{11k^6h^6}{226800}\right)\\
&b_{101}=-\frac{h^8k^6}{1814400},\quad b_{100}=h^2\left(1-\frac{k^2h^2}{12}+\frac{k^4h^4}{360}-\frac{11k^6h^6}{226800}\right),\ c_{101}=\frac{1}{12}\left(\frac{h^6k^8}{151200}\right)\\
&c_{100}=\frac{h^4}{12}\left(1-\frac{h^4k^2}{30}-\frac{11h^4k^4}{18900}\right),\ d_{100}=\frac{h^8k^2}{181400},\ d_{101}=\frac{h^6}{360}\left(1-\frac{h^2k^2}{630}\right)\\
&e_{100}=\frac{11h^8}{226800},\quad e_{101}=\frac{h^8}{1814400}
\end{aligned}\right\}$$

3.3 TEST EXAMPLE

The accuracy and performance of the schemes obtained in Section 3.2 have been shown with an example using MATLAB code. All calculations are performed for a uniform mesh with size h on the unit interval [0, 1], and with conditions, considered on boundary in the unit interval.

The e^{2-} norm vector, i.e., $(e_1, e_2, ... e_m)$ as $\|e\|_2 = \sum_{i=1}^{M} |e|_i^2$, is used to compare the solution.

PROBLEM

Consider

$u'' + k^2u = 3\pi \sin x \;\; 0 < x < 1$ with $u(0) = u(1) = 0$, of which the exact solution is given by

$$u(x) = \left(\frac{3\pi}{k^2-1}\right)(\sin x - \sin kx)$$

The eigenvalues of the problem (homogeneous) are $n\pi,\; n = 0,1,2,3....$

In case one of the eigenvalues is equal to k^2, the solution does not exist. Thus, we obtain the numerical solution for the eigenvalues, which are close to k.

The e^{2-} error norms are shown in Table 3.1 for the numerical calculation of the problem with $O(h^6)$, $O(h^8)$, and $O(h^{10})$, and at $k = 1, 10, 20, 30, 40, 50, 60, 70, 80$, comparing the degree of accuracy of these schemes.

In general, the accuracy of $O(h^{10})$ over $O(h^8)$ and that of $O(h^8)$ over $O(h^6)$ improve significantly, and the inaccuracy e^2 in these schemes decreases as the number of nodes increases. If $k = 1$, on the other hand, $O(h^{10})$ and $O(h^8)$ decrease as N increases.

The e^{2-} norm of the errors for the $O(h^6)$, $O(h^8)$, and $O(h^{10})$ schemes at $k = 1$ and for $N = 16, 32, 64, 128, 256$, and 512 is shown in Figure 3.1. Figures 3.2–3.4 demonstrate the effect of increasing the number of nodes at different fixed wave numbers $k = 10, 30, 50$. Further, Figures 3.5–3.7 show the comparison of the e^{2-} norm of the errors for $O(h^6)$, $O(h^8)$, and $O(h^{10})$ schemes for $k = 1, 10$, and 50, respectively. The goal of these calculations is to see how the approximate solution behaves when it gets close to the eigenvalue.

3.4 NUMERICAL RESULTS

TABLE 3.1
Comparison of the Order of Accuracy of the Schemes

K	*N*	$O(h)^6$	$O(h)^8$	$O(h)^{10}$
1	16	4.2386e–008	2.9566e–012	1.2832e–016
	32	2.6491e–009	4.6197e–014	5.0126e–o19
	64	1.6557e–010	7.2182e–016	1.9581e–021
	128	1.0348e–011	1.1278e–017	7.6487e–024
	256	6.4675e–011	1.7623e–019	2.9878e–026
	512	4.0422e–014	2.7535e–021	1.1671e–028
10	16	4.2386e–006	2.9566e–008	1.2832e–010
	32	2.6491e–007	4.6197e–10	1.2832e–010
	64	1.6557e–008	7.2182e–012	1.9581e–015
	128	1.0348e–009	1.1278e–013	7.6487e–018
	256	6.4675e–011	1.7623e–015	2.9878e–020
	512	4.0422e–012	2.7535e–017	1.1671e–022
20	16	1.6954e–005	4.7305e–007	8.2127e–010
	32	1.0596e–006	7.3914e–009	8.2127e–009
	64	6.6227e–008	1.1549e–010	1.2532e–013
	128	4.1392e–009	1.8046e–012	4.8952e–016
	256	2.5870e–010	2.8196e–013	1.9122e–018
	512	1.6167e–011	4.4056e–016	7.4694e–021
30	16	3.8147e–005	2.3948e–006	9.3548e–008
	32	2.3842e–006	3.7419e–008	3.6542e–010
	64	1.4910e–007	5.8468e–010	1.4274e–012
	128	9.3132e–009	9.1355e–012	5.5759e–015
	256	5.8208e–010	1.4273e–013	2.1781e–017
	512	3.6381e–011	2.2304e–015	8.5081e–020
40	16	6.7817e–005	7.5638e–006	5.2561e–007
	32	4.2386e–006	1.1826e–007	2.0532e–009
	64	2.6491e–007	1.8479e–009	8.0202e–012
	128	1.6557e–009	2.8873e–011	3.1329e–014
	256	1.0348e–009	4.5114e–013	1.2238e–016
	512	6.4675e–011	7.0490e–015	4.7804e–019
50	16	1.0596e–004	1.8479e–005	2.0051e–006
	32	6.6227e–006	2.8873e–007	7.8323e–009
	64	4.1390e–007	4.5114e–009	3.0595e–011
	128	2.5870e–008	7.o490e–011	1.1951e–013
	256	1.6169e–009	1.1014e–012	4.6684e–016
	512	1.0105e–010	1.7210e–014	1.8236e–018
60	16	1.5259e–004	3.8317e–005	5.9871e–006
	32	9.5367e–006	5.9871e–007	2.3387e–008
	64	5.9605e–007	9.3514e–009	9.1355e–011

(Continued)

TABLE 3.1 (*Continued*)
Comparison of the Order of Accuracy of the Schemes

K	*N*	$O(h)^6$	$O(h)^8$	$O(h)^{10}$
	128	3.7253e–008	1.467e–011	3.5686e–013
	256	2.3283e–009	2.2839e–012	1.3940e–015
	512	1.4552e–010	3.5686e–014	5.4452e–018
70	16	2.0769e–004	7.0889e–005	1.5097e–006
	32	1.2918e–005	1.1092e–006	2.3387e–008
	64	8.1129e–007	1.7331e–008	2.3036e–010
	128	5.0705e–008	2.7080e–010	8.9980e–013
	256	3.1691e–009	4.2312e–012	3.5151e–015
	512	1.9807e–010	6.6112e–014	1.3731e–017
80	16	2.7127e–004	1.2110e–004	3.3639e–005
	32	1.6954e–005	1.1092e–007	1.3140e–007
	64	1.0596e–006	2.9566e–008	5.1330e–010
	128	6.6227e–008	4.6197e–010	2.0051e–012
	256	4.1397e–009	7,2182e–012	7.8323e–015
	512	2.5820e–010	1.1278e–013	3.0595e–017

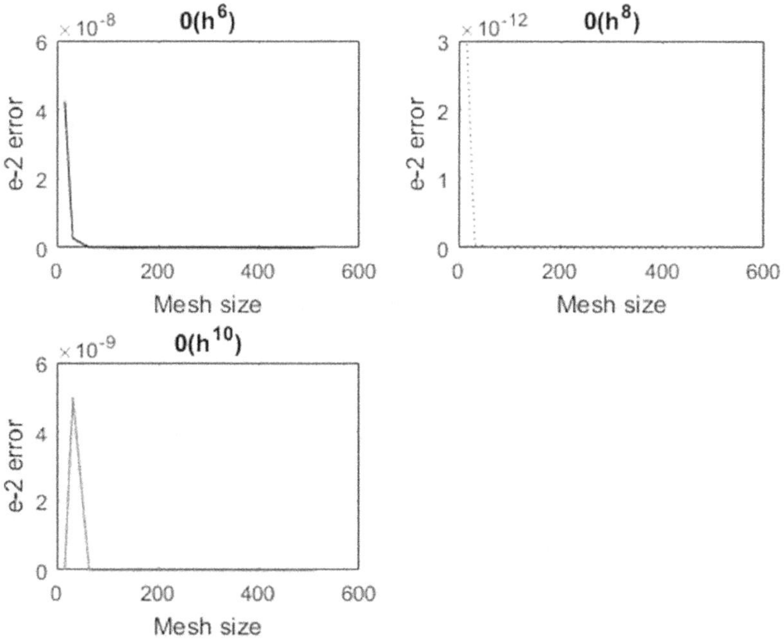

FIGURE 3.1 The impact of increasing the number of nodes at a constant wave number ($k = 1$).

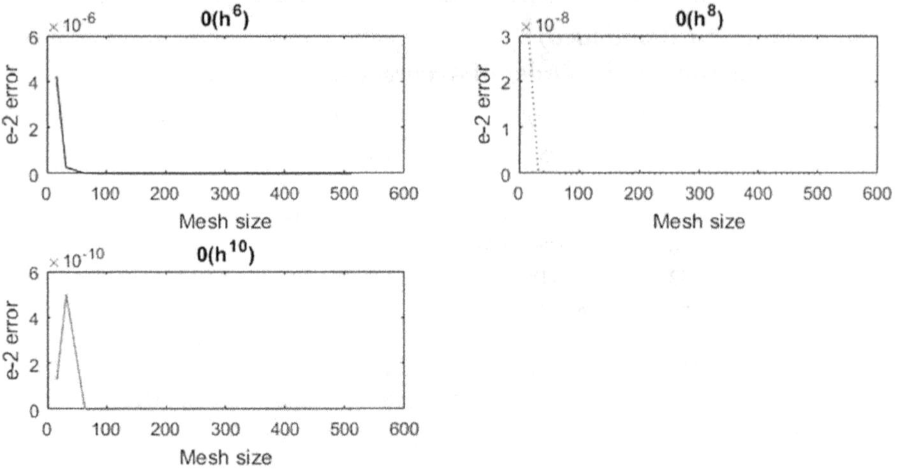

FIGURE 3.2 The impact of increasing the number of nodes at a constant wave number ($k = 10$).

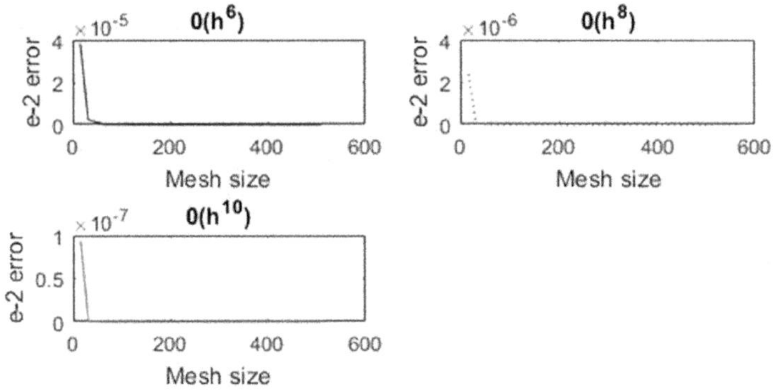

FIGURE 3.3 The impact of increasing the number of nodes at a constant wave number ($k = 30$).

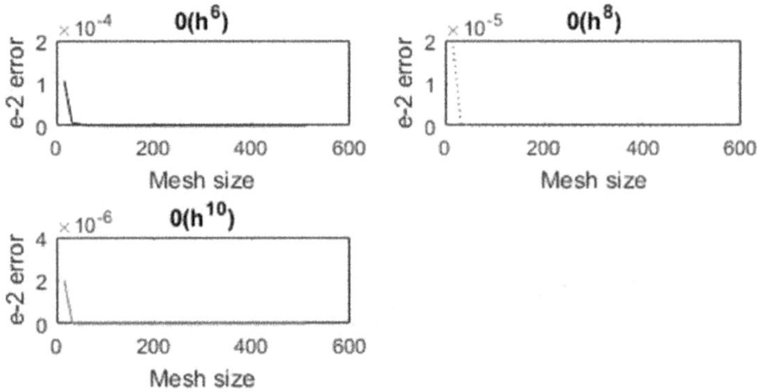

FIGURE 3.4 The impact of increasing the number of nodes at a constant wave number ($k = 50$).

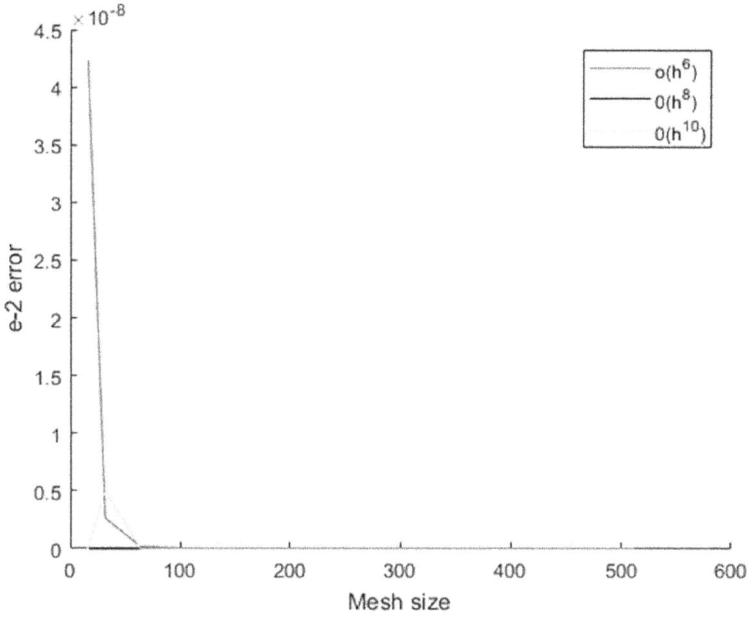

FIGURE 3.5 Comparison of the order of accuracy of the schemes at a constant wave number ($k = 1$).

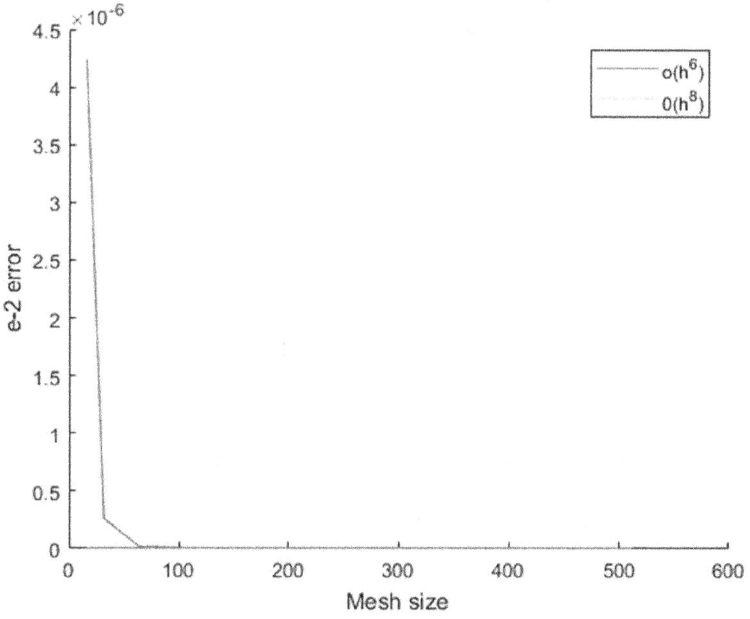

FIGURE 3.6 Comparison of the order of accuracy of the schemes at a constant wave number ($k = 10$).

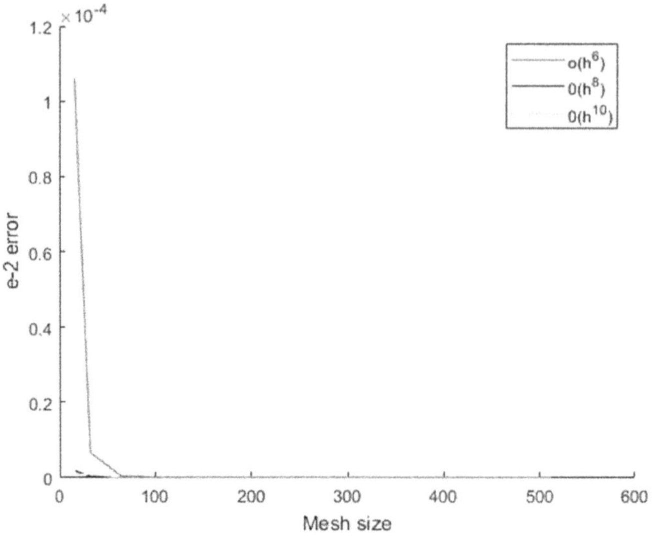

FIGURE 3.7 Comparison of the order of accuracy of the schemes at a constant wave number ($k = 50$).

3.5 CONVERGENCE ANALYSIS

A convergence analysis for Schemes (3.16), (3.23), and (3.31) with Dirichlet boundary conditions has been performed in this section. To begin, we'll introduce the following norm notation, which will be utilized in the subsequent sections. For the vector $u = \{u_i\}_{i=1}^{J} \in \Re^{J-1}$,

$$\|u\| = \left(\sum_{i=1}^{J-1} |u_i|^2 h \right)^{\frac{1}{2}}, \qquad \|u\|_\infty = \sup_{1 \le i \le J-1} |u_i|,$$

and for the vector $u_i = \{u_1, u_2, ..., u_{J-1}\}^T$ with $1 \le J \le J-1$, we have

$$\|u\| = \left(\sum_{i=1}^{J-1} \|u_i\|^2 h \right)^{\frac{1}{2}}$$

Similar to the derivations of the sixth-order, eighth-order, and tenth-order schemes and with their corresponding truncation errors $T = 0(h^6), 0(h^8)$, and $0(h^{10})$ in the previous section, the simple matrix is of the form

$$DU = \vec{b},$$

where D is a symmetric tri-diagonal matrix. When $h \to 0$ and $kh \to 0$, the coefficient matrix D tends to the matrix

$$\begin{bmatrix} 2 & -1 & 0 & \dots 0 & 0 \\ -1 & 2 & -1 & \dots 0 & 0 \\ \dots & \dots & \dots & \dots \dots & .. \\ 0 & 0 & 0 & \dots 2 & -1 \\ 0 & 0 & 0 & \dots -1 & 2 \end{bmatrix}_{(J-1)x(J-1)} \tag{3.33}$$

The fact that matrix (3.33) is positive definite is well known. As a result, when kh is small enough, the coefficient matrix D is also positive definite. So, given a suitably tiny condition (3.16), (3.23), and (3.31), each has a unique solution.

Assume that v_i is the value of the exact solution of (1) at the grid points $x_i = ih, (i = 0, 1, \dots J)$ and let $e_i = u_i - v_i$. Then the error vector $E = \{e_i\}$ satisfies

$$\delta_x^2 u_i = u_i^{(2)}\left(\frac{h^2}{12} + \frac{h^4}{360}\delta_x^2\right)\left(-k^2 u_i^{(2)} + f_i^{(2)}\right) + 0\left(h^6\right) \tag{3.34}$$

We rewrite (3.33) as follows:

$$\delta_x^2 u_i + k^2\left(1 - \frac{k^2h^2}{12}\left(1 + \frac{h^2}{30}\delta_x^2\right)\right)u_i = \left(1 - \frac{k^2h^2}{12}\left(1 + \frac{h^2}{30}\delta_x^2\right)\right)f_i = \frac{h^2}{12}\left(1 + \frac{h^2}{30}\delta_x^2\right)f'$$

$$\delta_x^2 u_i = u_i^{(2)} + \frac{h^2 u_i^{(4)}}{12} + \frac{h^2 u_i^{(6)}}{360} + \left(\frac{h^6}{20160} + \frac{h^8\delta_x^2}{1814400}\right)\left(-k^2 u_i^{(6)} + f_i^{(6)}\right) + O\left(h^{10}\right)$$

It is well known that matrices (3.16), (3.23), and (3.31) have the eigenvalues

$$\lambda_j = 2 - 2\cos\frac{j\pi}{J} = 4\sin^2\frac{j\pi h}{2}$$

and the corresponding eigenvectors

$$\eta_j = (\eta, \eta_2, \dots, \eta_{j-1}, \eta_j),\ \eta_{m,j} = \sin\frac{mj\pi}{J}, m = 1,\ J-1$$

for $j = 1, \dots, J-1$. When h is sufficiently small, the smallest eigenvalue behaves like

$$\min_{1 \le J-1} \lambda_j = 4\sin^2\frac{mj\pi}{J} \approx \pi^2 h^2 \tag{3.35}$$

Since the eigenvectors set $\{\eta_j\}$ constitutes an orthogonal basis of $\Re^{J-1}$, we can write the error vector

$$E = \sum_{j=1}^{J-1} a_j \eta_j, \quad \|E\|^2 = \sum_{j=1}^{J-1} \left|a_j\right|^2 \left\|u_j\right\|^2.$$

Now consider the inner product $(DE,E)=(h^2T,E)$. When $kh\to 0$, we have

$$(DE,E)=\left(D\sum_{j=1}^{J-1}a_j\eta_j,\sum_{j=1}^{J-1}a_j\eta_j)\to(\sum_{j=1}^{J-1}a_j\lambda_j\eta_j,\sum_{j=1}^{J-1}a_j\eta_j\right)$$
$$=\sum_{j=1}^{J-1}\lambda_j|a_j|^2\,\|\eta_j\|^2\geq C\pi^2h^2\sum_{j=1}^{J-1}|a_j|^2\,\|\eta_j\|^2=C\pi^2h^2\|E\|^2. \tag{3.36}$$

where C is a generic constant that can take on different values depending on the context. On the other hand, there's

$$(h^2T,E)\leq h^2\|T\|\,\|E\| \tag{3.37}$$

Combining (3.36) and (3.37), we have

$$\|E\|\leq C\|T\|$$

which leads to the following result.

Theorem 3.1

Assume kh is small enough, where k is the wave number and h is the grid length. Then, (3.16), (3.23), and (3.31) have a one-of-a-kind solution. Moreover, for the sufficiently smooth solution u_i of (1) and the approximate solution u_h of schemes (3.16), (3.23), and (3.31), with Dirichlet boundary conditions, there exist the following error estimates.

$$\|u-u_h\|\leq Ch^6,\quad \|u-u_h\|\leq Ch^8,\text{ and } \|u-u_h\|\leq Ch^{10},$$

respectively, where C is constant dependent on k, u, and f.

For mixed boundary condition $u_x|_{x=0}=a,\ u|_{x=1},=b$, the limiting matrix corresponding to (3.32) is of the same form, but with the size $J\times J$, which has the eigenvalues

$$\lambda_j=2-2\cos\frac{(2_j-)\pi}{2J}$$

and the corresponding eigenvectors

$$\eta_j=(\eta_{1,j},\eta_{2,j},\ldots,\eta_{J,j})^T,\ \eta_{m,j}=\cos\frac{(m-1)(2_j-1)\pi}{2J},\quad m=1,\ldots,J.$$

for $j=1,\ldots,J$.

3.6 CONCLUSIONS

The Dirichlet boundary conditions were used to derive the sixth-order, eighth-order, and tenth-order compact finite difference schemes for the one-dimensional Helmholtz problem. To verify the validity, accuracy, efficiency, and robustness of the schemes, a numerical experiment was conducted. In this project, MATLAB software was used. The novel compact finite difference schemes are substantially more efficient than classic second-order methods, according to the computer experiment. No further numerical boundary conditions are required because the system is compact. The results show that there is a significant improvement in the accuracy of $O(h^{10})$ over $O(h^8)$ and that of $O(h^8)$ over $O(h^6)$; also, the error decreases as the number of nodes increases in each of these schemes.

The proposed schemes can be used in mathematical physics and computational finance with the Dirichlet boundary condition because the schemes are more versatile in terms of application to complicated geometries and boundary conditions than previous methods such as non-compact finite difference schemes.

REFERENCES

Al-Shibani, F. S., Ismail, A. I. M., & Abdullah, F. A. (2013). Compact finite difference methods for the solution of one dimensional anomalous sub-diffusion equation. *General Mathematical Notes, 18*(2), 104–119.

Amini, N., Shin, C., & Lee, J. (2021). Second-order implicit finite-difference schemes for the acoustic wave equation in the time-space domain. *Geophysics, 86*(5), T421–T437.

Babuska, I. M., & Sauter, S. A. (1997). Is the pollution effect of the FEM avoidable for the Helmholtz equation considering high wave numbers? *SIAM Journal on Numerical Analysis, 34*(6), 2392–2423.

Cocquet, P. H., Gander, M. J., & Xiang, X. (2021). Closed form dispersion corrections including a real shifted wavenumber for finite difference discretizations of 2d constant coefficient Helmholtz problems. *SIAM Journal on Scientific Computing, 43*(1), A278–A308.

Córdova, L. J., Rojas, O., Otero, B., & Castillo, J. (2014). A comparative study of two compact finite differences methods: standard vs. mimetic. In *Ingeniería y ciencias aplicadas: modelos matemáticos y computacionales: memorias del XII Congreso Internacional de Métodos Numéricos en Ingeniería y Ciencias Aplicadas: CIMENICS'2014: Isla de Margarita, Venezuela, 24 al 26 de marzo de 2014* (pp. 1–6). Sociedad Venezolana de Métodos Numéricos en Ingeniería.

Córdova, L. J., Rojas, O., Otero, B., & Castillo, J. (2016). Compact finite difference modeling of 2-D acoustic wave propagation. *Journal of Computational and Applied Mathematics, 295*, 83–91.

Dastour, H., & Liao, W. (2021). A generalized optimal fourth-order finite difference scheme for a 2D Helmholtz equation with the perfectly matched layer boundary condition. *Journal of Computational and Applied Mathematics, 394*, 113544.

Dastour, H., & Liao, W. (2021). An optimal 13-point finite difference scheme for a 2D Helmholtz equation with a perfectly matched layer boundary condition. *Numerical Algorithms, 86*(3), 1109–1141.

Duressa, G. F., Bullo, T. A., & Kiltu, G. G. (2016). Fourth order compact finite difference method for solving one dimensional wave equation. *International Journal of Engineering and Applied Sciences, 8*(4), 30–39.

Elconsul, A. M., & Lagha, E. O. (2013). Compact finite difference schemes for one-dimensional Helmholtz equation. *University Bulletin, 2*(15), 5–22.

Erlangga, Y., & Turkel, E. (2012). Iterative schemes for high order compact discretizations to the exterior Helmholtz equation*. *ESAIM: Mathematical Modelling and Numerical Analysis*, *46*(3), 647–660.

Erturk, E., & Gökçöl, C. (2006). Fourth-order compact formulation of Navier–Stokes equations and driven cavity flow at high Reynolds numbers. *International Journal for Numerical Methods in Fluids*, *50*(4), 421–436.

Fang, Q., Nicholls, D. P., & Shen, J. (2007). A stable, high-order method for three-dimensional, bounded-obstacle, acoustic scattering. *Journal of Computational Physics*, *224*(2), 1145–1169.

Fu, Y. (2008). Compact fourth-order finite difference schemes for Helmholtz equation with high wave numbers. *Journal of Computational Mathematics*, 98–111.

Harari, I., & Turkel, E. (1995). Accurate finite difference methods for time-harmonic wave propagation. *Journal of Computational Physics*, *119*(2), 252–270.

Hermanns, M., & Hernandez, J. A. (2008). Stable high-order finite-difference methods based on non-uniform grid point distributions. *International Journal for Numerical Methods in Fluids*, *56*(3), 233–255.

Hirsh, R. S. (1975). Higher order accurate difference solutions of fluid mechanics problems by a compact differencing technique. *Journal of Computational Physics*, *19*(1), 90–109.

Ito, K., Qiao, Z., & Toivanen, J. (2008). A domain decomposition solver for acoustic scattering by elastic objects in layered media. *Journal of Computational Physics*, *227*(19), 8685–8698.

Jiang, Y., & Ge, Y. (2020). An explicit fourth-order compact difference scheme for solving the 2D wave equation. *Advances in Difference Equations*, *2020*(1), 1–14.

Lele, S. K. (1992). Compact finite difference schemes with spectral-like resolution. *Journal of computational physics*, *103*(1), 16–42.

Li, L., Jiang, Z., & Yin, Z. (2018). Fourth-order compact finite difference method for solving two-dimensional convection–diffusion equation. *Advances in Difference Equations*, *2018*(1), 1–24.

Li, M., & Tang, T. (2001). A compact fourth-order finite difference scheme for unsteady viscous incompressible flows. *Journal of Scientific Computing*, *16*(1), 29–45.

Li, M., Tang, T., & Fornberg, B. (1995). A compact fourth-order finite difference scheme for the steady incompressible Navier-Stokes equations. *International Journal for Numerical Methods in Fluids*, *20*(10), 1137–1151.

Li, S., Wang, J., & Luo, Y. (2015). A fourth-order conservative compact finite difference scheme for the generalized RLW equation. *Mathematical Problems in Engineering*, *2015*.

Liu, D., Kuang, W., & Tangborn, A. (2009). High-order compact implicit difference methods for parabolic equations in geodynamo simulation. *Advances in Mathematical Physics*, *2009*.

Nabavi, M., Siddiqui, M. K., & Dargahi, J. (2007). A new 9-point sixth-order accurate compact finite-difference method for the Helmholtz equation. *Journal of Sound and Vibration*, *307*(3–5), 972–982.

Okoro, F. M., & Owoloko, E. A. (2010). A one-way dissection of high-order compact scheme for the solution of 2D Poisson equation. *International Journal of Physical Sciences*, *5*(8), 1277–1283.

Oyakhire, F.I. (2017). Compact finite difference schemes for 1D, 2D and 3D Helmholtz equations sing Pade approximations. *IOSR Journal of mathematics (IOSR-JM)*, 15 (6), 10–19.

Oyakhire, F. I. (2019). High order compact finite difference techniques for stochastic advection diffusion equation. *International Journal of Mathematics and Statistics Studies*, 7(3), 1–11.

Oyakhire, F. I., Ibina E. O., Okoro U. U. (2019). Modified asymmetric for Black-Scholes equation of European options pricing system, *International Journal of Mathematics and Statistics Studies*, *7*(3), 12–21.

Sanyasiraju, Y. V. S. S., & Manjula, V. (2005). Higher order semi compact scheme to solve transient incompressible Navier-Stokes equations. *Computational Mechanics*, *35*(6), 441–448.

Shen, J., & Wang, L. L. (2005). Spectral approximation of the Helmholtz equation with high wave numbers. *SIAM Journal on Numerical Analysis*, *43*(2), 623–644.

Singer, I., & Turkel, E. (2006). Sixth-order accurate finite difference schemes for the Helmholtz equation. *Journal of Computational Acoustics*, *14*(03), 339–351.

Singer, I., & Turkel, E. (1998). High-order finite difference methods for the Helmholtz equation. *Computer Methods in Applied Mechanics and Engineering*, *163*(1–4), 343–358.

Spotz, W. F. (1995). *High-order compact finite difference schemes for computational mechanics* (Doctoral dissertation, The University of Texas at Austin).

Spotz, W. F., & Carey, G. F. (1995). High-order compact finite difference methods. In *Preliminary Proceedings International Conference on Spectral and High Order Methods*, Houston, TX, pp. 397–408.

Spotz, W. F., & Carey, G. F. (2001). Extension of high-order compact schemes to time-dependent problems. *Numerical Methods for Partial Differential Equations: An International Journal*, *17*(6), 657–672.

Sutmann, G. (2007). Compact finite difference schemes of sixth order for the Helmholtz equation. *Journal of Computational and Applied Mathematics*, *203*(1), 15–31.

Tian, Z., & Ge, Y. (2003). A fourth-order compact finite difference scheme for the steady stream function–vorticity formulation of the Navier–Stokes/Boussinesq equations. *International Journal for Numerical Methods in Fluids*, *41*(5), 495–518.

Turkel, E., Gordon, D., Gordon, R., & Tsynkov, S. (2013). Compact 2D and 3D sixth order schemes for the Helmholtz equation with variable wave number. *Journal of Computational Physics*, *232*(1), 272–287.

Wang, F., Pan, M., & Wang, Y. (2017). Construction of compact finite difference schemes by classic differential quadrature. *Applied Sciences*, *7*(3), 284.

Wong, Y. S., & Li, G. (2011). Exact finite difference schemes for solving Helmholtz equation at any wavenumber. *International Journal of Numerical Analysis and Modeling, Series B*, *2*(1), 91–108.

Xiu, D., & Shen, J. (2007). An efficient spectral method for acoustic scattering from rough surfaces. *Communication Computer Physics*, *2*(1), 54–72.

Zhang, K. K., Shotorban, B., Minkowycz, W. J., & Mashayek, F. (2006). A compact finite difference method on staggered grid for Navier–Stokes flows. *International Journal for Numerical Methods in Fluids*, *52*(8), 867–881.

4 Study of Second Harmonic (SH) Wave Conversion from Fundamental Wave with Plane Wave Approximation

Madhu, Naman Mathur, and Prashant Povel Dwivedi
Manipal University, Jaipur

CONTENTS

4.1 INTRODUCTION

One of the foremost fundamental wave conversion techniques is second harmonic generation, and it serves as the basis for several useful applications. The effect of second harmonic generation (SHG) was first observed in 1961 [1] and theoretically analyzed in 1962 [2]. Since then, it's been of excessive interest in both applied and fundamental physics, but this effect is usually weak as in most nonlinear optical phenomena, and really intense light waves are required. There are many efforts that are aimed at searching ways to extend the nonlinear response. One such approach is using the surface plasmon resonance (SPR) phenomenon in metal nanoparticles (NPs) [3,4]. These NPs are known to greatly enhance the electrical field of an incident light wave at the SPR that significantly strengthens a nonlinear response [5]. In the literature, there's overwhelming written evidence, based on both experimental and theoretical studies, to refute the likelihood of generating an optical second harmonic on the passage of light through isotropic media [6].

DOI: 10.1201/9781003222255-4

It is documented that there are two mechanisms that will cause the depletion of the fundamental wave within the intracavity SHG process: The primary one is the energy conversion from the fundamental wave to the SHG wave within the SHG nonlinear process, resulting in the depletion of the fundamental wave, and the second process includes the results of the primary process. The circulating fundamental power within the laser cavity is reduced, thanks to the extra loss of the fundamental wave caused by the SHG nonlinear process [7]. In this chapter, we are using the tactic of energy conversion from fundamental wave to second harmonic generation wave within the nonlinear process.

In this chapter, we are using the method of energy conversion from fundamental wave to second harmonic generation wave in the nonlinear process. We present a model that comprehensively and intuitively models the depletion of the fundamental wave, and with the help of second harmonic intensity and the intensity of the fundamental wave (input wave), we have to calculate the intensity of the fundamental wave. Due to the mechanisms of input intensity I_0, it considers the depletion of the fundamental wave as a function of input intensity, at the condition of phase matching.

4.2 THEORY OF SECOND HARMONIC GENERATION (SHG)

In this chapter, we are considering that the fundamental wave incident on a nonlinear crystal (PPLN) will generate a second harmonic wave. The two types of losses, i.e., linear and nonlinear losses, in the crystal are considered in the model. The linear loss is a result of imperfect coatings at the surface of the nonlinear crystal, while the nonlinear loss is caused by the depletion of the fundamental wave (due to the energy conversion from the fundamental wave to the SHG wave).

The fundamental wave depletion inside the nonlinear crystal is accounted for within the nonlinear coupled wave equations, and then, the calculated nonlinear coefficient includes both the nonlinear and linear depletion. To explain the second-order nonlinear processes, first, we discuss the process of polarization. We all know that optics is the study of the interaction of electromagnetic radiation and matter. When an electromagnetic wave interacts with a dielectric material, it induces a polarization – the dipole moment per unit volume. The dipole moment primarily arises due to the displacement of the valence electrons from their stationary orbits. The direction and magnitude of the polarization are dependent on the magnitude, the direction of propagation, and the polarization direction of the applied electric field inside the medium. The induced polarization, P, can be expressed as a power series of the applied electric field in the form [8]:

$$\vec{P} = \chi^{(1)}\vec{E} + \chi^{(2)}\vec{E}\ \vec{E} + \chi^{(3)}\vec{E}\ \vec{E}\ \vec{E} + \cdots$$

$$\vec{P} = \vec{P}^{L} + \vec{P}^{NL} \tag{4.1}$$

where ε is the permittivity of the vacuum, $\bar{E}$ is the electric field component of the electromagnetic wave, and χ is the susceptibility tensor.

The linear polarization is defined as:

$$\vec{P}^L = \varepsilon\chi^{(1)}\vec{E} \tag{4.2}$$

The nonlinear polarization is represented as:

$$\vec{P}^{NL} = \varepsilon(\chi^{(2)}\vec{E}\ \vec{E} + \chi^{(3)}\vec{E}\ \vec{E}\vec{E} + \ldots) \tag{4.3}$$

The second-order nonlinear polarization can be represented as follows:

$$\vec{P}^{NL} = \varepsilon\chi^{(2)}\vec{E}\ \vec{E} \tag{4.4}$$

where $\chi^{(2)}$ is the second-order nonlinear susceptibility tensor.

Let us consider the optical field incident upon a second-order nonlinear optical $\chi^{(2)}$ medium, which consists of two distinct frequency components as:

$$\vec{E}(t) = \vec{E}_1 e^{i\omega_1 t} + \vec{E}_2 e^{i\omega_2 t} \tag{4.5}$$

The second-order contribution to the nonlinear polarization is of the form:

$$\vec{P}^2(t) = \varepsilon\chi^{(2)}\ \vec{E}(t)^2 \tag{4.6}$$

$$\vec{E}(t) = \frac{1}{2}\left(\vec{E}_1 e^{i\omega_1 t} + \vec{E}_2 e^{i\omega_2 t} + \vec{E}_1^* e^{-i\omega_1 t} + \vec{E}_2^* e^{-i\omega_2 t}\right) = \frac{1}{2}\left(\vec{E}_1 e^{i\omega_1 t} + \vec{E}_2 e^{i\omega_2 t} + CC\right) \tag{4.7}$$

Since this is a nonlinear relation, the optical field should be written in the real form. Substituting equation (4.7) in equation (4.6), we get

$$P^2(t) = \varepsilon\chi^{(2)}\frac{1}{4}\left(\vec{E}_1 e^{i\omega_1 t} + \vec{E}_2 e^{i\omega_2 t} + \vec{E}_1^* e^{-i\omega_1 t} + \vec{E}_2^* e^{-i\omega_2 t}\right)^2$$

$$= \varepsilon\chi^{(2)}\frac{1}{4}\left(E_1^2 e^{2i\omega_1 t} + E_2^2 e^{2i\omega_2 t} + 2E_1E_2 e^{i(\omega_1+\omega_2)t} + 2E_1E_2^* e^{i(\omega_1-\omega_2)t} + C.C\right)$$

$$+\ldots\varepsilon\chi^{(2)}\frac{1}{2}\left(E_1E_1^* + E_2E_2^*\right) \tag{4.8}$$

In equation (4.8), the first two terms on the right-hand side represent the physical process described as second harmonic generation (SHG), the third and fourth terms represent the physical processes called sum frequency generation (SFG) and difference frequency generation (DFG), respectively, and the last term is known as the optical rectification (OR). We do not study the complex conjugates (c.c.) of equation (4.8), as they do not lead to any extra processes other than the above-mentioned processes.

To understand further the generation of the new frequencies and the transfer of energy among the interactive waves, we have to solve the Maxwell's electromagnetic equations in the medium [4.8].

$$\nabla^2 \vec{E} + \frac{n^2}{c^2}\frac{\partial^2 \vec{E}}{\partial t^2} = -\mu_0 \frac{\partial^2 \vec{P}_{NL}}{\partial t^2} \tag{4.9}$$

For interactions (linear and nonlinear), where the amplitudes of the fields change slowly on the scale of the wavelength in space and the optical period in time, one can invoke the slowly varying envelope approximation (SVEA) [9,10], which simplifies our second-order wave equation (4.9) to a first-order derivative in the propagation direction as [8]:

$$\frac{\partial E(\omega)}{\partial z} = \frac{i\mu_0 c\sigma}{2n} P^{NL}(\omega) \tag{4.10}$$

For the process of second harmonic generation ($\omega_1 = \omega_2 = \omega$ and $\omega_3 = 2\,\omega$), there are only two interacting waves, ω and 2ω. The coupled wave equations for SHG can then be represented as [8]:

$$\frac{\partial E_{2\omega}}{\partial z} = ik\frac{2\omega}{n_{2\omega}c} d_{NL} E_{\omega}^2 \exp(i\Delta kz) \tag{4.11}$$

$$\frac{\partial E_{\omega}}{\partial z} = ik\frac{2\omega}{n_{\omega}c} d_{NL} E_{\omega}^{*} E_{2\omega} \exp(-i\Delta kz) \tag{4.12}$$

Equation (4.11) represents the generated second harmonic wave, and equation (4.12) represents the evolution of the fundamental wave, where E_ω and $E_{2\omega}$ are the field amplitudes of the fundamental and second harmonic waves, respectively, d_{NL} is the nonlinear coefficient of the nonlinear crystal along the propagation direction of the fundamental wave, ω and 2ω are the angular frequencies of the fundamental and second harmonic waves, respectively, and n_ω and $n_{2\omega}$ are the refractive indices of the medium at ω and 2ω, respectively. At low conversion efficiency, where the intensity of fundamental wave can be assumed to be constant, the above two equations (4.11) and (4.12) can be reduced to one linear first-order equation and that equation can be solved analytically in plane wave approximation. For this simplified case, the second harmonic intensity is given by [9,10]:

$$I_{2\omega} = \frac{\varepsilon n_{2\omega} c}{2}\left|E_{2\omega}^2\right| = \frac{2\omega^2 d_{NL}^2 L^2 I_{\omega}^2(0)}{\varepsilon n_{\omega}^2 n_{2\omega} c^3} \sin c^2\left(\frac{\Delta kL}{2}\right) \tag{4.13}$$

where L is the length of the nonlinear material and I_ω is the fundamental wave intensity.

$$\eta_0(SHG) = \frac{I_{2\omega}L}{I_{\omega}} \tag{4.14}$$

where $\eta_{0\,(SHG)}$ is the SHG efficiency for depleted wave approximation and can be represented as above.

4.3 SHG – PLANE WAVE

The next step is to determine the equation of second harmonic generation using plane wave (SHG – plane wave) [11].

$$\frac{dA_{2\omega}}{dz} = i\left(\frac{\omega d_{eff}}{n_2\ c}\right)A_\omega^2 e^{-i\Delta kz} \tag{4.15}$$

where $\Delta k = k_{2\omega} - 2k_\omega$

$$\frac{dA_\omega}{dz} = i\left(\frac{\omega d_{eff}}{n_1 c}\right)A_{2\omega}A^* e^{i\Delta kz} \tag{4.16}$$

In the equations, A_ω and $A_{2\omega}$ are the field amplitudes of the fundamental and second harmonic waves, respectively, and d_{eff} is the nonlinear coefficient of the nonlinear crystal along the propagation direction of the fundamental wave.

After normalization,

$$z' = \frac{Z}{L} \quad B_{2\omega}(z) = \frac{\sqrt{n_2}\,A_{2\omega}\ (z)}{\sqrt{n_1}\,|A_\omega(0)|} \text{ and } B_\omega(z) = \frac{\sqrt{n_1}\,A_\omega(z)}{\sqrt{n_2}\,|A_\omega(0)|} \tag{4.17}$$

With the help of equations (4.16) and (4.17), we get the following equations:

$$\frac{dB_2}{dz} = i\Gamma_L B_1^2 e^{-i\Delta k_l z} \tag{4.18}$$

$$\frac{dB_1}{dz} = i\Gamma_L B_2 B_1^* e^{i\Delta k_L z} \tag{4.19}$$

where $\Delta k_L = \Delta kL$; phase mismatch $\Gamma_L = \Gamma L$; $\Gamma = \frac{\omega d_{eff}}{c\sqrt{n_1 n_2}}|A_\omega(0)|$; and Γ_L is the sample length of crystal.

$$I_\omega = \frac{1}{2}\varepsilon_0 c n_\omega |A_\omega(0)| \text{ is the input (fundamental wave) intensity} \tag{4.20}$$

We have used equations (4.18) and (4.19) for our simulation analysis for conversion from fundamental to second harmonic wave.

4.4 RESULTS AND DISCUSSION

At the phase matching, the fundamental wave intensity $I_\omega = 3.3462 * 10^{13}\ W/m^2$, which has been calculated by equation (4.20); in Figure 4.1(a), along the normalized length of the crystal at z=0, the fundamental wave intensity (I_ω) is maximum. At the condition of phase matching (Δk=0), the fundamental wave decreases and the second harmonic wave increases up to the normalized length of 0.6, indicating energy transfer from fundamental wave to second harmonic wave, and at the normalized length of 0.6, both the fundamental and second harmonic waves have equal amounts of energy. In the normalized length range from 0.6 up to 1, the intensity of the second harmonic wave goes to maximum. At the value of normalized length of 1, we get 80% energy conversion in the second harmonic wave. In the complete convergence process from fundamental to second harmonic wave, the total energy is conserved as shown in Figure 4.1a.

In Figure 4.1b, at the condition of phase mismatching ($\Delta k = \pi$), there is a transfer of energy from fundamental wave to second harmonic wave up to the normalized length of 0.7 and the conversion efficiency is approximated as 38%. Above the normalized length of 0.7 of the crystal, the conversion efficiency decreases and reaches 20% at the end of the crystal. And also we observe that at phase mismatching, there is no overlap of energy between fundamental and second harmonic waves.

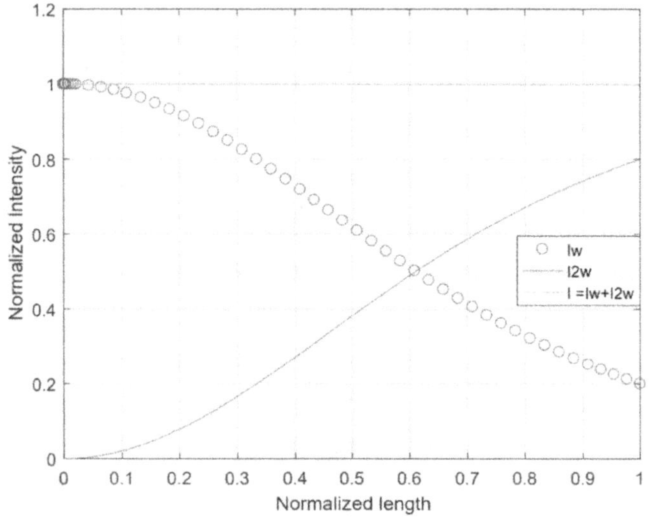

(a)

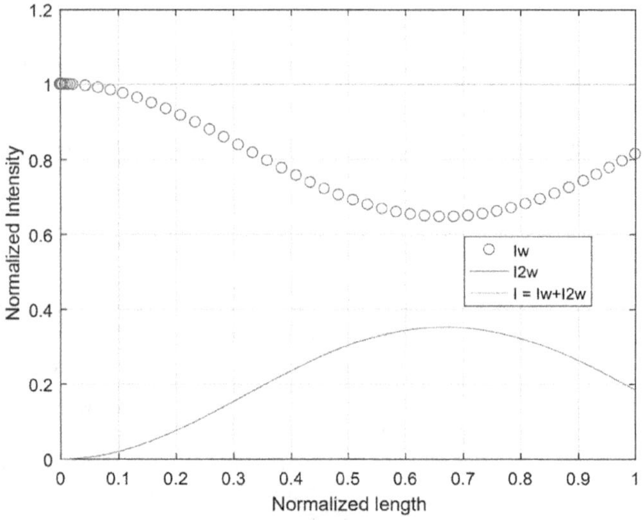

(b)

FIGURE 4.1 Fundamental and second harmonic wave intensities along the sample length with (a) phase matching and (b) phase mismatching.

In Figure 4.2a, when the input intensity is increased fourfold ($4I_\omega$), at the condition of phase matching ($\Delta k = 0$), the fundamental wave decreases and the second harmonic wave increases up to the normalized length of 0.5, indicating energy transfer from fundamental wave to second harmonic wave, and at the normalized length of 0.5, both the fundamental and second harmonic waves have equal amounts of energy. In the range of the normalized length between 0.5 and 1, the intensity of the second harmonic wave becomes maximum. At the value of the normalized length

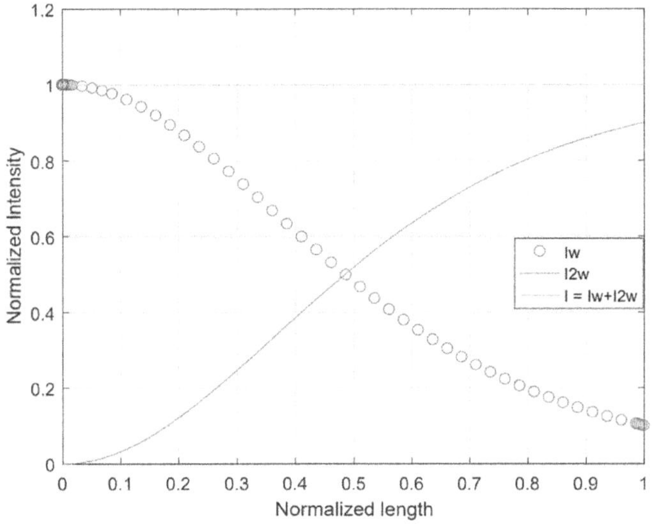

(a)

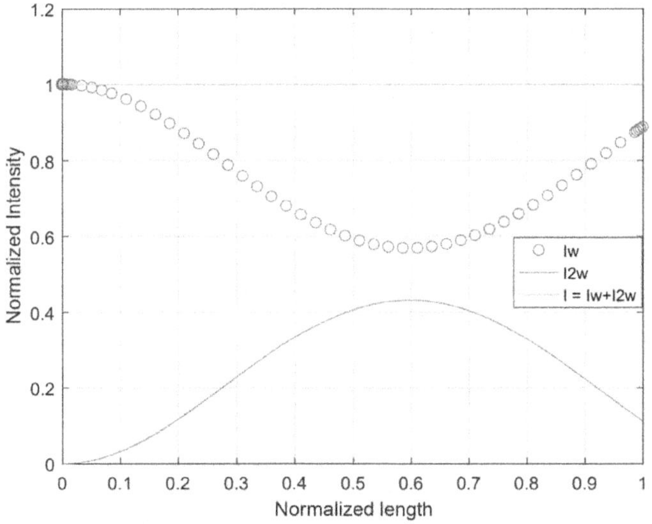

(b)

FIGURE 4.2 Fundamental $(4I_\omega)$ and second harmonic wave intensities along the sample length with (a) phase matching and (b) phase mismatching.

of 1, we get 90% energy conversion in the second harmonic wave. In the complete convergence process from fundamental to second harmonic wave, the total energy is conserved as shown in Figure 4.2a.

In Figure 4.2b, at the condition of phase mismatching ($\Delta k = \pi$), there is a transfer of energy from fundamental to second harmonic wave up to the normalized length of 0.6 and the conversion efficiency is approximated 41%. Above the normalized length of the crystal of 0.6, the conversion efficiency decreases and reaches 10% at the end of the crystal. And also we observe that at phase mismatching, there is no overlap of energy between the fundamental and second harmonic waves.

4.5 CONCLUSIONS

In conclusion, the energy is transferred from the depleted fundamental wave to second harmonic wave in the nonlinear process. The simulation results of the normalized intensity vs the normalized length of crystal are determined. At the phase-matching condition, we conclude that the energy transfer from fundamental to second harmonic wave is maximized and, at mismatching, it is minimized. So we can say that phase matching plays a crucial role in the conversion from fundamental to second harmonic wave. One more observation from our simulation results is that if the intensity of the fundamental wave is increased, then the position of normalized length of the crystal decreases, where the conversion of energy is 50% and also the conversion efficiency is increased. So we can say that for depleted fundamental wave, the conversion efficiency depends upon the phase matching between the fundamental and second harmonic waves, and the fundamental wave intensity.

REFERENCES

1. Franken, P. A. (1961). Generation of optical harmonics. *Physics Review Letters*.
2. Bloembergen, N., & Pershan, P. S. (1962). Light waves at the boundary of nonlinear media. *Physics Review.*
3. Maier, S. P. (2007). *Fundamentals and Applications.* Springer: Bath.
4. Shahbazyan, T. V., & Stockman, M. I. (2013). *Plasmonics: Theory and Applications.* Plasmonics: Theory and Applications.
5. Lipovskii, S. A. (2018). Understanding the second-harmonic generation enhancement and behavior in Metal Core–dielectric shell nanoparticles. *The Journal of Physical Chemistry.*
6. David, L., Andrews, P. A. (1995). Theory of second harmonic generation in randomly oriented species. *Chemical Physics.*
7. Liam Flannigan, T. K.-Q. (March 2021). Study of fundamental wave depletion in intracavity second harmonic generation. *Optics Express.*
8. Samanta, G. K. (2009). *High-Power, Continuous-Wave Optical Parametric Oscillators from Visible to Near-Infrared*, Ph.D. Dissertation, Universitat Politecnica de Catalunya Barcelona.
9. Sutherland, R. L. (1996). *Handbook of Nonlinear Optics. Second Edition.* Marcel Dekker, Inc.
10. Boyd, R. W. (2003). *Nonlinear Optics. Second Edition.* Academic Press.
11. Povel, D. P. (2015). *Fabrication of Periodically Poled LiNbo3 Crystal and Evaluation of Domain Randomness by diffraction Noise Measurement*, Ph. D. Dissertation, Pusan National University.

5 Mathematical Analysis for Small Signal Stability of Electric Drives

Anubhav Agrawal
BML Munjal University

CONTENTS

5.1 INTRODUCTION: BACKGROUND AND DRIVING FORCES

By far, DC machines have been used in industries for variable speed operation because of their easy speed control, but such drives suffer from limited overloading capacity and frequent maintenance. The introduction of power electronic devices has made it possible to achieve precise speed control and less maintenance of squirrel cage IMD [1]. The supply frequency is directly related to the speed of the rotating magnetic field. Hence, variable frequency and voltage magnitude offers desired speed and torque [2,3]. The variable frequency and voltage magnitude can be achieved through power electronic converters, either two stage (AC–DC–AC) or single stage (AC–AC).

In this study, a single-stage (AC–AC) power electronic converter known as 'matrix converter' (MC) is used [4–8]. This direct AC–AC converter offers several significant advantages over the other existing AC–AC commercial converters, such as adjustable power factor, four-quadrant operation, absence of large DC-link capacitor, and capability of regeneration and high-quality sinusoidal input/output waveforms. The AC drive saturates in open-loop operation; hence, it is mandatory to study the closed-loop operation of the drive [9–11].

DOI: 10.1201/9781003222255-5

The induction machines had been operated using a scalar control technique known as V/F control, at constant flux [12]. This control strategy is easy to implement and cost-effective and allows the induction machine to operate at knee point, but when compared to the vector control, the machine drive offers poor performance. In this study, the drive system is modeled using the vector control scheme, which has superior properties [13].

Mathematics is a crisp language with well-defined rules and helps us to manipulate and formulate ideas and underlying assumptions [14]. The mathematical modeling of the induction motor drive is done in d–q reference frame. The optimal design of PI controller is necessary for stable performance of the drive system. The controller parameters strongly affect the response of the system, and therefore, an accurate design of the PI controller parameters is of utmost importance. Various mathematical techniques have been proposed for PI controller design. The conventional Ziegler–Nichols tuning for PID controller is discussed in [15], and an efficient automatic tuning algorithm is presented in [16]. In yet another study [17], a fuzzy neural network-based algorithm is illustrated to design the PID controllers.

The D-partition technique is used to determine the boundaries of a stable PI controller parameter plane. This technique, in contrast to conventional methods, is capable of coordinating parameters in parametric plane irrespective of the system order [18–20]. The D-partition technique is applied to locate the stability domain in controller parameters [21]. The authors in [22] presented the closed-loop DC drive performance based on parameter plane synthesis. In [23,24], the stable gain parameters of CSI-fed IMD are calculated through the same technique.

In this paper, a novel mathematical technique is proposed to design the PI controller parameters of an AC drive system. The analysis is done in continuous time domain. The characteristic equation of control loop is developed through small perturbations about a steady-state operating point. The probable stable region obtained from the D-partitioning technique is checked by frequency scanning technique to validate its stability. Thereafter, the response of control loop is determined at the different controller parameters, selected from the stable region. The best suitable parameters are selected from the step variation in reference value.

5.2 SYSTEM DESCRIPTION

The power electronic converter taken up in this study is MC. This converter system is an array of 'm*n' switches structured such that it connects 'm' input phases to 'n' output phases [25–27]. In particular, it is a three-phase to three-phase forced commutated cycloconverter with bidirectional switches (BS) that connect each output phase to each input phase without using DC-link elements, as shown in Figure 5.1. The output voltage is constructed by proper cutting of the input voltages using an appropriate switching algorithm.

5.3 MATHEMATICAL MODELING OF THE DRIVE

The mathematical block diagram of the drive is shown in Figure 5.2. It comprises of a PI controller, a power electronic converter, a 3-Φ induction machine, and a feedback

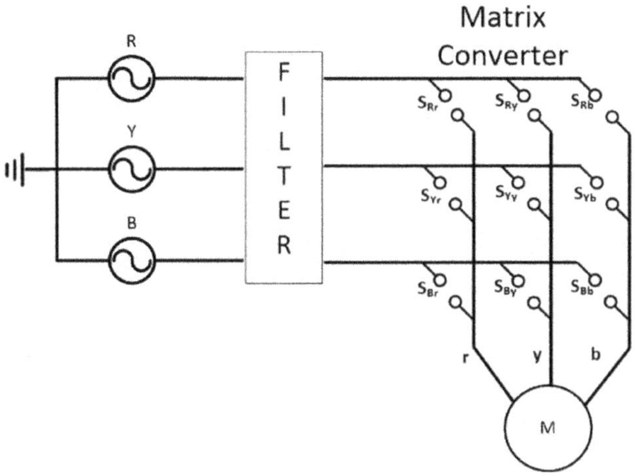

FIGURE 5.1 Basic power circuit of MC.

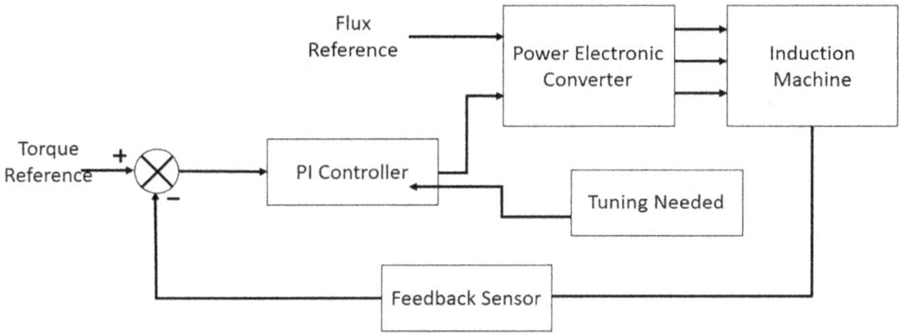

FIGURE 5.2 Block diagram of the proposed drive scheme.

sensor. The AC–AC power electronic converter is operated using the modified Venturini algorithm. The torque component of the motor current (q-axis) is sensed and used as a feedback signal after being filtered in a low-pass filter. The current feedback signal is compared with the torque reference to generate the error signal. The error is handled through the PI controller, which determines the reference voltage signals for the 'd'-axis and 'q'-axis voltages (Vds* and Vqs*). Here, Vqs* is a function of iqs*. The controlled output voltage of MC is fed to the motor for desired performance.

The matrix converter output voltages can be written as [9]:

$$V_{as} = qV_{im}\cos(\omega_o t) - \frac{q}{6}V_{im}\cos(3\omega_o t) + \frac{1}{4}V_{im}\cos(3\omega_i t) + \text{harmonics} \tag{5.1}$$

$$V_{bs} = qV_{im}\cos\left(\omega_o t \frac{2\pi}{3}\right) - \frac{q}{6}V_{im}\cos(3\omega_o t) + \frac{1}{4}V_{im}\cos(3\omega_i t) + \text{harmonics} \tag{5.2}$$

$$V_{cs} = qV_{im}\cos\left(\omega_o t + \frac{4\pi}{3}\right) - \frac{q}{6}V_{im}\cos(3\omega_o t) + \frac{1}{4}V_{im}\cos(3\omega_i t) + \text{harmonics} \tag{5.3}$$

These voltages can be transformed to $d^e - q^e$ reference frame rotating at a synchronous electrical angular velocity of the fundamental component of voltage applied to the machine stator (i.e., ω_e). For convenience, the d^e-axis is aligned with the 'r'-axis of the input supply connected at stator terminal at t=0 $(\theta = \omega_e t)$. Therefore, the output frequency ω_o is the same as ω_e. The transformed voltages can be written as:

$$V_{qs} = V_o\left[1 - \frac{1}{6}\cos(\omega_e t)\cos(3\omega_e t) + \frac{1}{4q}\cos(\omega_e t)\cos(3\omega_i t) + \text{harmonics}\right] \quad (5.4)$$

$$V_{ds} = V_o\left[-\frac{1}{6}\cos(\omega_e t)\cos(3\omega_e t) + \frac{1}{4q}\cos(\omega_e t)\cos(3\omega_i t) + \text{harmonics}\right] \quad (5.5)$$

5.3.1 Current Transducer

The output of the current transducer is directly proportional to the current of the motor. Therefore, the gain is constant. A low-pass filter is used to reduce the ripples and harmonics from the current. The speed and current feedback signal are the following:

$$i_{qsf} = \frac{1}{1 + pT_f} i_{qs} \quad (5.6)$$

5.3.2 Current Controller

The PI controller is used to achieve the current control since this ensures quick response with zero steady-state error and is given as:

$$V_{qs} = \frac{k_1(1 + pT_1)}{pT_1}\left(i_{qs}^* - i_{qsf}\right) \quad (5.7)$$

5.3.3 Induction Motor

The IM is modeled in d–q reference frame and can be described by the following fourth-order matrix equation in $d^e - q^e$ reference frame [2]:

$$\begin{bmatrix} V_{qs}^e \\ V_{ds}^e \\ V_{qr}^e \\ V_{dr}^e \end{bmatrix} = \begin{bmatrix} R_s + \frac{p}{\omega_b}X_s & \frac{\omega_e}{\omega_b}X_s & \frac{p}{\omega_b}X_m & \frac{\omega_e}{\omega_b}X_m \\ -\frac{\omega_e}{\omega_b}X_s & R_s + \frac{p}{\omega_b}X_s & -\omega_e\frac{X_m}{\omega_b} & \frac{p}{\omega_b}X_m \\ \frac{p}{\omega_b}X_m & \frac{\omega_{sl}}{\omega_b}X_m & R_r + \frac{p}{\omega_b}X_r & \frac{\omega_{sl}}{\omega_b}X_r \\ -\frac{\omega_{sl}}{\omega_b}X_m & \frac{p}{\omega_b}X_m & -\frac{\omega_{sl}}{\omega_b}X_r & R_r + \frac{p}{\omega_b}X_r \end{bmatrix} \begin{bmatrix} i_{qs}^e \\ i_{ds}^e \\ i_{qr}^e \\ i_{dr}^e \end{bmatrix} \quad (5.8)$$

The rotor flux linkages for the 'd'-axis and 'q'-axis, keeping in mind that the flux is aligned with the 'd'-axis, are given as:

$$i_{qr}^{e} = -\frac{L_m}{L_r} i_{qs}^{e} \tag{5.9}$$

$$i_{dr}^{e} = -\frac{\varphi_{dr}}{L_r} - \frac{L_m}{L_r} i_{ds}^{e} \tag{5.10}$$

Using equations (5.8), (5.9), and (5.10), the 'q'-axis stator voltage can be rewritten as:

$$V_{qs}^{e} = \left(R_s + L_a p\right) i_{qs}^{e} + w_e L_s i_{ds}^{e} \tag{5.11}$$

where $L_a = \sigma L_s$ and $\sigma = \left(1 - \frac{L_m^2}{L_r L_s}\right)$.

Also, we know that $w_e = w_r + w_{sl} = w_r + \frac{i_{qs}}{i_{ds}} \frac{R_r}{L_r}$; substituting in equation (5.11) and rearranging, we get:

$$i_{qs}^{e} = \frac{V_{qs}^{e} - w_r L_s i_{ds}^{e}}{R_s + \frac{R_r L_s}{L_r} + L_a p} = \frac{K_{a1}}{\left(1 + sT_{a1}\right)} \left(V_{qs}^{e} - w_r L_s i_{ds}^{e}\right) \tag{5.12}$$

where $R_a = R_s + \frac{R_r L_s}{L_r}$, $K_{a1} = \frac{1}{R_a}$, and $T_{a1} = \frac{L_a}{R_a}$.

5.3.4 Power Converter

The MC output voltage depends upon the reference voltage generated by the current controller. The three-phase input AC voltage is chopped in accordance with the reference signals. Although the output is proportional to these reference values, the switching pulses are not instantaneous. For a simpler analysis, the delay in the

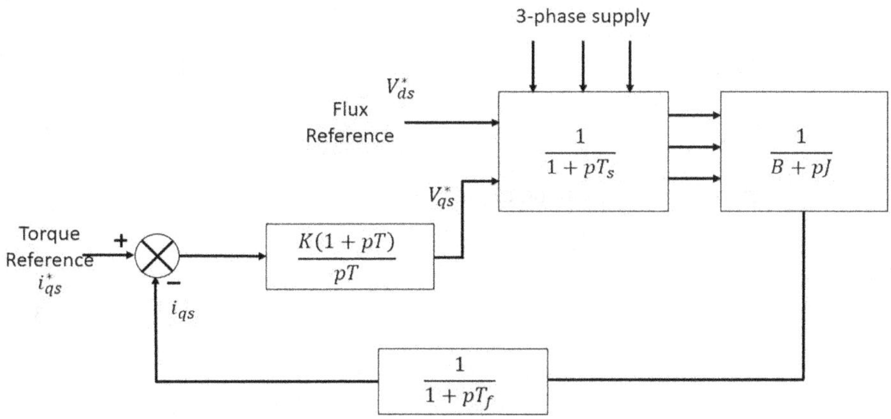

FIGURE 5.3 Mathematical model of the system.

switching pulses is approximated by a simple first-order time lag with a time constant equal to the reciprocal of the switching frequency [23]. Hence, the MC is treated as a linear continuous element, although it is nonlinear sampled data in real time. The output of MC is given by:

$$V_{out} = \frac{1}{1+pT_s} V_{ds}^{*} \tag{5.13}$$

The detailed mathematical model of the drive scheme is shown in Figure 5.3.

5.4 PARAMETER PLANE SYNTHESIS TECHNIQUE FOR CONTROLLER DESIGN

The parameter plane synthesis method (D-partition technique) is used for finding the region in parameter plane as continuous data system, which ensures certain desired qualities of transient response, such as damping ratio and degree of relative stability.

5.4.1 Design of Proportional Integral (PI) Controller

The small perturbations about steady-state operating point are considered to develop the characteristic equations of speed and current loops separately. The controllers are designed on the basis of both system relative stability and response of the drive.

The system of equations (5.8) can be rewritten as:

$$\begin{aligned} w_{sl}V_{qs}^{e} &= w_{sl}\left\{R_s + \frac{p}{w_b}\left(X_s - \frac{X_m^2}{X_r}\right)\right\}i_{qs}^{e} + \left(w_{sl} * w_e\frac{X_m}{w_b} + p\frac{X_m R_r}{X_r} + p^2\frac{X_m}{w_b}\right)i_{dr}^{e} \\ 0 &= w_{sl}\left(-\frac{X_m R_r}{X_r}\right)i_{qs}^{e} + \left\{w_{sl}\frac{X_r^2}{w_b} + \frac{w_b * R_r^2}{X_r} + p(2R_r) + p^2\frac{X_r}{w_b}\right\}i_{dr}^{e} \end{aligned} \tag{5.14}$$

To design the q-axis current controller parameters (K and T), motor flux and speed are assumed to be constant because mechanical time constants of motor are higher than the electrical time constants. Furthermore, slip speed and synchronous speed are also considered constant.

The perturbed quantities can be expressed as:

$$\begin{aligned} V_{qs}^{e} &= V_{qso}^{e} + \Delta V_{qs}^{e} \\ i_{qs}^{e} &= i_{qso}^{e} + \Delta i_{qs}^{e} \\ i_{dr}^{e} &= i_{dro}^{e} + \Delta i_{qs}^{e} \end{aligned} \tag{5.15}$$

Substituting and rearranging using equations (5.14) and (5.15), the characteristic equation is written as:

$$\begin{bmatrix} \omega_{slo}\left\{ \begin{matrix} R_s + \dfrac{p}{w_b}\left(X_s - \dfrac{X_m^2}{X_r}\right) + \\ \dfrac{k(1+pT)}{pT} * \dfrac{1}{1+pT_s} * \dfrac{K_{a1}}{(1+sT_{a1})} * \dfrac{1}{1+pT_f} \end{matrix} \right\} & \left\{ \begin{matrix} \omega_{slo} * \omega_e \dfrac{X_m}{w_b} + \\ p\dfrac{X_m R_r}{X_r} + p^2 \dfrac{X_m}{\omega_b} \end{matrix} \right\} \\ -\omega_{slo}\left(\dfrac{X_m R_r}{X_r}\right) & \left\{ \begin{matrix} \omega_{slo}\dfrac{X_r^2}{\omega_b} + \dfrac{\omega_b * R_r^2}{X_r} + \\ p(2R_r) + p^2 \dfrac{X_r}{\omega_b} \end{matrix} \right\} \end{bmatrix} \begin{bmatrix} \Delta i_{qs}^e \\ \Delta i_{ds}^e \end{bmatrix} = 0 \tag{5.16}$$

The characteristic equation of the current loop is given as $D(p) = 0$, which is rewritten as:

$$\alpha p F_2(p) + \beta F_1(p) + p F_1(p) = 0 \tag{5.17}$$

where $\alpha = 1/K$ and $\beta = 1/T$.

Putting $p = -\sigma + j\omega$ and varying ω from $-\infty$ to $+\infty$ for a constant σ. The D-partition boundary curve in α–β plane for $\sigma = 0$ is shown in Figure 5.4. The curve is shaded according to the sign of Δ, which is given as:

$$\Delta = I_m\left[F_1(p)\right] * R_e\left[pF_2(p)\right] - I_m\left[pF_2(p)\right] * R_e\left[F_1(p)\right] \tag{5.18}$$

Moving into the direction of increasing ω, the boundary curve is shaded twice on left-hand side if $\Delta > 0$ and on the right-hand side if $\Delta < 0$. The innermost region is the most probable stable region. An operating point ($K_1 = 1/15$, $T_1 = 1/150$) is taken from the probable stable region and is substituted in the characteristic equation to study the stability. The locus of the vector D(p) is plotted in the real and imaginary planes with variation in ω as shown in Figure 5.5. The characteristic curve is shaded once on its left-hand side as ω increases from $-\infty$ to $+\infty$. The characteristic curve encloses the

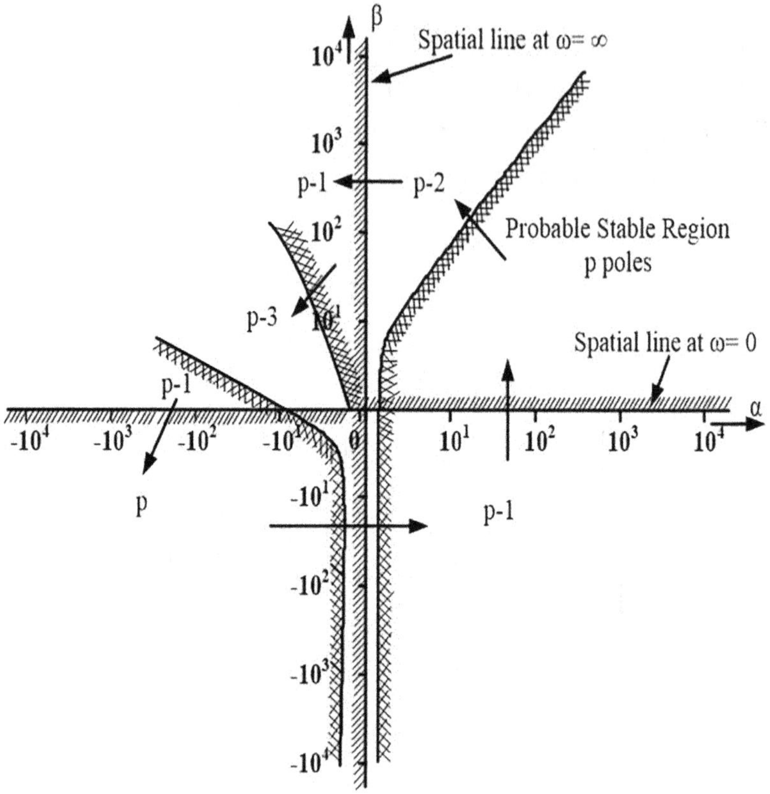

FIGURE 5.4 D-partition boundary for torque controller design for absolute stability (σ=0).

origin, thus a stable response. To strengthen the results, the author has considered few more points ((1/20, 1/200), (1/25, 1/250)) lying inside the closed region and all these points show the similar behavior. Another random point (K12 = 1.0, T12 = 1/500) lying outside this region is considered, and the locus of the characteristic vector D(p) is shown in Figure 5.5. The characteristic curve does not enclose the origin; hence, the region outside region is unstable.

To ensure the maximum relative stability, the D-partition curves are plotted for different values of σ and varying ω from $-\infty$ to $+\infty$. As σ is increased, the stable region with higher degree of stability shrinks. The D-partition curves for variation in σ are shown in Figure 5.6. It is found that the region with higher degree of stability goes on decreasing as σ is increased.

The final selection of current controller parameters from the most stable operating region of the parameter plane is done on the basis of transient response of torque current control loop. The transient response of the control loop is obtained for step variation in reference torque current from 1.0 to 1.05 p.u. of rated current value using MATLAB/Simulink blocksets. The transient response of the current loop is determined assuming that motor is running at constant slip.

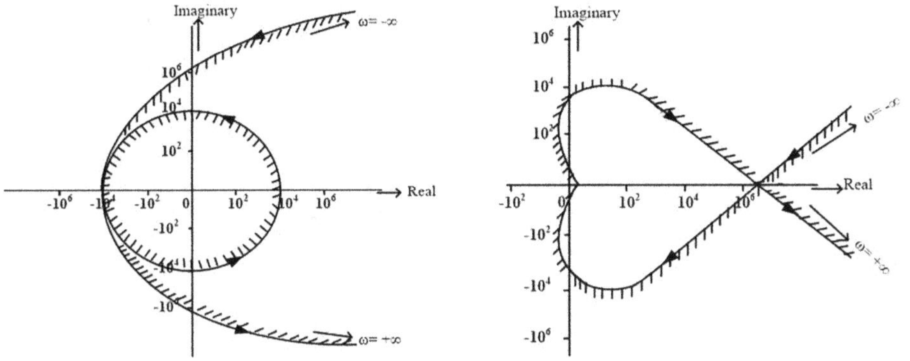

FIGURE 5.5 Frequency scanning technique for torque controller loop parameters (σ constant) with $\sigma=0.0$, $\alpha=15.0$, $\beta=150.0$ (stable) and $\sigma=0.0$, $\alpha=1.0$, $\beta=500.0$ (unstable).

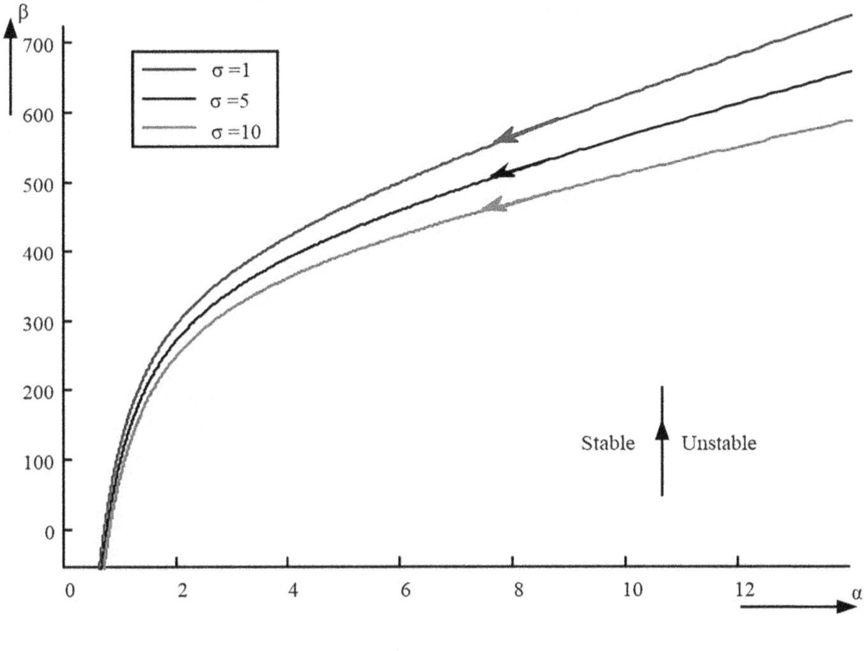

FIGURE 5.6 D-partition boundary for torque controller design for different values of σ.

The percentage overshoot and settling time for a large dataset of α and β are summarized in Table 5.1. The wide variation in PI parameters is studied, and it is clear that the point ($K=1/15$, $T=1/150$) gives the best result.

The response of the drive for some representative points close to the best fit value is listed in Table 5.2. The corresponding transient response curves are shown in Figure 5.7. The comparison of various transient response curves shows that for $K_1=1/15$ and $T_1=1/150$, the settling time is 0.04 seconds with 0.28% peak overshoot. Therefore, the transfer function of the torque current controller is selected as $\dfrac{10(1+0.0067p)}{p}$.

TABLE 5.1
Overshoot (%) and Settling Time (Sec) for Different Sets of *K* and *T*

β \ α	100	150	200	250	300
10	(0.71, 0.76)	(–, 0.104)	(–, 0.065)	(0.18, 0.06)	(0.51, 0.62)
15	(2.67, 0.94)	(0.28, 0.04)	(0.92, 0.06)	(1.14, 0.062)	(2.12, 0.87)
20	(1.93, 0.82)	(–, 0.061)	(0.41, 0.058)	(0.95, 0.058)	(1.45, 0.72)
25	(1.44, 0.78)	(–, 0.075)	(0.12, 0.05)	(0.57, 0.063)	(1.02, 0.68)
30	(1.2, 0.86)	(–, 0.09)	(0.2, 0.05)	(0.33, 0.057)	(0.72, 0.67)

TABLE 5.2
Overshoots and Settling Times for Step Change in Reference Torque

Case	Gain (K_1)	Time Constant (T_1)	Overshoot (%)	Settling Time (sec)
1	0.05263	0.00667	-	0.047
2	0.06667	0.007143	0.195	0.048
3	0.06667	0.00667	0.28	0.04
4	0.09090	0.00667	0.952	0.042
5	0.07143	0.01	0.475	0.05

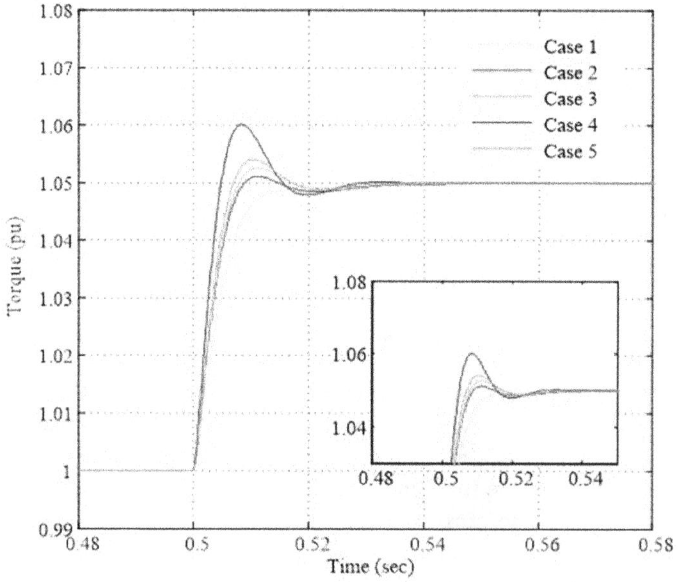

FIGURE 5.7 Transient response of torque controller loop.

5.5 CONCLUSIONS

This chapter deals with the small signal stability of an indirect vector-controlled induction motor fed by matrix converter. The mathematical model of the drive system has been developed, and the characteristic equation of the model has been derived. The system nonlinearity is linearized by using small perturbation algorithm around a stable operating point. The linearized characteristic equation for current controller is developed in terms of PI controller parameters. The transient response of the system shows that the D-partition technique is an efficient technique for the choice of the design of PI controller parameters. Furthermore, this technique is capable of coordinating parameters in parametric plane irrespective of the system order. For fastest transient response, the D-partition technique provides a possible range of controller parameters as discussed in the chapter. Finally, the results of frequency scanning technique validate the stability of the region identified by plane partition method.

REFERENCES

1. Bose, B. K., & Bose, B. K. (Eds.) (1997). *Power Electronics and Variable Frequency Drives: Technology and Applications* (Vol. 996). Piscataway, NJ: IEEE Press.
2. Bose Bimal, K. (2002). Modern Power Electronics and AC Drives. New Jersey: Prentice Hall.
3. Leonhard, W. (2001). Control of electrical drives. Springer Science & Business Media.
4. Wheeler, P. W., Zhang, H., & Grant, D. A. (1994, September). A theoretical and practical consideration of optimised input filter design for a low loss matrix converter. *In Ninth International Conference on Electromagnetic Compatibility*, 1994. (Conf. Publ. No. 396) (pp. 138–142). IET.
5. Wheeler, P. W., Rodriguez, J., Clare, J. C., Empringham, L., & Weinstein, A. (2002). Matrix converters: A technology review. *IEEE Transactions on Industrial Electronics*, 49(2), 276–288.
6. Dubey, G. K. (2001). Fundamentals of electrical drives. Alpha Science Int'l Ltd.
7. Venturini, M., & Alesina, A. (1980, June). The generalised transformer: A new bidirectional, sinusoidal waveform frequency converter with continuously adjustable input power factor. In *1980 IEEE Power Electronics Specialists Conference* (pp. 242–252). IEEE.
8. Venturini, M. (1980). A new sine wave in sine wave out, conversion technique which eliminates reactive elements. *Proceedings Powercon* 7.
9. Altun, H., & Sünter, S. (2003). Matrix converter induction motor drive: modeling, simulation and control. *Electrical Engineering*, 86(1), 25–33.
10. She, H., Lin, H., Wang, X., & Yue, L. (2010, September). Vector control of induction motor based on output voltage compensation of matrix converter. *In 2010 IEEE Energy Conversion Congress and Exposition* (pp. 1851–1858). IEEE.
11. Casadei, D., Serra, G., & Tani, A. (2001). The use of matrix converters in direct torque control of induction machines. *IEEE Transactions on Industrial Electronics*, 48(6), 1057–1064.
12. Krishnan, R. (2001). Electric Motor Drives: Modeling, Analysis, and Control. Pearson.
13. Haghgoeian, F., Ouhrouche, M. A., & Thongam, J. S. (2005, November). Speed estimation using neural network in vector controlled induction motor drive. In Proceedings of WSEAS International Conference on Dynamical Systems and Control (pp. 2–4).
14. Nieniewski, M. J., & Marleau, R. S. (1987). Mathematical modeling of a digital current control loop for electrical drives. *IEEE Transactions on Industrial Electronics*, (1), 107–114.

15. O'dwyer, A. (2009). Handbook of PI and PID Controller Tuning Rules. World Scientific.
16. Yu, C. C. (2006). Autotuning of PID Controllers: A Relay Feedback Approach. Springer Science & Business Media.
17. Lee, C. H. (2004). A survey of PID controller design based on gain and phase margins. *International Journal of Computational Cognition*, 2(3), 63–100.
18. Siljak, D. D. (1964). Analysis and synthesis of feedback control systems in the parameter plane i-linear continuous systems. *IEEE Transactions on Applications and Industry*, 83(75), 449–458.
19. Siljak, D. D. (1969). Nonlinear Systems-The Parameter Analysis and Design.
20. Hwang, C., Hwang, L. F., & Hwang, J. H. (2010). Robust D-partition. *Journal of the Chinese Institute of Engineers*, 33(6), 811–821.
21. Hwang, C., & Hwang, J. H. (2004). Stabilisation of first-order plus dead-time unstable processes using PID controllers. *IEEE Proceedings-Control Theory and Applications*, 151(1), 89–94.
22. Ahmad, U., DR Kohli, S. (1999). Design and analysis of pulse-width modulated closed-loop DC drive torsional system using parameter plane technique. *Electric Machines &Power Systems*, 27(8), 833–848.
23. Agarwal, P., & Verma, V. K. (2001). Parameter plane synthesis of a current source inverter fed induction motor drive. *Journal of the Institution of Engineers (India): Electrical Engineering Division*.
24. Agarwal, P., & Verma, V. K. (1992). Synthesis and performance of digitally controlled current source. *Electric Machines & Power Systems*, 20(2), 149–160.
25. Zuckerberger, A., Weinstock, D., & Alexandrovitz, A. (1997). Single-phase matrix converter. *IEE Proceedings-Electric Power Applications*, 144(4), 235–240.
26. Wang, H., Chen, X., Su, M., Liang, X., Dan, H., Zhang, G., & Wheeler, P. (2021). Three-level indirect matrix converter with neutral-point potential balance scheme for adjustable speed drives. *IEEE Transactions on Transportation Electrification*, 99, 1.
27. Singh, A. K., & Pattnaik, S. (2021). An indirect modulation strategy of matrix converters for unbalanced input supply with reduced commutation losses. *International Journal of Electronics*, 1–22.

NOMENCLATURE

Vas, Vbs, Vcs	Induction motor stator phase voltages
ias, ibs, ics	Induction motor stator phase currents
Vab	Line voltage of motor, rise 'a' over 'b'
iar, ibr, icr	Induction motor rotor phase currents
Xs, rs	Stator reactance and resistance per phase
Xr, rr	Rotor reactance and resistance per phase
Xm	Mutual reactance per phase
Ls, Lr	Stator and rotor self-inductances
Lm	Mutual inductance between stator and rotor
Σ	Degree of relative stability
P	Pole pairs
Te	Electromagnetic torque
Tl	Load torque

Ωo	Output frequency of converter
Ωi	Input frequency of converter
Ωe	Synchronous frequency
Ωr	Rotor speed of induction motor
ωsl	Slip speed of induction motor
Ωs	Synchronous speed of induction motor
Θe	Angular position of synchronous reference
Ωb	Base angular frequency in rad/sec
Vdse, idse	Voltage and current in d-axis stator winding in synchronously rotating reference frame
Vqse, iqse	Voltage and current in q-axis stator winding in synchronously rotating reference frame
idse, iqse	Currents in d-axis and q-axis rotor winding in stator rotating reference frame
idre, iqre	Currents in d-axis and q-axis rotor winding in synchronously rotating reference frame
Suffix o	Represents a variable in steady-state operation
Suffix b	Represents base quantity
K1, T1, K2, T2	Gain and time constant of current PI controller in continuous time domain
Kp1, KI1, Kp2, KI2	Proportional and integral gains of current controller in discrete time domain
Tf	Current filter time constant
P	Differential operator d/dt or complex frequency
Z	Damping ratio

6 COVID-19 Outbreak
A Predictive Mathematical Study

Preeti Deolia and Anuraj Singh
ABV-Indian Institute of Information Technology and Management, Gwalior

Pradeep Malik
Shree Guru Gobind Singh Tricentenary University, Gurugram

CONTENTS

6.1 INTRODUCTION

The world continued without a hitch, and all countries were occupied in revamping their gross domestic product (GDP). Suddenly, an epidemic starts over, which overwhelmed the whole world in a brief period and drastically took the face of a pandemic called the novel coronavirus disease (COVID-19). Coronaviruses are highly transmissible and are morbific viruses. According to a report of the International Committee on Taxonomy of Viruses (ICTV), coronaviruses belong to the subfamily Coronavirinae, a member of the family Coronaviridae and the order Nidovirales. The subfamily Coronavirinae consists of four biological groups: α-, β-, γ-, and δ-coronavirus [1].

DOI: 10.1201/9781003222255-6

Similar to other respiratory diseases, the COVID-19 virus can be spread in the form of tiny particles exposed from an infected individual's nose and mouth. Also, the T-zone tissue areas of the face, including the eyes, nose, and mouth, are the main entry points of the virus into the human body. The infection resulted in hyposmia (decreased sense of smell)/anosmia (inability to smell anything), loss of the sense of taste, and poor appetite. In many cases, the COVID virus carriers may not show symptoms such as cough and fever, although they have hyposmia/anosmia and low appetite symptoms.

Nowadays, several research papers and online resources are available, analyzing the spread of COVID-19 [2,3]. To understand the dynamics of a disease spread in a population, mathematical models play a crucial role and have a positive history of applications to help humans, such as in dengue [4,5] and malaria [6,7].

An Susceptible, Exposed, Infectious, Removed (SEIR)-type model is the most fitted model to explore the dynamics of coronavirus disease [8–11]. Therefore, in this chapter, a modified SEIR model is accounted to discuss the transmission dynamics of the COVID-19 disease incorporating infection in incubation period and the effect of pathogen concentration on the susceptible in the environment. Lastly, numerical findings including some pre-assumed scenarios are represented and analyzed, along with concluding remarks of the study.

6.2 MODEL FORMULATION

In this work, a mathematical model is considered under the following assumptions:

1. The transmission dynamics of COVID-19 disease is similar to the Susceptible, Exposed, Infectious, Hospitalized, Recovered, Pathogen (SEIHRP)-model.
2. The role of the exposed and infected individuals in the transmission of the disease directly as well as indirectly is considered.
3. Natural recovery of asymptomatic and symptomatic individuals having strong immunity.
4. Only symptomatic and COVID-19-confirmed individuals are hospitalized.
5. A fraction of COVID-19 pathogen is removed from the environment due to sanitization.

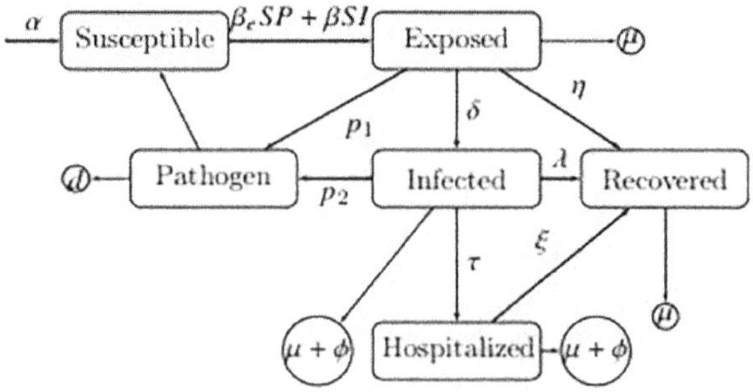

FIGURE 6.1 SEIHRP-model.

TABLE 6.1
Variables of the Model

Variables	Description of Variables
S	Susceptible
E	Exposed
I	Infected
H	Hospitalized
R	Recovered
P	Pathogen

The flow diagram for the transmission of COVID-19 virus is shown in Figure 6.1, and the description of variables is given in Table 6.1.

The mathematical model is represented as follows:

$$\begin{aligned}
\frac{dS}{dt} &= \alpha - \beta SI - \beta_e SP - \mu S \\
\frac{dE}{dt} &= \beta SI + \beta_e SP - (\mu + \eta + \sigma\delta)E \\
\frac{dI}{dt} &= \sigma\delta E - (\tau + \lambda + \mu + \phi)I \\
\frac{dH}{dt} &= \tau I - (\xi + \mu + \phi)H \\
\frac{dR}{dt} &= \eta E + \xi H + \lambda I - \mu R \\
\frac{dP}{dt} &= p_1 E + p_2 I - dP
\end{aligned} \tag{6.1}$$

with the following non-negative initial conditions:

$$S(0) = S_0, E(0) = E_0, I(0) = I_0, H(0) = H_0, R(0) = R_0, P(0) = P_0.$$

All model parameters are defined in Table 6.2 and are assumed to be non-negative for the existence of the SEIHRP-model (6.1).

for all $t \geq 0$.

6.3 PRELIMINARIES

In this section, the qualitative nature of the solutions to system (6.1) is discussed. Let $R_+^6 = \{(S,E,H,I,R,P) \in R^6 \mid S > 0, E > 0, H > 0, I > 0, R > 0, P > 0\}$ be the positive cone.

TABLE 6.2
Parameters of the Model

Parameters	Description of Parameters
A	Birth rate of the susceptible individuals
M	Natural death rate
B	Direct transmission rate of virus
β_e	Indirect transmission rate of COVID-19 virus
Δ	Rate of transfer of exposed to the infected class
H	Natural recovery rate of exposed
Λ	Natural recovery rate of infected
$\frac{1}{\sigma}$	Incubation period
T	Transfer rate of infected individuals to the hospital
Φ	Death rate due to COVID-19
Ξ	Recovery rate of hospitalized individuals
p_1	Shedding rate of virus from the exposed class to the environment
p_2	Shedding rate of virus from the infected class to the environment
D	Removal rate of the COVID-19 virus from the environment due to sanitization

6.3.1 Boundedness of the Solution

In this subsection, a positively invariant feasible region is determined. The solutions to system (6.1) are bounded in this region.

Theorem 6.3.1

Solutions to system (6.1) with specified initial conditions are bounded in the domain

$$\Lambda = \left\{ (S,E,I,H,R,P) \in R^6 \geq 0 : 0 \leq (S+E+I+H+R+P) = N(t) \leq \frac{\alpha}{\mu}, \right.$$

$$\left. 0 \leq P(t) \leq \frac{(p_1 + p_2)\alpha}{\mu d} \right\}$$

Proof: The total population $N(t)$ at time t is

$$N = S + E + I + H + R + P.$$

Then,

$$\frac{dN}{dt} \leq \alpha - \mu N.$$

Using Gronwall's inequality, it can be concluded that

$$0 \le N(t) \le \frac{\alpha}{\mu}.$$

Now, for the pathogen parameter $P(t)$:

$$\frac{dP}{dt} \le (p_1 + p_2)\frac{\alpha}{\mu} - dP. \tag{6.2}$$

Again applying Gronwall's inequality in equation (6.2) leads to

$$0 \le P(t) \le \frac{(p_1 + p_2)\alpha}{\mu d}.$$

Hence, any solution to system (6.1) is bounded in Λ.

6.4 MODEL ANALYSIS

6.4.1 Equilibrium Points and Basic Reproduction Number

System (6.1) has the following equilibrium points:

- Disease-free equilibrium (DFE) point $M_0 = \left(\frac{\alpha}{\mu}, 0, 0, 0, 0, 0\right)$.
- The endemic equilibrium point $M^* = \left(S^*, E^*, I^*, H^*, R^*, P^*\right)$ exists where

$$S^* = \frac{d(\tau+\lambda+\mu+\phi)(\mu+\eta+\sigma\delta)E^*}{\beta\sigma\delta d + \beta_e\left(p_1 + (\tau+\lambda+\mu+\phi) + p_2\sigma\delta\right)},$$

$$E^* = \frac{1}{(\mu+\eta+\sigma\delta)}\left[\alpha - \frac{\mu d(\tau+\lambda+\mu+\phi)(\mu+\eta+\sigma\delta)}{\beta\sigma\delta d + \beta_e\left(p_1(\tau+\lambda+\mu+\phi) + p_2\sigma\delta\right)}\right],$$

$$I^* = \frac{\sigma\delta E^*}{(\tau+\lambda+\mu+\phi)}, H^* = \frac{\tau\sigma\delta E^*}{(\xi+\mu+\phi)(\tau+\lambda+\mu+\phi)},$$

$$R^* = \frac{1}{\mu}\left[\eta + \frac{\xi\tau\sigma\delta}{(\xi+\mu+\phi)(\tau+\lambda+\mu+\phi)} + \frac{\lambda\sigma\delta}{(\tau+\lambda+\mu+\phi)}\right]E^*,$$

$$P^* = \frac{1}{d}\left[p_1 + \frac{p_2\sigma\delta}{(\tau+\lambda+\mu+\phi)}\right]E^*.$$

6.4.1.1 Basic Reproduction Number

It is defined as the average number of secondary cases that would be generated by a primary case in a totally susceptible population. Using the notation $Y = (E, I, P)$ for the model, the following vector functions are obtained:

$$F_i(Y)=\begin{pmatrix} \beta SI+\beta_e SP \\ 0 \\ 0 \end{pmatrix} \text{ and } V_i(Y)=\begin{pmatrix} (\mu+\eta+\sigma\delta)E \\ -\sigma\delta E+(\tau+\lambda+\mu+\varphi)I \\ -p_1E-p_2I+dP \end{pmatrix}$$

Then, the next-generation matrix FV^{-1} is given by

$$FV^{-1}=\begin{pmatrix} \dfrac{\alpha\beta\sigma\delta d+\beta_e\alpha\left(\sigma\delta p_2+p_1(\tau+\lambda+\mu+\phi)\right)}{\mu d(\tau+\lambda+\mu+\phi)(\mu+\eta+\sigma\delta)} & \dfrac{\alpha\beta d+\beta_e p_2\alpha}{\mu(\tau+\lambda+\mu+\phi)} & \dfrac{\alpha\beta_e}{\mu d} \\ 0 & 0 & 0 \\ 0 & 0 & 0 \end{pmatrix}$$

Therefore, the basic reproduction number is

$$R_0=\rho\left(FV^{-1}\right)=\frac{\alpha\beta\sigma\delta d+\beta_e\alpha\left(\alpha\delta p_2+p_1(\tau+\lambda+\mu+\phi)\right)}{\mu d(\tau+\lambda+\mu+\phi)(\mu+\eta+\sigma\delta)},$$

where ρ is the spectral radius.

6.4.2 Stability Analysis

The variational matrix J at DFE (M_0) is given by

$$J=\begin{pmatrix} -\mu & 0 & \dfrac{-\beta\alpha}{\mu} & 0 & 0 & \dfrac{-\beta_e\alpha}{\mu} \\ 0 & -(\mu+\eta+\sigma\delta) & \dfrac{-\beta\alpha}{\mu} & 0 & 0 & \dfrac{-\beta_e\alpha}{\mu} \\ 0 & \sigma\delta & -(\tau+\lambda+\mu+\phi) & 0 & 0 & 0 \\ 0 & 0 & \tau & -(\xi+\mu+\phi) & 0 & 0 \\ 0 & \eta & \lambda & \xi & -\mu & 0 \\ 0 & p_1 & p_2 & 0 & 0 & -d \end{pmatrix} \tag{6.3}$$

6.4.3 Local Stability at Disease-Free Equilibria

Theorem 6.4.3.1

The disease-free state of system (6.1) is locally stable for $R_0<1$ and unstable for $R_0>1$ provided $(\tau+\lambda+\mu+\phi)>d$.

Proof: The variational matrix at M_0 gives the following characteristic equation,

$$(\lambda'+\mu)^2(\lambda'+(\xi+\mu+\phi))\big(\lambda'^3+(d+\tau+\lambda+2\mu+\phi+\eta+\sigma\delta)\big)\lambda'^2$$
$$+\left(d(\tau+\lambda+2\mu+\phi+\eta+\sigma\delta)+(\mu+\eta+\sigma\delta)(\tau+\lambda+\mu+\phi)-\frac{\alpha\beta\sigma\delta}{\mu}-\frac{\alpha\beta_e p_1}{\mu}\right)\lambda'$$
$$+\left(d(\mu+\eta+\sigma\delta)(\tau+\lambda+\mu+\phi)-\frac{\alpha\beta\sigma\delta d}{\mu}-\frac{\alpha\beta_e(\sigma\delta p_2+(\tau+\lambda+\mu+\phi)p_1)}{\mu}\right)=0 \tag{6.4}$$

Clearly, the linear factors of equation (6.4) are negative and the other linear factors of the remained cubic equation are negative following the Routh–Hurwitz criterion as follows:

The cubic part of equation (6.4) can be rewritten as

$$\lambda'^3+B_1\lambda'^2+B_2\lambda'+B_3=0,$$

where

$$B_1=d+\tau+\lambda+2\mu+\phi+\eta+\sigma\delta$$

$$B_2=d(\tau+\lambda+2\mu+\phi+\eta+\sigma\delta)+(\mu+\eta+\sigma\delta)(\tau+\lambda+\mu+\phi)-\frac{\alpha\beta\sigma\delta}{\mu}-\frac{\alpha\beta_e p_1}{\mu}$$

$$B_3=d(\mu+\eta+\sigma\delta)(\tau+\lambda+\mu+\phi)-\frac{\alpha\beta\sigma\delta d}{\mu}-\frac{\alpha\beta_e(\sigma\delta p_2+(\tau+\lambda+\mu+\phi)p_1)}{\mu}.$$

According to Routh–Hurwitz criterion, system (6.1) is locally stable if and only if $B_1,B_2,B_3>0$ and $B_1B_2-B_3>0$.

Now, for $R_0<1$,

$$B_1=d+\tau+\lambda+2\mu+\phi+\eta+\sigma\delta>0$$

$$B_2=d(\tau+\lambda+2\mu+\varphi+\eta+\sigma\delta)+(\mu+\eta+\sigma\delta)(\tau+\lambda+\mu+\varphi)(1-R_0)>0.$$

$$B_3=d(\mu+\eta+\sigma\delta)(\tau+\lambda+\mu+\varphi)(1-R_0)>0.$$

Now, since $B_2>d(\tau+\lambda+2\mu+\phi+\eta+\sigma\delta)+(\mu+\eta+\sigma\delta)(\tau+\lambda+\mu+\phi)$ and $B_3<d(\mu+\eta+\sigma\delta)(\tau+\lambda+\mu+\phi)$,

$$B_1B_2-B_3>d^2(\tau+\lambda+2\mu+\phi+\eta+\sigma\delta)+d(\tau+\lambda+2\mu+\phi+\eta+\sigma\delta)^2+$$
$$(\tau+\lambda+2\mu+\phi+\eta+\sigma\delta)(\mu+\eta+\sigma\delta)(\tau+\lambda+\mu+\phi)>0$$

Therefore, the disease-free state (M_0) of system (6.1) is locally stable.

6.4.4 Global Stability

In this subsection, the results regarding the global stability of disease-free state are established.

Theorem 6.4.4.1

The disease-free equilibria M_0 are globally stable iff $R_0 < 1$.

Proof: Consider a real-valued Lyapunov function defined as

$$V = \int_{\frac{\alpha}{\mu}}^{S(t)} \left(1 - \frac{\alpha}{\mu z}\right) dz + E + \frac{(\mu + \eta + \sigma\delta) I}{\sigma\delta}. \tag{6.5}$$

Differentiating equation (6.5) for all $t \geq 0$, it is obtained that

$$V' = -\frac{(\alpha - \mu S)^2}{\mu S} + \frac{\alpha(\beta I + \beta_e P)}{\mu} - \frac{(\mu + \eta + \sigma\delta)(\tau + \lambda + \mu + \varphi) I}{\sigma\delta}$$

$$= -\frac{(\alpha - \mu S)^2}{\mu S} - \frac{(\mu + \eta + \sigma\delta)(\tau + \lambda + \mu + \varphi) I}{\sigma\delta}(1 - R_0) \leq 0.$$

The equality holds if and only if $S(t) = S_0, I(t) = 0$, and $E(t) = 0$. The largest invariant set in the region $\left\{(S(t), E(t), I(t)) \in R_+^6 \mid \frac{dV}{dt} = 0\right\}$ is $\left\{\left(\frac{\alpha}{\mu}, 0, 0\right)\right\}$. Therefore, by using LaSalle's invariance principle, M_0 is globally stable.

6.5 NUMERICAL SIMULATION

In order to substantiate the obtained findings for system (6.1), the numerical analysis is performed for different sets of parameter values. Some parameters are taken from the literature [12,13].

In Figure 6.2, a time series graph is plotted to investigate the global stability of infection-free equilibria. For these values of parameters, the basic reproduction number $R_0 = 0.97279 < 1$; thus, the condition for the global stability of the DFE M_0 is satisfied. The time series drawn for the parameters S, E, I, and P in Figure 6.2 clearly reflects that both the infected and pathogen classes converge to zero.

6.6 CONCLUSIONS

In this work, the transmission dynamics of COVID-19 disease is represented by a modified SEIR model. The total human population is dissected into five portions: Susceptible, Exposed, Infected, Hospitalized, and Recovered along with Pathogen compartment. The primary characteristics of the model, including positivity and boundedness, are investigated. The infection-free equilibrium point is locally and

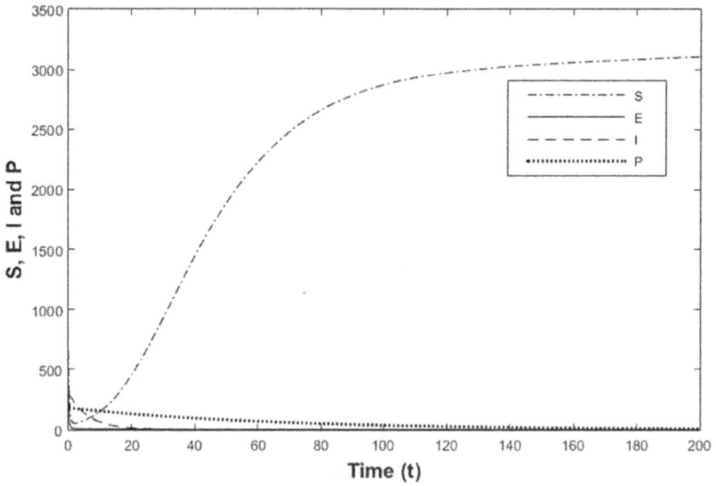

FIGURE 6.2 Time series plot of $S(t), E(t), I(t)$, and $P(t)$ at $b = 0.00001231, E(t), I(t)$, and $P(t)$ converges to disease-free state $(R_0 = 0.97279)$.

globally stable if the basic reproduction number R_0 is less than one, otherwise unstable. The aforementioned numerical simulation gives evidence for our analytical results.

REFERENCES

1. Woo, P. C., Lau, S. K., Huang, Y., & Yuen, K. Y. (2009). Coronavirus diversity, phylogeny and interspecies jumping. *Experimental Biology and Medicine*, *234*(10), 1117–1127.
2. Khan, M. A., & Atangana, A. (2020). Modeling the dynamics of novel coronavirus (2019-nCov) with fractional derivative. *Alexandria Engineering Journal*, *59*(4), 2379–2389.
3. Guan, W. J., Ni, Z. Y., Hu, Y., Liang, W. H., Ou, C. Q., He, J. X. & Zhong, N. S. (2020). Clinical characteristics of 2019 novel coronavirus infection in China. *MedRxiv.*
4. Abidemi, A., Abd Aziz, M. I., & Ahmad, R. (2020). Vaccination and vector control effect on dengue virus transmission dynamics: Modelling and simulation. *Chaos, Solitons & Fractals*, *133*, 109648.
5. Ganegoda, N., Götz, T., & Wijaya, K. P. (2021). An age-dependent model for dengue transmission: Analysis and comparison to field data. *Applied Mathematics and Computation*, *388*, 125538.
6. Mohammed-Awel, J., & Gumel, A. B. (2019). Mathematics of an epidemiology-genetics model for assessing the role of insecticides resistance on malaria transmission dynamics. *Mathematical Biosciences*, *312*, 33–49.
7. Agusto, F. B., Del Valle, S. Y., Blayneh, K. W., Ngonghala, C. N., Goncalves, M. J., Li, N. & Gong, H. (2013). The impact of bed-net use on malaria prevalence. *Journal of Theoretical Biology*, *320*, 58–65.
8. Cooke, K. L., & Van Den Driessche, P. (1996). Analysis of an SEIRS epidemic model with two delays. *Journal of Mathematical Biology*, *35*(2), 240–260.
9. Abta, A., Kaddar, A., & Alaoui, H. T. (2012). Global stability for delay SIR and SEIR epidemic models with saturated incidence rates. *Electronic Journal of Differential Equations*, *2012*(23), 1–13.

10. Han, S., & Lei, C. (2019). Global stability of equilibria of a diffusive SEIR epidemic model with nonlinear incidence. *Applied Mathematics Letters*, *98*, 114–120.
11. Kassa, S. M., Njagarah, J. B., & Terefe, Y. A. (2020). Analysis of the mitigation strategies for COVID-19: from mathematical modelling perspective. *Chaos, Solitons & Fractals*, *138*, 109968.
12. Khan, M. A., & Atangana, A.. (2020). Modeling the dynamics of novel coronavirus (2019-nCov) with fractional derivative. *Alexandria Engineering Journal*, *59*(4), 2379–2389.
13. Sardar, T., Nadim, S. S., Rana, S., & Chattopadhyay, J. (2020). Assessment of lockdown effect in some states and overall India: A predictive mathematical study on COVID-19 outbreak. *Chaos, Solitons & Fractals*, *139*, 110078.

7 Modeling the Role of Rain in Cleaning the Environment

Sandeep Sharma and Fateh Singh
DIT University, Dehradun

CONTENTS

NOMENCLATURE

T	Density of the raindrops
G	Concentration of the gaseous pollutants
P_1	Concentration of the particulate matter formed naturally
P_2	Concentration of the particulate matter generated through various anthropogenic activities
G_a	Concentration of the gaseous pollutants in the absorbed phase
r	Rate of raindrops generation
α	Natural depletion rate of the raindrops
θ	Reversible rate coefficients (lies between 0 and 1)
π	Fraction of the absorbed phase going to the gaseous phase
p_1, p_2	Rate at which the two particulate matters are emitted into the environment
θ_1	Rate of raindrops density interaction with gaseous pollutants G
θ_2, θ_3	Rate of raindrops density interaction with two particulate matters P_1, P_2
g	Emission rate of gaseous pollutants in the environment
γ_1	Natural depletion rate for the gaseous pollutants G
γ_2, γ_3	Natural depletion rates of the particulate matter P_1 and P_2
Λ	Rate of removal of gaseous pollutants in the absorbed phase
δ_1	Rate of conversion of gaseous pollutants to the absorbed phase

DOI: 10.1201/9781003222255-7

δ_2, δ_3	Removal rates of particulate matter from the environment due to rain
δ_4	Removal rate of absorbed phase due to rain

7.1 INTRODUCTION

The need for a conducive environmental condition for the survival of living organisms is a well-established fact. In particular, the human population highly needs clean and safe water and air for its survival. In the absence of these factors, it will be very difficult to live a healthy life. The contaminated water and air caused a number of adverse effects on the human body. In particular, air pollution is one of the major causes of mortality across the globe. In the recent past, the concentration of a variety of air pollutants has increased at an exponential rate. Along with natural activities, anthropogenic activities are the major source of air pollution (Kampa and Castanas, 2008). Air pollutants are mainly classified into four categories, namely gaseous pollutants, persistent organic pollutants, particulate matters, and heavy metals (Kampa and Castanas, 2008). The increasing concentration of air pollutants not only disturbs the climatic conditions, but also causes many adverse impacts on the human body (Manisalidis et al., 2020). Regular exposure to air pollution leads to a malfunctioning immune system. Subsequently, it increases the susceptibility of an individual to getting infections (Glencross et al., 2020). A detailed study on the impacts of air pollution on the immune system is carried out (Glencross et al., 2020). Thus, air pollution also contributes to the spread of infectious diseases. On account of this, a number of studies investigated the role of environmental pollution on the spread of infectious diseases (Lafferty and Holt, 2003, Sharma and Kumari, 2017, Kumari and Sharma, 2018, Sharma and Kumari, 2019). Lafferty and Holt (2003) used a generic SIR model to demonstrate the various possible impacts of stress on the dynamics of the disease. Kumari and Sharma (2018) modified a generic SIS model by introducing a separate compartment of the stressed individuals. During the modeling process, the transmission rate is modified by incorporating two pollution parameters. The study finds that pollution certainly plays a pivotal role in the spread of infectious diseases. Sharma and Kumari (2019) formulated a new compartmental model to investigate the impact of environmental pollution on waterborne diseases. Through numerical simulations, the authors demonstrated that environmental pollution increases the size of the epidemics once they invade the population. A mathematical model with multiple transmission pathways to comprehend the role of pollution on the spread of cholera is used by Sharma and Kumari (2017). The study finds the direct involvement of pollution parameters in the dynamics of the disease.

It is a well-established fact that the concentration of the air pollutants significantly reduces in the rainy season. The water droplets absorb a number of pollutants and thus clean the environment. Due to the importance of the problem, a number of mathematical models have been proposed to analyze the role of rain in removing air pollutants from the environment (Shukla et al., 2008a and b, Naresh et al., 2007, Sundar et al., 2009, Naresh, 2003).

Naresh (2003) proposed a mathematical model consisting of three differential equations to gather insight into the removal of pollutants by rain. A differential equations-based model was proposed to study the removal of particulate matters and gaseous pollutants by Naresh et al. (2007). The authors demonstrated that if the rate of formation of raindrops is sufficiently large, the pollutants can be completely removed from the environment. Shukla et al. (2008a) implemented a mathematical model to study the removal of two particulate matters and a gaseous pollutant by rain from the environment. The study found the conditions under which air pollutants can be removed from the environment. Shukla et al. (2008b) proposed a nonlinear mathematical model to investigate the role of cloud density in the removal of air pollutants by rain. The model also considered gravitational settling and reversible reaction processes. The analysis of the model reveals that air pollutants can be removed when the rates of formation of cloud droplets and raindrops are sufficiently large. The interaction of hot gases with cloud droplets and raindrops with the help of a mathematical model is investigated by Sundar et al. (2009). With the help of dynamical analysis of the proposed model, the authors found that if gases are too hot, then they reduce the rate of formation of raindrops and thus prevent the rain.

Recently, some stochastic differential equations have been proposed to study the impact of random fluctuation on the formation of raindrops and, subsequently, the removal of air pollutants by rain (Misra and Tripathi, 2018, 2019). Misra and Tripathi (2018) considered the cloud droplets, raindrops, and concentration of particulate matters as the interacting phases. The model assumed that the conversion rate of cloud droplets into raindrops is not the same as the rate of formation of cloud droplets. The model is then subjected to a detailed dynamical analysis to derive reasonable conditions for the removal of particulate matters by rain. A stochastic model was proposed by Misra and Tripathi (2019) to study the removal of air pollutants from the environment from the rain-deficient regions using aerosols. The analysis of the stochastic version of the stochastic mathematical provided more realistic results than the deterministic mathematical model. The stability conditions of the stochastic model mainly depend upon the intensity of the white noise.

Motivated from the above, in this work we propose a five-dimensional mathematical model consisting of raindrops, gaseous pollutants, absorbed phase of particulate matters, and two particulate matters. Out of the two particulate matters, one is produced by the gaseous pollutants and the other is generated by different anthropogenic activities.

7.2 MATHEMATICAL MODEL

During the formulation of the mathematical model, we consider the environment of a polluted city consisting of gaseous pollutants and particulate matters. It is a well-established fact that anthropogenic activities have a major contribution to the concentration of particulate matter (Kampa and Castanas, 2008). On account of this, we consider a separate equation pertaining to the particulate matters generated through human activities. The resulting model system is governed by the following set of differential equations:

$$\frac{dT}{dt} = r - \alpha T - \theta_1 GT - \theta_2 P_1 T - \theta_3 P_2 T$$
$$\frac{dG}{dt} = g - \gamma_1 G - \delta_1 GT + \theta \Lambda G_a + \pi \delta_4 G_a T$$
$$\frac{dP_1}{dt} = p_1 - \gamma_2 P_1 - \delta_2 P_1 T \tag{7.1}$$
$$\frac{dP_2}{dt} = p_2 - \gamma_3 P_2 - \delta_3 P_2 T$$
$$\frac{dG_a}{dt} = \delta_1 GT - \Lambda G_a - \delta_4 TG_a$$

In model system (7.1), T represents the density of the raindrops, while the concentration of the gaseous pollutants is denoted by G. P_1 is the concentration of the particulate matter formed naturally, and P_2 is the concentration of the particulate matter generated through various anthropogenic activities. The concentration of the gaseous pollutants in the absorbed phase is represented by G_a. It is assumed that raindrops are generated at a constant rate r. α is the natural depletion rate of the raindrops. p_1 and p_2 are the rates at which the two particulate matters are emitted into the environment. Further, γ_2 and γ_3 are the natural depletion rates of the particulate matters P_1 and P_2, respectively. In the modeling process, it is assumed that the interactions between two phases have been governed by the law of mass action. Further, the raindrops density also decreases due to the interaction with the gaseous pollutants G and with the two particulate matters P_1, P_2 at rates θ_1, θ_2, and θ_3, respectively. g represents the emission rate of gaseous pollutants into the environment, while γ_1 is the natural depletion rate for the gaseous pollutants. The concentration of the gaseous pollutants also decreases due to their conversion to the absorbed phase at a rate δ_1. The gaseous pollutants in the absorbed phase are removed at a rate Λ. It is also assumed that a fraction $\theta \Lambda G_a$ of the same re-enter the environment due to the recycling process. The concentration of the absorbed phase of gaseous pollutants decreases at a rate $\delta_4 TG_a$. The model also considers a reversible process due to which a fraction $\pi \delta_4 TG_a$ may re-enter the environment. Further, δ_2 and δ_3 are the removal rates at which the particulate matter are removed from the environment by rain.

The proposed model system is investigated under the following initial conditions:

$$T(0) \geq 0, G \geq 0, P_1 \geq 0, P_2 \geq 0, G_a \geq 0.$$

Further, to obtain the bounds for the system variables, we have the following region of attraction (Freedman and So, 1985):

$$\Omega = \left\{ 0 < T \leq \frac{r}{\alpha}, 0 < G + G_a \leq \frac{g}{\gamma_m}, 0 < P_1 \leq \frac{p_1}{\gamma_2}, 0 < P_2 \leq \frac{p_2}{\gamma_3} \right\}$$

where $\gamma_m = min\{(\gamma_1+\delta_1),(1-\theta)\Lambda\}$ to investigate the dynamics of model system (7.1).

7.3 EQUILIBRIUM ANALYSIS

The proposed model system (7.1) has a unique non-trivial equilibrium point $E^*(T^*,G^*,P_1^*,P_2^*,G_a^*)$. The components of E^* can be obtained by solving the following system of equations:

$$r-\alpha T-\theta_1 GT-\theta_2 P_1 T-\theta_3 P_2 T=0$$

$$g-\gamma_1 G-\delta_1 GT+\theta\Lambda G_a+\pi\delta_4 G_a T=0$$

$$p_1-\gamma_2 P_1-\delta_2 P_1 T=0$$

$$p_2-\gamma_3 P_2-\delta_3 P_2 T=0$$

$$\delta_1 GT-\Lambda G_a-\delta_4 TG_a=0$$

as follows:

$$T=\frac{r}{\alpha+\theta_1 G+\theta_2 P_1+\theta_3 P_2}$$

$$G=\frac{g(\Lambda+\delta_4 T)}{\left[\Lambda\gamma_1+(\delta_4\gamma_1+(1-\theta)\Lambda\delta_1)T+(1-\pi)\delta_4\delta_1 T^2\right]}=f_1(T)$$

$$P_1=\frac{p_1}{(\gamma_2+\delta_2 T)}=f_2(T) \tag{7.2}$$

$$P_2=\frac{p_2}{(\gamma_3+\delta_3 T)}=f_3(T)$$

$$G_a=\frac{\delta_1 GT}{(\Lambda+\delta_4 T)}$$

Next, using f_1, f_2, and f_3 in the first equation of the system, we obtain

$$F(T)=r-\alpha T-\theta_1 f_1 T-\theta_2 f_2 T-\theta_3 f_3 T \tag{7.3}$$

It can be noted that $F(0)=r>0$ and $F\left(\frac{r}{\alpha}\right)<0$. Further,

$$F'(T)=-\alpha-\theta_1\left\{f_1+Tf_1'\right\}-\theta_2\left\{f_2+Tf_2'\right\}-\theta_3\left\{f_3+Tf_3'\right\}<0.$$

From the above calculation, we conclude that $F(T)=0$ has a unique root T^* between 0 and $\frac{r}{\alpha}$. We can find the other component of E^* using T^*.

7.4 STABILITY ANALYSIS

In this section, we use the Lyapunov function approach to obtain the conditions on the local and global stability of the equilibrium point E^*. The stability conditions help us to predict the long-term behavior of the system. We begin with the local stability and then derive the conditions for global stability of the equilibrium point E^*.

7.4.1 Local Stability

The following theorem ascertains the local stability of the equilibrium point E^*.

Theorem 7.1

The equilibrium point E^* is locally asymptotically stable under the following conditions

$$\frac{3(\theta\pi+\pi\delta_4T^*)^2}{\left(\gamma_1+\delta_1T^*\right)}<(\pi+\delta_4T^*)^2 min\left\{\frac{4}{7}\frac{T\left(\pi+\delta_4T^*\right)}{T^*(\delta_1G^*-\delta_4G_a^*)^2},\frac{\left(\gamma_1+\delta_1T^*\right)}{(\delta_1T^*)^2}\right\}$$

$$\frac{T}{T^*}>\frac{7}{\left(\gamma_1+\delta_1T^*\right)}max\{(\theta_1T^*)^2,(\delta_1G^*-\pi\delta_4G_a^*)^2\} \tag{7.4}$$

Proof. To prove the above theorem, we consider the following Lyapunov function:

$$V_1=\frac{1}{2}\left(T_1^2+k_1G_1^2+k_2P_{11}^2+k_3P_{21}^2+k_4G_{a1}^2\right)$$

Here, k_i $(i=1,2,3,4)$ are constants to be determined and T_1, G_1, P_{11}, P_{21}, and G_{a1} represent the small perturbations as follows:

$$T=T^*+T_1, G=G^*+G_1, P_1=P_1^*+P_{11}, P_2=P_2^*+P_{21}, G_a=G_a^*+G_{a1}.$$

On differentiating the function V_1, we get

$$V_1'=T_1T_1'+k_1G_1G_1'+k_2P_{11}P_{11}'+k_3P_{21}P_{21}'+k_4G_{a1}G_{a1}' \tag{7.5}$$

The linearization of model system (7.1) around the equilibrium point E^* gives the following system of equations:

$$\begin{aligned}
T_1'&=-\frac{T}{T^*}T_1-\theta_1T^*G_1-\theta_2T^*P_{11}-\theta_3T^*P_{21}\\
G_1'&=-\left(\delta_1G^*-\pi\delta_4G_a^*\right)T_1-\left(\gamma_1+\delta_1T^*\right)G_1+\left(\theta\pi+\pi\delta_4T^*\right)G_{a1}\\
P_{11}'&=-\delta_2P_1^*T_1-\left(\gamma_2+\delta_2T^*\right)P_{11}\\
P_{21}'&=-\delta_3P_2^*T_1-\left(\gamma_3+\delta_3T^*\right)P_{21}\\
G_{a1}'&=\left(\delta_1G^*-\delta_4G_a^*\right)T_1+\delta_1T^*G_1-\left(\pi+\delta_4T^*\right)G_{a1}.
\end{aligned} \tag{7.6}$$

Using equation (7.6) in equation (7.5), we obtain

$$V_1' = -\frac{T}{T^*}T_1^2 - k_1\left(\gamma_1 + \delta_1 T^*\right)G_1^2 - k_2\frac{p_1}{P_1^*}P_{11}^2 - k_3\frac{p_2}{P_2^*}p_{21}^2 - k_4\left(\pi + \delta_4 T^*\right)G_{a1}^2$$

$$-\theta_1 T^* T_1 G_1 - \theta_2 T^* T_1 P_{11} - \theta_3 T^* T_1 P_{21} - k_1\left(\delta_1 G^* - \pi\delta_4 G_a^*\right)T_1 G_1$$

$$+k_1\left(\theta\pi + \pi\delta_4 T^*\right)G_1 G_{a1} - k_2\delta_2 P_1^* P_{11}T_1 - k_3\delta_3 P_2^* P_{21}T_1 + k_4\left(\delta_1 G^* - \delta_4 G_a^*\right)T_1 G_{a1}$$

$$+k_4\delta_1 T^* G_{a1} G_1$$

From the above relation, we can infer that V_1' is negative definite provided the following conditions are satisfied:

$$\frac{k_1}{7}\left(\frac{T}{T^*}\right)\left(\gamma_1 + \delta_1 T^*\right) > (\theta_1 T^*)^2$$

$$\frac{2k_2}{7}\left(\frac{T}{T^*}\right)\left(\gamma_2 + \delta_2 T^*\right) > (\theta_2 T^*)^2$$

$$\frac{2k_3}{7}\left(\frac{T}{T^*}\right)\left(\gamma_3 + \delta_3 T^*\right) > (\theta_3 T^*)^2$$

$$\frac{1}{7}\left(\frac{T}{T^*}\right)\left(\gamma_1 + \delta_1 T^*\right) > k_1(\delta_1 g^* - \pi\delta_4 g_2^*)^2$$

$$\frac{k_4}{3}\left(\gamma_1 + \delta_1 T^*\right)\left(\pi + \delta_4 T^*\right) > k_1(\theta\pi + \delta_4\pi)^2 \tag{7.7}$$

$$\frac{2}{7}\frac{T}{T^*}\left(\gamma_2 + \delta_2 T^*\right) > k_2(\delta_2 p_1^*)^2$$

$$\frac{2}{7}\frac{T}{T^*}\left(\gamma_3 + \delta_3 T^*\right) > k_3(\delta_3 p_2^*)^2$$

$$\frac{2}{21}\frac{T}{T^*}\left(\pi + \delta_4 T^*\right) > k_4(\delta_1 g^* - \delta_4 g_a^*)^2$$

$$\frac{k_1}{3}\left(\gamma_1 + \delta_1 T^*\right)\left(\pi + \delta_4 T^*\right) > k_4(\delta_1 T^*)^2$$

Now, if we select

$$k_1 = 1, \left(\frac{7T^*}{2T}\right)\frac{(\theta_1 T^*)^2}{\left(\gamma_2 + \delta_2 T^*\right)} < k_2 < \left(\frac{2T}{7T^*}\right)\frac{\left(\gamma_2 + \delta_2 T^*\right)}{(\delta_2 p_1^*)^2} \quad and$$

$$\left(\frac{7T^*}{2T}\right)\frac{(\theta_3 T^*)^2}{\left(\gamma_3 + \delta_3 T^*\right)} < k_3 < \left(\frac{2T}{7T^*}\right)\frac{\left(\gamma_3 + \delta_3 T^*\right)}{(\delta_3 p_2^*)^2}$$

then the conditions for the local stability of the equilibrium point E^* are satisfied.

7.4.2 Global Stability

Next, we will study the global stability of the equilibrium point E^*. The conditions given in the following theorem establish the global stability of E^*.

Theorem 7.2

The equilibrium point $E*$ is globally asymptotically stable if the following conditions are satisfied

$$\frac{3(\theta\pi + \pi\delta_4 T)^2}{(\gamma_1 + \delta_1 T)(\pi + \delta_4 g^*)} < \left(\frac{1}{21}\right)(\pi + \delta_4 G^*)min\begin{Bmatrix} \left(\frac{4T}{T^*}\right)\left(\frac{1}{(\delta_1 G - \delta_4 G_a)^2}\right), \\ \frac{7(\gamma_1 + \delta_1 T)}{(\delta_1 T^*)} \end{Bmatrix} \tag{7.8}$$

$$\left(\frac{T}{T^*}\right) > \frac{7}{(\gamma_1 + \delta_1 T)} max\{(\theta_1 T)^2, (\delta_1 G^* - \pi\delta_4 G_a^*)^2\}$$

Proof. To obtain the above result, we consider the following Lyapunov function:

$$V_2 = \frac{1}{2}\left[(T - T^*)^2 + m_1(G - G^*)^2 + m_2(p_1 - p_1^*)^2 + m_3(p_2 - p_2^*)^2 + m_4(G_a - G_a^*)^2\right]$$

In the above expression of the function V_2, $m_i s$ are the constants that we have to determine.

The derivative of the function V_2 yields the following expression:

$$V_2' = (T - T^*)T' + m_1(G - G^*)G' + m_2(P_1 - P_1^*)P_1' + m_3(P_2 - P_2^*)P_2' + m_4(G_a - G_a^*)G_a' \tag{7.9}$$

Using model system (7.1), we obtain the following expression of V_2':

$$\begin{aligned} V_2' = &(T - T^*)(T - \alpha T - \theta_1 GT - \theta_2 P_1 T - \theta_3 P_2 T) \\ &+ m_1(G - G^*)(g - \gamma_1 G - \delta_1 GT + \theta\Lambda G_a + \pi\delta_4 G_a T) \\ &+ m_2(P_1 - P_1^*)(p_1 - \gamma_2 P_1 - \delta_2 P_1 T) + m_3(P_2 - P_2^*)(p_2 - \gamma_3 P_2 - \delta_3 P_2 T) \\ &+ m_4(G_a - G_a^*)(\delta_1 GT - \Lambda G_a - \delta_4 TG_a) \end{aligned} \tag{7.10}$$

After performing some rearrangements of the terms, we can write the above expression in the following form:

$$
\begin{aligned}
V_2' = &-\left(\frac{T}{T^*}\right)(T-T^*)^2 - m_1\left(\gamma_1+\delta_1 T^*\right)(G-G^*)^2 \\
&- m_2\left(\gamma_2+\delta_2 T^*\right)(P_1-P_1^*)^2 - m_3\left(\gamma_3+\delta_3 T^*\right)(P_2-P_2^*)^2 \\
&- m_4\left(\pi+\delta_4 G^*\right)(G_a-G_a^*)^2 - \theta_1 T\left(T-T^*\right)\left(G-G^*\right) \\
&- \theta_2 T\left(T-T^*\right)\left(P_1-P_1^*\right) - \theta_3 T\left(T-T^*\right)\left(P_2-P_2^*\right) \\
&- m_1\left(\delta_1 G^* - \pi\delta_4 G_a^*\right)\left(T-T^*\right)\left(G-G^*\right) \\
&+ m_1\left(\theta\pi+\pi\delta_4 T\right)\left(G_a-G_a^*\right)\left(G-G^*\right) - m_2\delta_2 P_1\left(T-T^*\right)\left(P_1-P_1^*\right) \\
&- m_3\delta_3 P_2\left(T-T^*\right)\left(P_2-P_2^*\right) + m_4\left(\delta_1 G-\delta_4 G_a\right)\left(T-T^*\right)\left(G_a-G_a^*\right) \\
&+ m_4 T^*\delta_1\left(G_a-G_a^*\right)\left(G-G^*\right)
\end{aligned}
\tag{7.11}
$$

Now, V_2' is negative definite provided the following inequalities are satisfied:

$$
\begin{aligned}
& m_1\left(\frac{T}{7T^*}\right)\left(\gamma_1+\delta_1 T\right) > (\theta_1 T)^2 \\
& m_2\left(\frac{2T}{7T^*}\right)\left(\gamma_2+\delta_2 T^*\right) > (\theta_2 T)^2 \\
& m_3\left(\frac{2T}{7T^*}\right)\left(\gamma_3+\delta_3 T^*\right) > (\theta_3 T)^2 \\
& \left(\frac{T}{7T^*}\right)\left(\gamma_1+\delta_1 T\right) > m_1(\delta_1 G^* - \pi\delta_4 G_a^*)^2 \\
& \left(\frac{m_4}{3}\right)\left(\gamma_1+\delta_1 T\right)\left(\pi+\delta_4 G^*\right) > m_1(\theta\pi+\pi\delta_4 T)^2 \\
& \left(\frac{T}{7T^*}\right)\left(\gamma+\delta_2 T^*\right) > m_2(\delta_2 P_1)^2 \\
& \left(\frac{T}{7T^*}\right)\left(\gamma_3+\delta_3 T^*\right) > m_3(\delta_3 P_2)^2 \\
& \left(\frac{4T}{21T^*}\right)\left(\pi+\delta_4 G^*\right) > m_4(\delta_1 G-\delta_4 G_a)^2 \\
& \left(\frac{m_1}{3}\right)\left(\gamma_1+\delta_1 T\right)\left(\pi+\delta_4 G^*\right) > m_4(\delta_1 T^*)^2 .
\end{aligned}
\tag{7.12}
$$

At last, if we select

$$m_1 = 1, \left(\frac{7T^*}{T}\right)\frac{(\theta_2 T)^2}{\left(\gamma_2 + \delta_2 T^*\right)} < m_2 < \left(\frac{T}{7T^*}\right)\frac{\left(\gamma_2 + \delta_2 T^*\right)}{(\delta_2 P_1)^2} \text{ and}$$

$$\left(\frac{7T^*}{T}\right)\frac{(\theta_3 T)^2}{\left(\left(\gamma_3 + \delta_3 T^*\right)\right)} < m_3 < \left(\frac{T}{7T^*}\right)\frac{\left(\gamma_3 + \delta_3 T^*\right)}{(\delta_3 P_2)^2}$$

then the conditions given in Theorem 7.2 are satisfied.

7.5 NUMERICAL SIMULATION

In this section, we perform a comprehensive numerical simulation to validate the analytical results. Further, we also investigate the impact of variation of the parameters on the system dynamics. To achieve this goal, we consider the following set of parameter values:

$$r = 12, \alpha = 0.3, \theta_1 = 0.002, \theta_2 = 0.004, g = 4.5, \gamma_1 = 0.3, \gamma_2 = 0.15, \gamma_3 = 0.7,$$
$$\Lambda = 0.65, \theta = 0.03, \theta_3 = 0.003, \delta_1 = 0.055, \delta_2 = 0.005, \delta_3 = 0.002, \delta_4 = 0.07,$$
$$\pi = 0.2, p_1 = 0.6, p_2 = 0.85$$

The above set of parameter values satisfy the local and global stability conditions for E^*. Further, the components of equilibrium point E^* are obtained as $E^* = (38.1284, 2.1978, 1.7614, 1.0950, 1.3887)$.

First, we study the global stability of the equilibrium point E^*. For this purpose, we plot different trajectories starting from different initial conditions. Figure 7.1

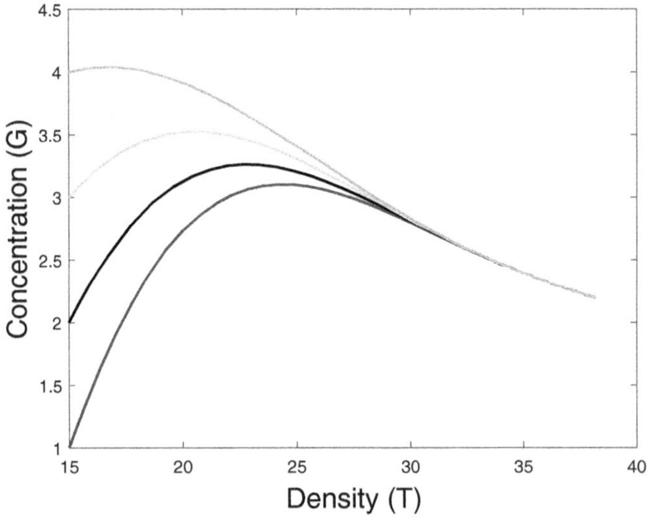

FIGURE 7.1 Nonlinear stability of E^* in $T-G$ plane.

depicts the global stability of E^* in $T-G$ plane, while Figure 7.2 shown the global stability of E^* in $G-G_a$ plane.

Next, we study the impact of the rate of formation of the raindrops (r) on the concentration of the air pollutants. In Figure 7.3, we vary the value of r and plot the variation in the concentration of the gaseous pollutants G against the same. It can be noted that the concentration of G reduces with the increase in the value of r.

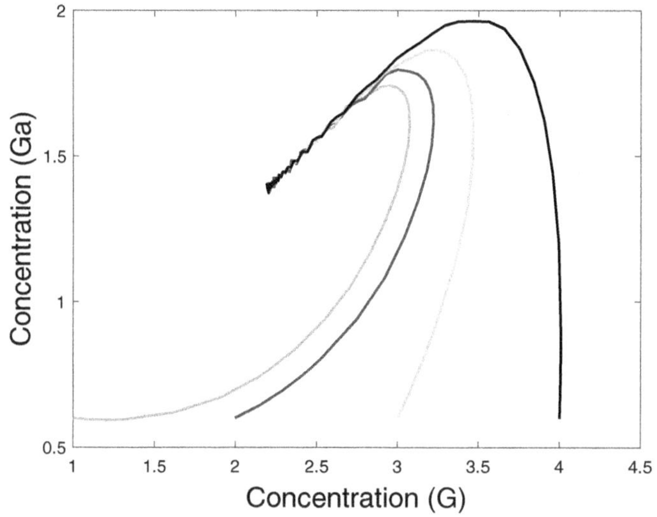

FIGURE 7.2 Nonlinear stability of E^* in $G-G_a$ plane.

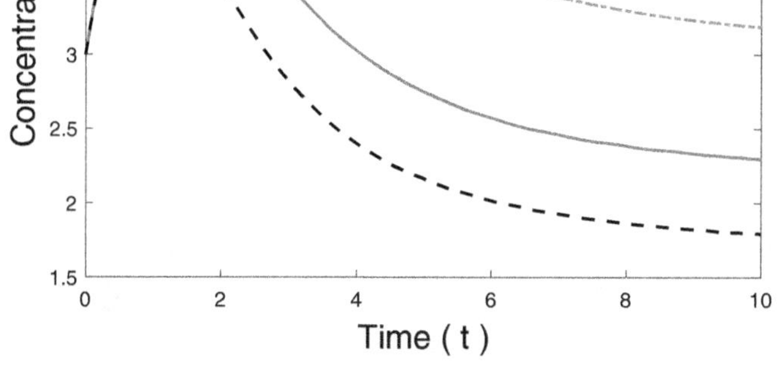

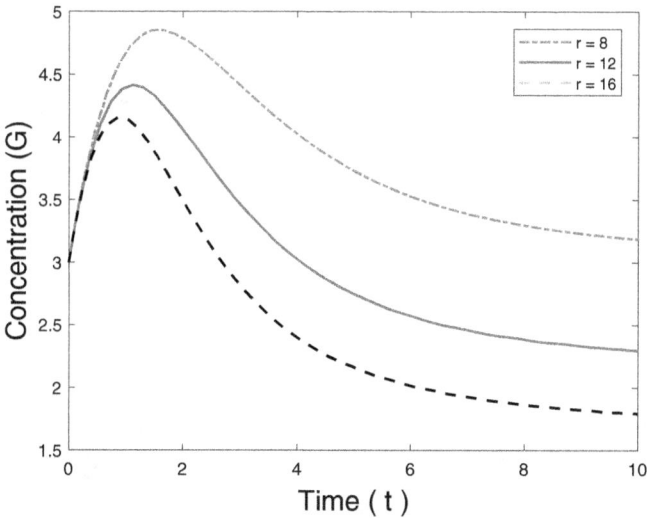

FIGURE 7.3 Variation in the concentration of G with respect to r.

In Figure 7.4, we study the variation in the concentration of P_1 with r. Again, we can observe that the concentration of the particulate matter P_1 also decreases as we increase the value of r. Similarly, we can conclude (from Figure 7.5) that the concentration of particulate matter P_2 also depleted with the increase in the rate of formation of raindrops. Further, we investigate the impact of δ_1, δ_2, and δ_3 on the concentrations of G, P_1, and P_2, respectively. In Figure 7.6, we obtain the variation in

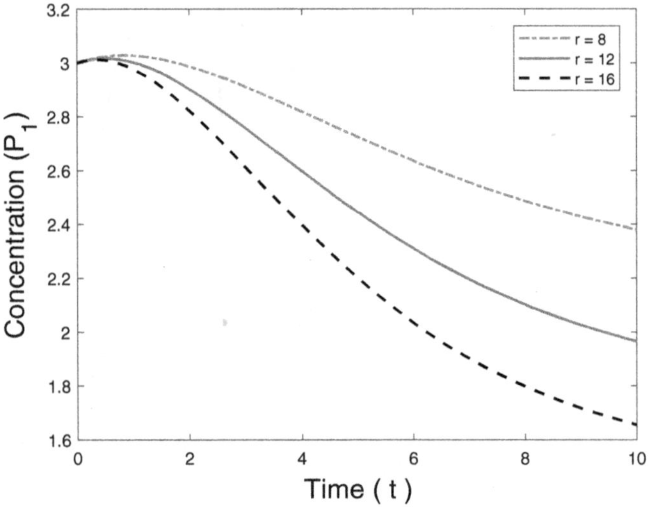

FIGURE 7.4 Variation in the concentration of P_1 with respect to r.

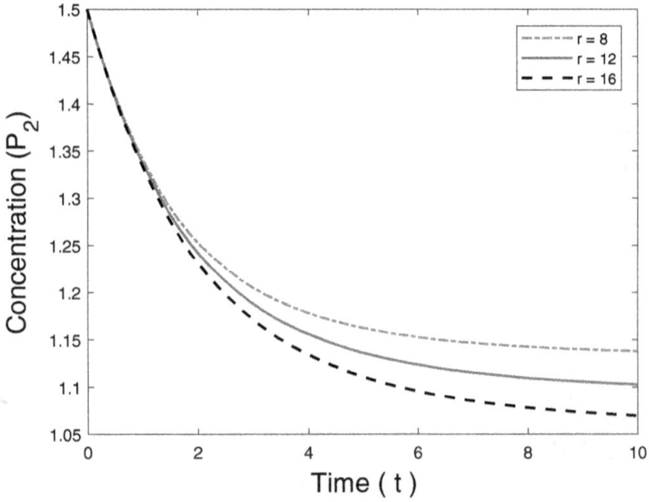

FIGURE 7.5 Variation in the concentration of P_2 with respect to r.

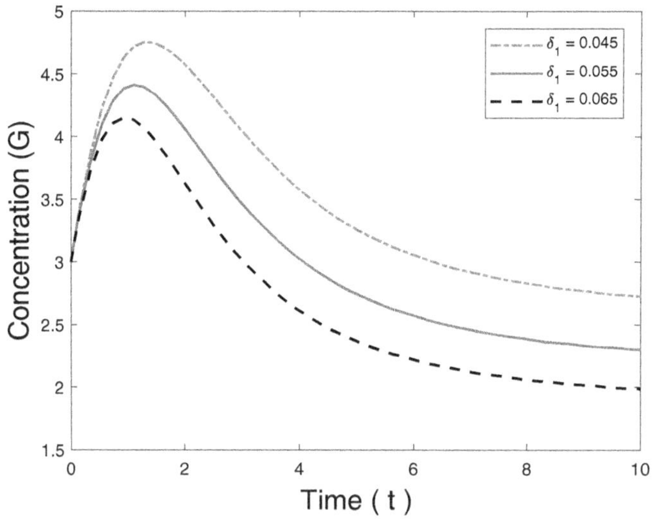

FIGURE 7.6 Variation in the concentration of G with respect to δ_1.

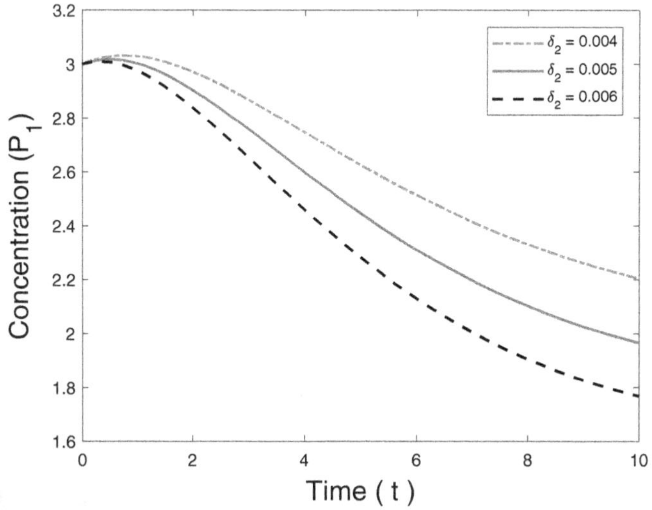

FIGURE 7.7 Variation in the concentration of P_1 with respect to δ_2.

the concentration of G for different values of δ_1. From the figure, it is clear that the concentration of G reduces significantly as we increase the value of δ_1. The change in the concentration of P_1 with respect to δ_2 is demonstrated in Figure 7.7. From the figure, we can observe that concentration of P_2 decreases with the increase in the value of δ_2. The same observation can be derived from Figure 7.8 for the second particulate matter.

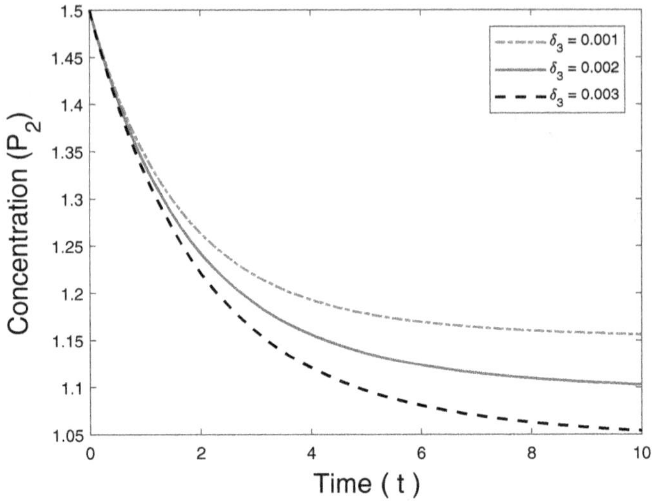

FIGURE 7.8 Variation in the concentration of P_2 with respect to δ_3.

7.6 CONCLUSIONS

The concentration of air pollutants is increasing at an alarming rate. The increase in the use of transport vehicles and industrial activities is main reason for the contamination of the environment. Air pollution has a number of hazardous impacts on the human body, resulting in cardiovascular and respiratory diseases. Therefore, controlling air pollution is one of the major research problems of the current time. On account of this, in this work, we formulated a mathematical model to study the removal of air pollutants (including air pollutants and particulate matters) due to rain. We consider gaseous pollutants and particulate matters as major air pollutants. Further, it is also assumed that the removal of pollutants takes place at a rate proportional to the product of the concentration of the particular pollutants and the density of the raindrops. The resultant model system (7.1) was subsequently subjected to dynamical analysis. It is observed that the resultant model system (7.1) has only one non-trivial equilibrium point E^*. Next, with the aid of the Lyapunov function, we obtain the conditions for local and global stability of E^*. The numerical results obtained in Section (7.5) also validate the analytical results. It is observed that the concentration of gaseous pollutants and particulate matters decreases significantly as the rate of formation of raindrops increases. In particular, if the air pollutants are emitted at a constant rate and the rate of formation of raindrops is sufficiently large, then air pollutants can be completely removed from the environment. On the other hand, in the absence of rain or low rainfall, the concentration of air pollutants keeps increasing.

Thus, from the above study, we conclude that rain is a significant tool to eliminate air pollutants from the environment. The cleaning effect of rain contributes to creating healthy living conditions.

REFERENCES

Freedman, H. I., & So, J. H. (1985). Global stability and persistence of simple food chains. *Mathematical biosciences*, *76*(1), 69–86.

Glencross, D. A., Ho, T. R., Camina, N., Hawrylowicz, C. M., & Pfeffer, P. E. (2020). Air pollution and its effects on the immune system. *Free Radical Biology and Medicine*, *151*, 56–68.

Kampa, M., & Castanas, E. (2008). Human health effects of air pollution. *Environmental pollution*, *151*(2), 362–367.

Kumari, N., & Sharma, S. (2018). Modeling the dynamics of infectious disease under the influence of environmental pollution. *International Journal of Applied and Computational Mathematics*, *4*(3), 1–24.

Lafferty, K. D., & Holt, R. D. (2003). How should environmental stress affect the population dynamics of disease?. *Ecology Letters*, *6*(7), 654–664.

Manisalidis, I., Stavropoulou, E., Stavropoulos, A., & Bezirtzoglou, E. (2020). Environmental and health impacts of air pollution: a review. *Frontiers in public health*, *8*, 14.

Misra, A. K., & Tripathi, A. (2018). A stochastic model for making artificial rain using aerosols. *Physica A: Statistical Mechanics and its Applications*, *505*, 1113–1126.

Misra, A. K., & Tripathi, A. (2019). Stochastic stability of aerosols-stimulated rainfall model. *Physica A: Statistical Mechanics and its Applications*, *527*, 121337.

Naresh, R. (2003). Qualitative analysis of a nonlinear model for removal of air pollutants. *International Journal of Nonlinear Sciences and Numerical Simulation*, *4*(4), 379–386.

Naresh, R., Sundar, S., & Shukla, J. B. (2007). Modeling the removal of gaseous pollutants and particulate matters from the atmosphere of a city. *Nonlinear Analysis: Real World Applications*, *8*(1), 337–344.

Sharma, S., & Kumari, N. (2017). Why to consider environmental pollution in cholera modeling?. *Mathematical Methods in the Applied Sciences*, *40*(18), 6348–6370.

Sharma, S., & Kumari, N. (2019). Dynamics of a waterborne pathogen model under the influence of environmental pollution. *Applied Mathematics and Computation*, *346*, 219–243.

Shukla, J. B., Misra, A. K., Sundar, S., & Naresh, R. (2008). Effect of rain on removal of a gaseous pollutant and two different particulate matters from the atmosphere of a city. *Mathematical and Computer Modelling*, *48*(5–6), 832–844.

Shukla, J. B., Sundar, S., Misra, A. K., & Naresh, R. (2008). Modelling the removal of gaseous pollutants and particulate matters from the atmosphere of a city by rain: Effect of cloud density. *Environmental Modeling & Assessment*, *13*(2), 255–263.

Sundar, S., Naresh, R., Misra, A. K., & Shukla, J. B. (2009). A nonlinear mathematical model to study the interactions of hot gases with cloud droplets and raindrops. *Applied Mathematical Modelling*, *33*(7), 3015–3024.

8 Image Watermarking with Polar Harmonic Moments

Sanoj Kumar and Manoj K. Singh
University of Petroleum and Energy Studies

Deepika Saini
Graphic Era deemed to be University

CONTENTS

8.1 INTRODUCTION: BACKGROUND AND DRIVING FORCES

Humans have always desired to build advanced and more advanced machines that follow their instruction at will. We have achieved a lot in this direction, but a lot of further improvements are essential. To achieve the goal, robust and very efficient mechanisms for object detection and pattern recognition are required. These mechanisms not only need to work efficiently in normal condition, but need to be robust enough to detect objects and patterns in various modified conditions, including rotation and scaling of the image space. Very useful artificial intelligence (AI) and machine learning (ML) algorithms are developed to investigate the objects or patterns from various critical conditions. In addition to training-based algorithms, features which are invariant to diverse conditions, including rotation and scaling, can be useful in determining the objects and patterns. Image moments are few such

DOI: 10.1201/9781003222255-8

invariant features which are quite useful in, but not limited to, object and pattern recognition.

The image moments have a wide range of applications, e.g., image watermarking (Singh et al., 2021, Ma et al., 2020 and Hosny and Darwish 2017), face recognition (Rahman et al., 2016, Farokhi et al., 2013), character recognition (Yadav and Purwar 2018), texture analysis (Ruberto et al., 2018) and edge detection (Wei and Zhao 2013). Nowadays, with the rapid growth in Internet technology and also network speed, the transfer of data has been increased (Cao 2017, Weihs and Ickstadt 2018). It includes many fields, e.g., satellite image data, medical, big data and Internet of things (IOT) (Kumar et al., 2019, Singh and Bhatnagar 2019, Singh et al., 2017, Bodkhe and Tanwar 2020, Hu et al., 2020). Due to the current situation, the medical field has been paid a lot of attention. The medical images are downloaded, uploaded and transferred for various applications. In some cases, the quality of these images are threatened. In this case, a user do not have the knowledge of the images whether they are attacked or not. So, the security of these images must be the primary concern. That's why the watermarks are embedded to these images (Swaraja et al., 2020, Khare and Srivastava 2021, Fan et al., 2019, Xia et al., 2019, Singh 2019). From the past decades, a lot of study has been investigated and various algorithms are generated for embedding the watermarks into medical as well as other images. With the detailed analysis of these methods, many of them play an effective role in protecting the quality and authenticity of these images from various attacks, including rotation, filtering, JPEG compression, cropping, the addition of noise and image flipping (Swaraja et al. 2020, Thakur et al. 2019, Zear et al., 2018, Parah et al., 2017). But, some methods are not so productive against these geometric attacks. To handle such attacks, the image watermarking algorithm must be geometric invariant. The image descriptors are the good solution in this case. Orthogonal moments are surely applied in those cases. The Zernike moments (ZM) and pseudo-Zernike moments (PZM) (Xin et al., 2007) are the good examples for the orthogonal family. It has been shown that the algorithm based on ZM and PZM performs great against image scaling, rotation and flipping attacks.

Some authors used polar harmonic transform (PHT) (Li et al., 2012) for image watermarking. PHT has also the caliber to gain the scaling invariance and image rotation in watermarking. It has the most relevance to insert the watermark rather than ZM and PZM. Radial harmonic Fourier moments (RHFMs) (Yang et al., 2015) and Fourier–Mellin moments (Shao et al., 2016) used in watermarking have the most caliber for strong image watermarking rather than ZM and PZM. With the whole study and comparison of the algorithms, one can conclude that the robustness of any watermarking algorithm based on moments purely depends on the accuracy of the estimated image moments. As the computed image moments are accurate, the watermarking algorithm is considered robust. The same thing is also justified in Hosny and Darwish (2017). They have estimated the PHT moments on a polar grid, and the accuracy of the estimated moments improved. The results of this show the robustness of the watermarking algorithm. In these moments-based algorithms, the bits of the watermark are inserted into the image moments. The watermarked image is regenerated by using these modified image moments. The bits of the watermark are extracted at the receiver end by estimating the image moments from the watermarked image. That's why the quality of the evoked watermarks is totally reliable on the accuracy of the estimated image moments.

In this chapter, an integration algorithm which uses the complete unit disk increases the accuracy of PHFT moments. The improved moments also contribute to the improved lustiness of the watermarking algorithm. The detailed explanation of the algorithm with the supported results are described in the below sections.

8.2 POLAR HARMONIC TRANSFORM

Polar harmonic transform (PHT) is a collection of three transformations, namely polar cosine transform (PCT), polar complex exponential transform (PCET) and polar sine transform (PST). In this chapter, we denote PCET with C_{uv}, where the index $u \in \mathbb{Z}$ denotes the order and the index $v \in \mathbb{Z}$ denotes the repetition, with $\mathbb{Z}$ representing the set of integers. The PCET of a function $h(r,\psi)$ with r representing the distance from the origin and ψ denotes the angular position of a point with respect to a fixed line. The coordinates (r,ψ) are restricted with $0 \le r \le 1$ and $0 \le \psi \le 2\pi$.

$$C_{uv} = \frac{1}{\pi}\int_0^{2\pi}\int_0^1 \overline{D_{uv}(r,\psi)h(r,\psi)}\, r\, dr\, d\psi \tag{8.1}$$

Here,
$$D_{uv}(r,\psi) = e^{i(2\pi ur^2+v\psi)}. \tag{8.2}$$

The functions $D_{uv}(r,\phi)$ are mutually orthogonal as shown in the following equation:

$$\int_0^{2\pi}\int_0^1 D_{u_a v_a}(r,\psi)\,\overline{D_{u_b v_b}(r,\psi)}\, h(r,\psi) r\, dr\, d\psi = \pi \text{ if } (u_a,v_a) = (u_b,v_b) \tag{8.3}$$

$$\int_0^{2\pi}\int_0^1 D_{u_a v_a}(r,\psi)\,\overline{D_{u_b v_b}(r,\psi)}\, h(r,\psi) r\, dr\, d\psi = 0 \text{ if } (u_a,v_a) \ne (u_b,v_b) \tag{8.4}$$

Similarly, PCT is denoted by $C_{uv}{}^c$, where the index $u \in \mathbb{Z}_+$ denotes the order and the index $v \in \mathbb{Z}$ denotes the repetition, with $\mathbb{Z}$ representing the set of integers. The PCT of a function $h(r,\psi)$ with r and ψ has its usual meaning as explained above.

$$C_{uv}^c = \Omega_u \int_0^{2\pi}\int_0^1 \overline{D_{uv}{}^c(r,\psi)h(r,\psi)}\, r\, dr\, d\psi \tag{8.5}$$

$$\text{Here } D_{uv}{}^c(r,\psi) = \cos(\pi ur^2)e^{iv\psi}, \tag{8.6}$$

$$\text{and } \Omega_u = \begin{cases} \dfrac{1}{\pi}, & \text{if order is zero} \\ \dfrac{2}{\pi}, & \text{if order is not zero} \end{cases} \tag{8.7}$$

Similar to PCT and PCET, PST is denoted by C_{uv}^s, where the index $u \in \mathbb{Z}_+$ denotes the order and the index $v \in \mathbb{Z}$ denotes the repetition, with $\mathbb{Z}$ representing the set of integers. PST is defined as:

$$C_{uv}^s = \Omega_u \int_0^{2\pi}\int_0^1 \overline{D_{uv}{}^s(r,\psi)h(r,\psi)}\, r\, dr\, d\psi \tag{8.8}$$

$$\text{Here } D_{uv}{}^{s}(r,\psi) = \sin(\pi u r^2)e^{iv\psi}. \tag{8.9}$$

Since the transformations are orthogonal, the function $h(r,\phi)$ can be reconstructed using

$$h(r,\phi) = \sum_{u=-\infty}^{\infty}\sum_{v=-\infty}^{\infty} C_{uv} \tag{8.10}$$

8.2.1 Moment Computations

Let $N \times N$ be the size of an image $I(i, j)$. Let us translate and rotate the image in such a position that the center of the image falls at the origin and the rows become parallel to the x-axis and the columns become parallel to the y-axis. The image is further scaled in such a manner that the leftmost column is at $x = -1$ and the rightmost column is at $x = 1$. Similarly, the lowermost row of the scaled image is associated with $y = -1$ and the uppermost row with $y = 1$. With these configurations of the image and coordinate system, the location of the pixels can be computed easily. Let us represent them as (x_i, y_k), $i,k = 0, 1, ..., N-1$. From computational perspective, PCET moments for an image $h(x_i, y_k)$ will be obtained in a simple form using equation 8.1 as

$$C_{uv} = \frac{2}{\pi}\sum_{i}\sum_{k} h(i,k)\overline{D_{uv}(x_i, y_k)}\ \delta x\ \delta y \tag{8.11}$$

Instead of taking $\overline{D_{uv}(x_i, y_k)\ \delta x\ \delta y}$ as single value by using the following integration, the moments can be assigned with better accuracy.

$$\overline{S_{uv}(x_i, y_k)\ \delta x\ \delta y} = \iint \overline{D_{uv}(r,\psi)}\ rdrd\psi \tag{8.12}$$

An image can be reconstructed from the moments by applying equation 8.13:

$$h(i,k) = \sum_{u=0}^{A}\sum_{v=0}^{B} C_{uv} D_{uv}(x_i, y_k) \tag{8.13}$$

Here A is an upper bound on u and B is an upper bound on v.

8.2.2 Watermarking Procedure

Let $n \times n$ be the size of a binary watermark. Therefore, n^2 bits should be embedded into the image. Let us denote the watermark image by W. Therefore, n^2 different image moments are computed for the watermark embedding. One of the moments is selected. The watermark bits are inserted into the absolute values of the complex-valued moments. For the present chapter, let us restrict the moment order between 0 and 25. In addition, let us also restrict the repetitions between 1 and 25. Let P_k for k as 1 to n^2 be the moments and $|P_k|$ be its absolute value. Here, the watermark bits are inserted into P_k. Let Δ be the quantization step and $P_k = r_k e^{\Theta k}$. Find two constants A_k and B_k satisfying division algorithm such that $|P_k| = A_k\Delta + B_k$. The values in $|P_k|$ are modified as (Singh et al. 2021):

$$|P_k^{'}| = \begin{cases} \left(A_k + \frac{3}{2}\right)\Delta, & \text{if } (W_k + A_k)\text{mod } 2 \text{ is } 1. \\ \left(A_k + \frac{1}{2}\right)\Delta, & \text{if } (W_k + A_k)\text{mod } 2 \text{ is } 0. \end{cases} \tag{8.14}$$

The image can then be reconstructed after inserting the modified moments into equation 8.15.

$$f(i,k) = 2\text{Re}\left(\sum_{j=0}^{n^2} |P_j^{'}|\, e^{\Theta j} D_j(x_i, y_k)\right) \tag{8.15}$$

Let $h(i,k)$ be the given image, $f(i,k)$ be the reconstructed image with unmodified moments, and $H(i,k)$ be the reconstructed image by using the modified moments. Also, $e(i,k)$ denotes the difference between $h(i,k)$ and $f(i,k)$. The watermarked image is constructed by adding $H(i,k)$ to $e(i,k)$. For more details on the algorithm, refer Singh et al. (2021). Let $|P_k'|$ be the absolute moment of the watermarked image. Find two constants A_k' and B_k' satisfying division algorithm such that $|P_k'| = A_k\Delta + B_k$. Then the following equation extracts the watermark bits.

$$W_k' = \begin{cases} 1, & \text{if } A_k \text{ mod } 2 \text{ is } 1. \\ 0, & \text{if } A_k \text{ mod } 2 \text{ is } 0. \end{cases} \tag{8.16}$$

for $k = 1, 2, \ldots, n^2$.

8.3 EXPERIMENTAL RESULTS AND DISCUSSIONS

To understand the capability and the functioning of the said algorithm, several experiments have been performed. For this purpose, we have used digital as well as medical images. Cameraman, jet plane, Lena, house, lake, mandrill, living room, pirate, peppers, walk bridge and woman (blonde & dark hair) are used as test images. These images are shown in Figure 8.1. The chest X-ray image database of Mendeley Data, V2 (Singh et al. 2021, Singh and Upneja 2012, Kermany et al., 2018) is used for medical images. This database is available publicly. This database is initially developed for the problem of classification in machine learning. It contains pneumonia and normal X-ray images. It contains 69 normal images of the first category. These images are used for the establishment of the said algorithm. The size of all the test images is 256×256, while the size of the watermark images is 16×16. However, the proposed algorithm is tested with all the test images, but here, the experimental results are described only for pepper image. The quantization step (Δ) plays a very important role. The high values of it give lower values of PSNR, while the lower

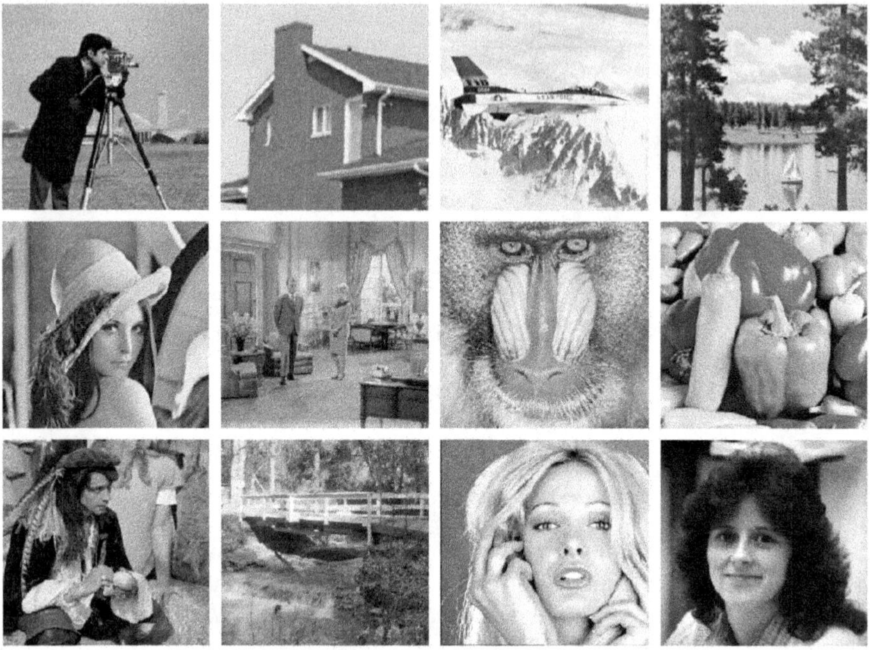

FIGURE 8.1 Test images.

values of it give high values of PSNR. But, low values of Δ yield luxurious exposure to different kinds of attacks and high values of Δ yield low exposure to these attacks. Hence, there must be a trade-off between the exposure of the algorithm and PSNR.

Based on various experiments, the value of Δ is chosen in such a way that the robustness of the algorithm should not be hampered and also the values of PSNR must be high (above 40). In the whole process, the value of Δ is set at 0.22. Generally, PSNR is selected to investigate the caliber of the watermarked image in terms of bit error ratio (BER), which is given as

$$BER = \frac{B}{N^2} \tag{8.17}$$

Here, B represents the number of those picture elements that do not coincide in the original and extracted watermarks. Both B and N are positive numbers and $B \le N^2$; hence, the value of BER always falls in [0,1]. The value of B is close to 0 if the original and extracted watermarks are identical, and the value of B is close to 1 if the original and extracted watermarks are different. So, for a good algorithm, BER must be very small (near to 0). The other metric that is selected for estimating the caliber of the watermarked image is PSNR and is defined as

$$PSNR = 10\ \log_{10} \frac{255^2}{MSE} \tag{8.18}$$

where MSE is measured between the watermarked and original images.

8.3.1 Rotational Invariance

The watermark is embedded into the test images. This watermark should be extracted from the watermarked images correctly. Therefore, it is essential that the moments must be evaluated accurately. Theoretically, PCET, PCT and PST are invariant to rotation. Due to this amazing exposure of the moments, when the watermark image is rotated, then the moments remain similar to that of the original image. This gives the error-free extracted watermark. The rotated watermarked images (counterclockwise) with the angles at 10°, 20°, 30°, 40°, 50°, 60°, 70°, 80° and 90° along with the evoked watermark are shown in Figure 8.2. The prominent quality of the outputs shows the lustiness of the proposed algorithm. The performances of the PCET, PCT and PST watermarking algorithms when the watermarked image (test images) is rotated counterclockwise at angles at 10°, 20°, 30°, 40°, 50°, 60°, 70°, 80° and 90° are shown in Table 8.1. The PSNR in the case of PST is higher than in the case of PCET and PCT; however, BER of PCET is smaller than that of PST and PCT.

8.3.2 Scaling Invariance

Another beautiful property of moments discussed in this chapter is that they are also scaling invariant. So, if they are evaluated accurately, the evoked watermark is similar to that of the original watermark. It can be verified easily from

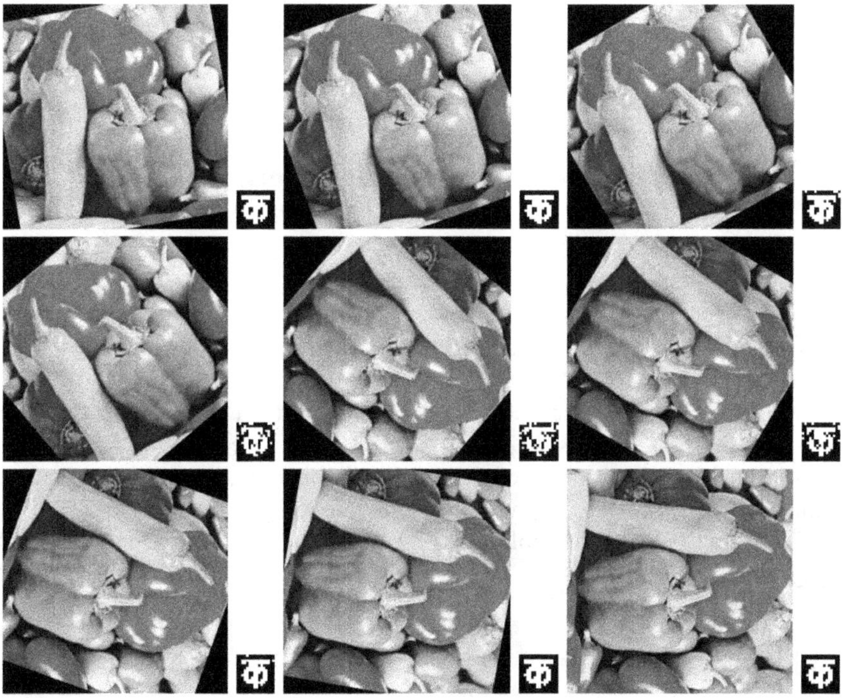

FIGURE 8.2 The rotated watermark images (counterclockwise) along with the evoked watermark.

TABLE 8.1
BER of the Evoked Watermark

Algorithm	PSNR	0°	10°	20°	30°	40°	50°	60°	70°	80°	90°
PCET	38.74	0.00	0.00	0.0065	0.0456	0.1003	0.0993	0.0485	0.0062	0.003	0.00
PCT	41.20	0.00	0.0016	0.0371	0.1396	0.2386	0.2363	0.1390	0.0394	0.0026	0.00
PST	41.50	0.00	0.00	0.0218	0.1016	0.2064	0.1953	0.1035	0.0251	0.0003	0.00

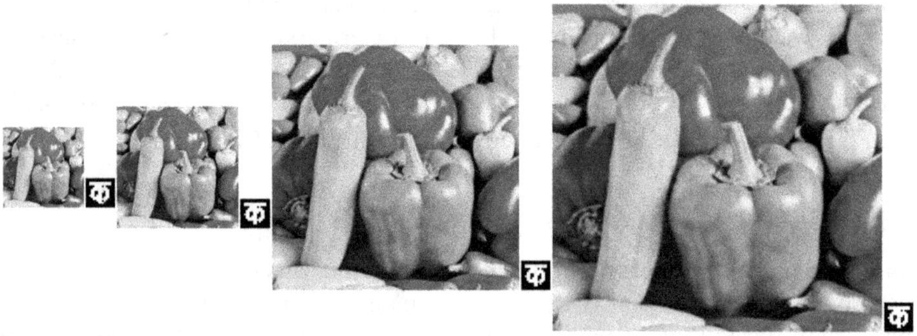

FIGURE 8.3 From left: the scaled watermarked images with a factor of 0.5, 0.75, 1.5 and 2 along with the corresponding evoked watermark.

TABLE 8.2
BER of Scaled Watermarked Image (Pepper)

Algorithm	0.5	0.75	1.5	2.0
PCET	0.0479	0.0019	0.0000	0.0000
PCT	0.0143	0.0003	0.0000	0.0000
PST	0.0068	0.0000	0.0000	0.0000

Figure 8.3. To estimate the robustness of the said methods, various scaling attacks are applied. The visual results for the same represent the robustness of the said algorithm. The values of BER of the evoked watermark are mentioned in Table 8.2 after scaling the watermarked image by a factor of 0.5, 0.75, 1.5 and 2.0, respectively. From Table 8.2, PST has a better performance than PCET and PCT.

8.3.3 Addition of Noise

Sometimes, the performance of a watermarking algorithm decreases in the presence of noises. In the literature, various types of noises are available, e.g., Gaussian, Poisson, salt-and-pepper and speckle noises. The success of the proposed techniques is tested on these noises. The watermarked images with the addition of these noises and the corresponding evoked watermarks are shown in Figure 8.4. The visual outputs show the robustness of these techniques. The value of BER when the watermark is extracted from the noisy watermarked image is mentioned in Table 8.3.

8.3.4 Image Filtering

The lustiness of the said algorithm is also examined using Gaussian, median and average filters. The aspect of the window is 5×5 and fixed for all the filters. The watermarked image is filtered using the above all filters. The values

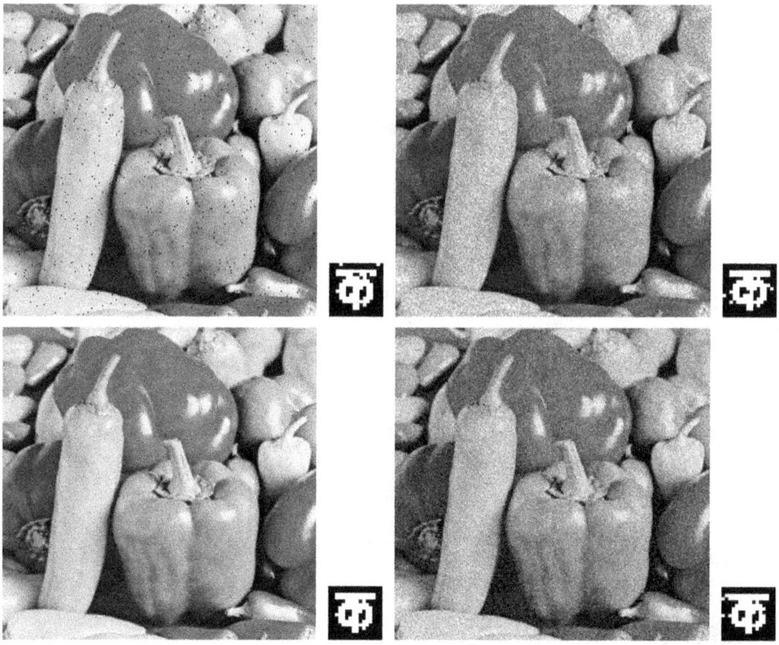

FIGURE 8.4 From top left in clockwise: the watermarked images after the addition of salt-and-pepper, speckle, Gaussian and Poisson noises along with the evoked watermark.

TABLE 8.3
BER When Noise is Added to Watermarked Image

Algorithm	Salt-and-Pepper	Speckle	Poisson	Gaussian
PCET	0.0250	0.0387	0.0000	0.0107
PCT	0.0075	0.4990	0.0000	0.0000
PST	0.0104	0.4844	0.0000	0.0000

TABLE 8.4
Image Filtering Attack Over Pepper Image

Algorithm	Median	Average	Gaussian
PCET	0.2689	0.3249	0.1663
PCT	0.3395	0.4404	0.1462
PST	0.3076	0.3975	0.1263

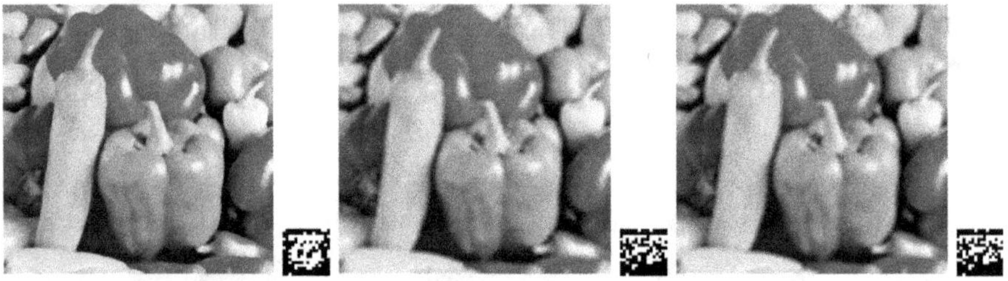

FIGURE 8.5 From left: the watermarked images and the evoked watermark against the median, average and Gaussian filters.

of BER of the watermark are mentioned in Table 8.4. AM and PZM do not perform well against these filtering attacks. BER of the said algorithm is at second position only to PST (Hosny and Darwish 2017) when the Gaussian and median filters are applied to the watermarked image. As we noted earlier, PST is not so much strong in the case of noise, scaling and rotation of the image. To analyze the effect of image filtering in the said algorithm, the aspect of the window is reduced to 3×3. In this case, the values of BER are 0. It shows the better performance of the said algorithm against filtering attacks. The same results are mentioned in Figure 8.5.

8.3.5 JPEG Compression

JPEG compression is another very common attack that is applied on the images. To test the lustiness of the said algorithm, JPEG compression attack is applied in all the test images with the character 25, 50, 75 and 90. The value of BER is 0.0078 for 25, while it is 0 for all the other values. Hence, one can assure that the said algorithm is strong against JPEG compression attack. The same results are shown in Figure 8.6 along with the evoked watermark image. The visual quality of the results as shown in Figure 8.6 suggests the lustiness of the said algorithm. A comparison has also been made with the other algorithms when the watermarked images is attacked with JPEG compression. The results of the same are mentioned in Table 8.5. From this table, one can measure that when the character is fixed at 50, 75 or 90, the value of BER in case of PCT (Hosny and Darwish 2017) is zero. But, when the character is at 25, the said algorithm has a better value of BER

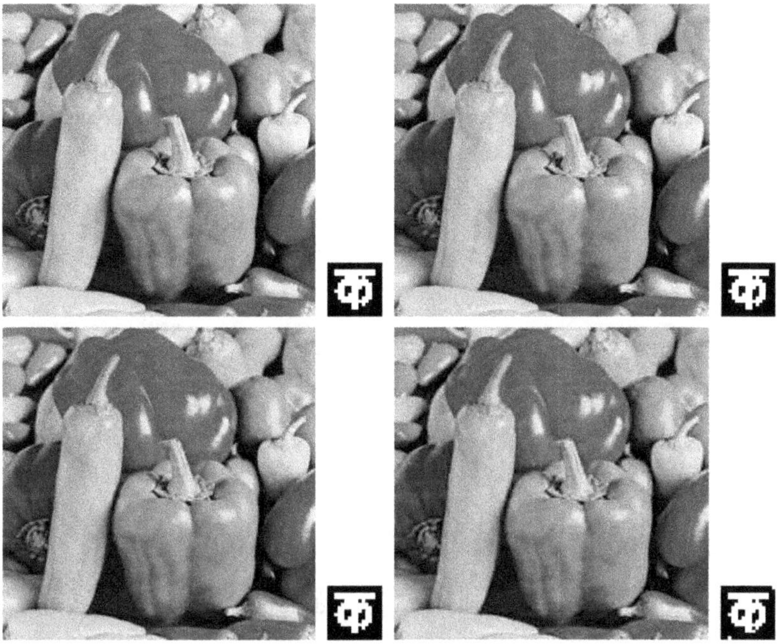

FIGURE 8.6 From top left in the direction of clock: the watermarked images attacked by JPEG compression attack with characters 25, 50, 75 and 90 along with the evoked watermark.

TABLE 8.5
JPEG Compression Attack on Pepper Image

Algorithm	$Q = 90$	$Q = 75$	$Q = 50$	$Q = 25$
PCET	0.0000	0.0000	0.0006	0.0140
PCT	0.0000	0.0000	0.0000	0.0104
PST	0.0000	0.0000	0.0000	0.0065

than PCT. The value of BER in case of PHFT (Ma et al., 2020) is comparable with the said algorithm at all the character indices.

8.3.6 Image Flipping

The robustness of the said algorithm is also tested with image flipping attacks. The horizontal, vertical and both-side flipping attacks are imposed. The value of BER is zero against the image flipping attacks. Hence, one can claim that the said algorithm is strong against image flipping attacks. Table 8.6 represents the values of BER of different methods. The value of BER in the case of PCT and PHFT are 0. But, the values of BER in case of the other methods are neither 0 nor negative. This suggests

TABLE 8.6
Image Flipping Attack on Pepper Image

Algorithm	Vertical	Horizontal	Both Flipping
PCET	0.4212	0.4212	0.0000
PCT	0.0000	0.0000	0.0000
PST	0.0000	0.0000	0.0000

FIGURE 8.7 From left: the watermarked images and the evoked watermark against vertical, horizontal and both-way flipping attacks.

that these methods are not so strong against the image flipping attacks. The watermarked images with the evoked watermark against the image flipping attacks are shown in Figure 8.7.

8.4 MOTION BLUR ATTACK

The said algorithm is also strong against the motion blur attack. The value of BER is zero for the horizontal and vertical motion blur attacks with kernel size 3. The value of BER is also evaluated with the kernel size 5 and is mentioned in Table 8.7. From this table, the value of BER in the case of PCT (Hosny and Darwish 2017) is smaller than the said method in the vertical motion. However, the value of BER is larger than that of the other methods except PHFT (Ma et al., 2020). Figure 8.8 represents the watermarked image and the corresponding evoked watermark in the case of horizontal and vertical motion blur attacks.

8.4.1 Gamma Modification Attack

The value of BER of the said algorithm is similar to that of PCT and PHFT in the case of gamma correction attack. But, the lustiness of the said algorithm is quite good compared to other algorithms as mentioned in Table 8.8. The values of BER are 0.0234 and 0.0312, respectively, against the gamma correction attack with $\gamma = 0.9$ and $\gamma = 1.1$. That indicates the great performance of the said algorithm in the case of gamma correction attack. The same is supported by the visual results shown in Figure 8.8.

TABLE 8.7
Motion Blur Attack on Pepper Image

	Chest X-ray	
Algorithm	**Vertical**	**Horizontal**
PCET	0.0166	0.0247
PCT	0.0003	0.0078
PST	0.0010	0.0042

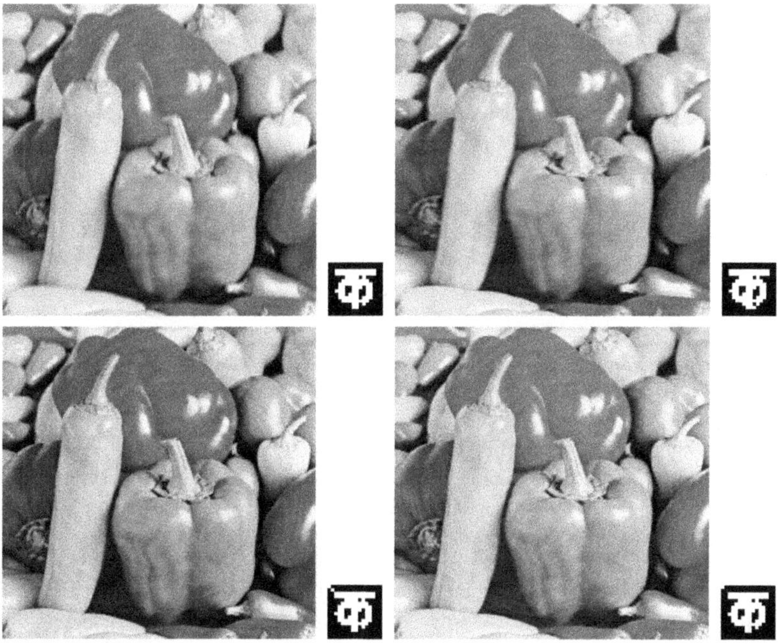

FIGURE 8.8 First row: the watermarked images along with the corresponding evoked watermark against the horizontal and vertical motion blur attacks. Second row: the watermarked images ($\gamma = 0.9; 1.1$) and the corresponding evoked watermark against gamma modification attack.

TABLE 8.8
Gamma Modification Attack on Pepper Image

	Chest X-ray	
Algorithm	$\gamma = \mathbf{0.9}$	$\gamma = \mathbf{1.1}$
PCET	0.0114	0.0107
PCT	0.0358	0.0351
PST	0.0469	0.0413

8.5 CONCLUSIONS

PHT includes image signifiers which are invariant in the case of scaling and rotation. The image moments are picked to generate the strong watermarking methods. PHFT moments are chosen for the said watermarking algorithm in this study. PST, PCT and PCET are also taken for image watermarking algorithms. The said methods are imposed to the ordinary as well as medical images to show the robustness of these methods against various attacks. However, the robustness of these methods can be improved by increasing the accuracy of the computed moments. A mixture of analytic as well as numerical methods is adopted in this chapter to improve the accuracy of PHFT moments. These accurate moments are adopted to the image watermarking. With the detailed analysis of the experiments, the said algorithm is found more accurate and robust compared to existing moments. There exist some algorithms that have a lower value of BER than the said algorithm in some attacks. Particularly, PCT (Hosny and Darwish 2017) and PHFT (Ma et al., 2020) obtained the robustness close to the said algorithm. However, in this study, only lower-order moments are selected. In further studies, higher-order moments can be selected and the effect of various attacks on the watermarking algorithms based on higher moments can be examined.

REFERENCES

Bodkhe, U., and Tanwar, S. 2020. Secure data dissemination techniques for IOT applications: Research challenges and opportunities. *Journal of Software: Practice and Experience.* 14(6): 563–571.

Cao, L. 2017. Data science: A comprehensive overview. *ACM Computing Surveys.* 50(3):1–42.

Fan, T.Y., Chao, H.C., and Chieu, B.C. 2109. Lossless medical image watermarking method based on significant difference of cellular automata transform coefficient. *Signal Processing: Image Communication.* 70: 174–183.

Farokhi, S., Shamsuddin, S. M., Flusser, J., Sheikh, U.U., Khansari, M., and Khouzani K.J. 2013. Rotation and noise invariant near-infrared face recognition by means of Zernike moments and spectral regression discriminant analysis. *Journal of Electronic Imaging,* 22(1):1–11.

Hosny, K.M. and Darwish, M. M. 2017. Invariant image watermarking using accurate polar harmonic transforms. *Computers & Electrical Engineering*, 62:429–447.

Hu, C., Pu, Y., Yang, F., Zhao, R., Alrawais, A. and Xiang, T. 2020. Secure and efficient data collection and storage of IOT in Smart Ocean. *IEEE Internet of Things Journal.* 7(10): 9980–9994.

Kermany, D., Zhang, K., and Goldbaum, M. 2018. Labeled optical coherence tomography (oct) and chest X-ray images for classification. http://web.archive.org/web/20080207010024/ http://www.808multimedia.com/winnt/kernel.htm, Mendeley Data, V2.

Khare, P. and Srivastava, V.K. 2021. A secured and robust medical image watermarking approach for protecting integrity of medical images. *Transactions on Emerging Telecommunications Technologies.* 32(7): https://doi.org/10.1002/ett.3918.

Kumar, S.A., Subramaniyaswamy, V., Kumar, V.V., Chilamkurti, N., and Logesh, R. 2019. SVD-based robust image steganographic scheme using RIWT and DCT for secure transmission of medical images. *Measurement*, 139:426–437.

Li, L., Li, S., Abraham A., and Pan, J. 2012. Geometrically invariant image watermarking using polar harmonic transforms. *Information Sciences*, 199: 1–19.

Ma, B., Chang, L., Wang, C., Li, J., Wang, X., and Shi, Y. 2020. Robust image watermarking using invariant accurate polar harmonic Fourier moments and chaotic mapping. *Signal Processing*, 172: 107544.

Parah, S.A., Sheikh, J.A., Nazir, F. A., Loan, A., and Bhat, G.M. 2017. Information hiding in medical images: A robust medical image watermarking system for e-healthcare. *Multimedia Tools and Applications*, 76(8): 10599–10633.

Rahman, S.M.M., Howlader, T., and Hatzinakos, D. 2016. On the selection of 2d krawtchouk moments for face recognition. *Pattern Recognition*, 54:83–93.

Ruberto, C.D., Putzu, L., and Rodriguez, G. 2018. Fast and accurate computation of orthogonal moments for texture analysis. *Pattern Recognition*, 83:498–510.

Shao, Z., Shang, Y., Zhang, Y., Liu, X., and Guo, D. 2016. Robust watermarking using orthogonal Fourier-Mellin moments and chaotic map for double images. *Signal Processing*, 120: 522–531.

Singh, A.K. 2019. Robust and distortion control dual watermarking in DWT domain using DCT and error correction code for color medical image. *Multimedia Tools and Applications*. 78(21):30523–30533.

Singh, C., and Upneja, R. 2012. A computational model for enhanced accuracy of radial harmonic Fourier moments. *In World Congress of Engineering*, London, UK, pp. 1189–1194.

Singh, M.K., Gautam, R., and Venkatachalam, P. 2017. Bayesian merging of misr and modis aerosol optical depth products using error distributions from aeronet. *IEEE Journal of Selected Topics in Applied Earth Observations and Remote Sensing*, 10(12): 5186–5200.

Singh, M.K., Kumar, Sanoj, Ali, M., and Saini, D. 2021. Application of a novel image moment computation in X-ray and MRI image watermarking. *IET Image Processing*, 15(3): 666–682.

Singh, S.P. and Bhatnagar, G. 2019. A novel biometric inspired robust security framework for medical images. *IEEE Transactions on Knowledge and Data Engineering*, 33(3):810–823.

Swaraja, K., Meenakshi, K., and Padmavathi, K. 2020. An optimized blind dual medical image watermarking framework for tamper localization and content authentication in secured telemedicine. *Biomedical Signal Processing and Control*. 55:101665.

Thakur, S., Singh, A.K., Ghrera, S.P., and Elhoseny, M. 2019. Multi-layer security of medical data through watermarking and chaotic encryption for tele-health applications. *Multimedia Tools and Applications*. 78(3): 3457–3470.

Wei, B.Z. and Zhao, Z.M. 2013. A sub-pixel edge detection algorithm based on Zernike moments. *The Imaging Science Journal*, 61(5):436–446.

Weihs, C. and Ickstadt, K. 2018. Data science: the impact of statistics. *International Journal of Data Science and Analytics*, 6(3):189–194.

Xia, Z., Wang, X., Zhou, W., Li, R., Wang, C., and Zhang, C. 2019. Color medical image lossless watermarking using chaotic system and accurate quaternion polar harmonic transforms. *Signal Processing*. 157: 108–118.

Xin, Y., Liao, S., and Pawlak, M. 2007. Circularly orthogonal moments for geometrically robust image watermarking. *Pattern Recognition*, 40(12): 3740–3752.

Yadav, M. and Purwar, R.K. 2018. Hindi handwritten character recognition using oriented gradients and Hu-geometric moments. *Journal of Electronic Imaging*, 27(5):1–11.

Yang, H., Wang, X., Wang, P., and Niu, P. 2015. Geometrically resilient digital watermarking scheme based on radial harmonic Fourier moments magnitude. *AEU - International Journal of Electronics and Communications*, 69(1): 389–399.

Zear, A., Singh, A.K., and Kumar, P. 2018. A proposed secure multiple watermarking technique based on DWT, DCT and SVD for application in medicine. *Multimedia Tools and Applications*. 77(4): 4863–4882.

9 Fractal Reptiles of the Plane with Holes using Reflections and Rotations

Akhlaq Husain
BML Munjal University

CONTENTS

9.1 INTRODUCTION

A compact set A with non-empty interior in the plane ($\mathbb{R}^2$ or $\mathbb{R}^3$) is called an *n-reptile* (or *n-rep tile* or simply *reptile*) if it can be tiled by n congruent tiles, each similar to A (see Croft et al., 1991). Reptiles are assumed to have simple topological structure in the sense that a reptile is a bounded tile that is equal to the closure of its interior (Croft et al., 1991, Falconer, 1990, Jordan and Ngai, 2005).

The simplest examples of reptiles are convex polygons such as parallelogram and the right-angled isosceles triangles with side lengths in the ratio $1 : \sqrt{2}$, and these are the only examples of 2-reptiles with this ratio (Croft et al., 1991, C17). Several authors have given general methods to construct n-reptiles using various construction strategies. Dekking (1982) gave a general method to generate fractal reptiles, space filling curves and nowhere differentiable functions using group endomorphisms and automorphisms. Bandt (1991) described a general method to create fractal reptiles using integer matrices.

John Conway posed the following question: Are there any reptiles with holes (i.e., non-simply connected reptile)? It was later corrected by Grunbaum as: Is there a connected open set whose closure is a reptile that is not simply connected? Croft, Falconer and Guy (1991) modified the same question as follows: What is the smallest integer n such that an n-reptile has a hole?

Jordan and Ngai (2005) presented a general geometric method to construct connected fractal reptiles of the plane with holes for any even integer $n \geq 4$. They also answered some questions concerning the geometry of fractal reptiles whose interiors

DOI: 10.1201/9781003222255-9

have infinitely many components and the closure of each component has infinitely many holes. The work of Jordan and Ngai oversimplified the original question of John Conway to: Are there 2- or 3-reptiles with holes?

For n equal to an odd integer, Jordan and Ngai (2005) also gave an example of a 9-reptile with holes and they claimed without proof that 9 is the smallest such integer. The method of Jordan and Ngai can be thought of as a generalization of the method by Bandt to construct reptiles using non-integer matrices. They also proved some important questions related to the topological structure of fractal reptiles in the plane. For example, they have shown that there exist reptiles in $\mathbb{R}^2$ with infinitely many components in the interior and the closure of some component is not a topological disk, and there exist reptiles such that the closure of every component of the interior of the reptile has a hole.

In this chapter, we present a concise summary of the article by Jordan and Ngai (2005) and construct new examples of n-reptile with holes for both even and odd values of n using rotations and elementary reflections. We also obtain reptiles without holes and examples of self-similar fractals as special cases of these reptiles. An example of a connected 9-reptile with holes using integer expanding matrix is constructed, and some open questions concerning reptiles with holes are stated in the end.

9.2 PRELIMINARIES AND SOME RESULTS

For any $A \subseteq \mathbb{R}^d$ $(d = 2,3)$, the interior, closure and boundary of A will be denoted by A°, $\bar{A}$ and ∂A, respectively.

Definition 9.1: Two sets A and B are said to be *essentially disjoint* if $\mu(A \cap B) = 0$, where μ is the d −dimensional measure on $\mathbb{R}^d$.

For a sequence of transformation $w_1, w_2, \ldots, w_k$ on $\mathbb{R}^d$, we denote the composition $w_1 \circ w_2 \circ \cdots \circ w_k$ simply by $w_1 w_2 \cdots w_k$. Let $\mathcal{F}, \mathcal{G}$ be two families of transformations on $\mathbb{R}^d$ and $A \subseteq \mathbb{R}^d$. We shall use the following notation:

$$\mathcal{F}\mathcal{G} = \{fg : f \in \mathcal{F},\ g \in \mathcal{G}\},\ \ \mathcal{F}(A) = \bigcap_{f \in \mathcal{F}} f(A) \tag{9.1}$$

Consider an iterated function system $\mathcal{F} = \{w_i\}_{i=1}^n$ to be denoted as IFS (in short) defined by

$$w_i(x) = \frac{1}{\sqrt[d]{n}} \rho_i(x) + t_i, \quad i = 1,2,\ldots,n$$

where $\rho_i(x)$ is an orthogonal transformation in $\mathbb{R}^d$ and $t_i \in \mathbb{R}^d$ are translation vectors. Note that each transformation in the IFS is a similitude and the contractivity factor of the IFS is $\frac{1}{\sqrt[d]{n}}$. For simplicity, we shall use stiff rotations and elementary reflections in hyperplanes through origin (as our orthogonal transformations) in the construction of reptiles as in Jordan and Ngai (2005). However, the construction using arbitrary reflections, rotations and improper rotations is also possible and we may obtain more interesting reptiles in those cases.

Let $A_0 \subseteq \mathbb{R}^d$ be non-empty, compact and invariant under $\mathcal{F}$ (i.e., $\mathcal{F}(A_0) \subseteq A_0$) such that

$$\mathring{A_0} \neq \varnothing,\ \overline{\mathring{A_0}} = A_0,\ \mu\left(\mathring{A_0}\right) = \mu(A_0) \tag{9.2}$$

Define

$$A_k = \mathcal{F}(A_{k-1}),\ A = \bigcap_{k=1}^{\infty} A_k \tag{9.3}$$

Then, A is the attractor of $\mathcal{F}$, which is a self-similar n-reptile and

$$A = \bigcap_{i=1}^{\infty} w_i(A) \tag{9.4}$$

It follows from equation (9.4) and the uniqueness of the attractor A that A is the closure of its interior and since the interior of A is non-empty, it is an n-reptile. Clearly, the fractal (similarity) dimension of A is d. Thus, the assumption that A has a non-empty interior is equivalent to the fact that the *open set condition* (Schief, 1994) is satisfied by the IFS.

We say that a reptile $A \subseteq \mathbb{R}^d$ is a reptile with a *hole* if the complement of the closure of some component of the interior of A has a bounded component. To answer the question posed by John Conway, Grunbaum gave an example of a 36-reptile in $\mathbb{R}^2$ which has a hole (see Croft et al., 1991, Figure C17]). Croft et al. (1991) asked the question: What is the least n for which an n-reptile in the plane has a hole?

Bandt and Wang (2001) and Luo et al. (2002) proved that if the interior of an n-reptile in $\mathbb{R}^2$ is connected, then the reptile is a topological disk. Therefore, the reptiles with holes are the reptiles having disconnected interiors. Jordan and Ngai (1991) proved a general result and described a geometric method to construct $2m$-reptiles with holes for every positive integer $m \geq 2$, which is given in the following theorem.

Theorem 9.1 (Jordan and Ngai, 2005, Theorem 1.1)

Let $n = 2m$ with $m \geq 2$, and let $\rho_{\frac{\pi}{2}}$ be the (anticlockwise) rotation by $\frac{\pi}{2}$ and σ_y be the reflection about the y-axis. Define an IFS by

$$w_i(x) = \begin{cases} \frac{1}{\sqrt{n}} \rho_{\frac{\pi}{2}}(x) + \left(i + \frac{1}{2}, 0\right), & -m \leq i \leq -1 \\ \frac{1}{\sqrt{n}} \rho_{-\frac{\pi}{2}}(x) + \left(i + \frac{1}{2}, 0\right), & 0 \leq i \leq m-2 \end{cases} \tag{9.5}$$

$$f(x) = \frac{1}{\sqrt{n}} \rho_{\frac{\pi}{2}}(x) \sigma_y(x) + \left(-m + \frac{1}{2}, 0\right) \tag{9.6}$$

and let A be the attractor of the IFS

$$\mathcal{F} = \{w_i\}_{-m}^{m-2} \cup \{f\}. \tag{9.7}$$

Then A is a connected n-reptile. Moreover, the interior of A is the union of the largest component, together with countably many geometrically similar sets, with the closure of each of these components containing infinitely many holes.

Thus, Theorem 9.1 reduces the question asked by Croft et al. (1991) to: Are there any 2−reptiles or 3-reptiles with holes? For the proof of Theorem 9.1, we refer to Jordan and Ngai (2005). Theorem 9.1 also affirms the following questions raised by Ngai and Tang (2005): If the interior of a reptile A in $\mathbb{R}^2$ has infinitely many components, is it possible that the closure of some component is not a topological disk, and is it possible that the closure of every component has a hole?

Our basis of construction of fractal reptiles with holes is Theorem 9.1. We shall revisit some examples considered by Jordan and Ngai (2005), and we will also construct new examples of connected fractal reptiles with holes. Pictures of some attractors are shown in Figure 9.1 for $m = 2,3,4,$ and 5 for the IFS defined by equation (9.7). The explicit mappings in the IFS are given below.

IFS for $m = 2$:

$$w_{-2}\begin{pmatrix} x \\ y \end{pmatrix} = \begin{pmatrix} 0 & -1/2 \\ 1/2 & 0 \end{pmatrix}\begin{pmatrix} x \\ y \end{pmatrix} + \begin{pmatrix} -3/2 \\ 0 \end{pmatrix},$$

$$w_{-1}\begin{pmatrix} x \\ y \end{pmatrix} = \begin{pmatrix} 0 & -1/2 \\ 1/2 & 0 \end{pmatrix}\begin{pmatrix} x \\ y \end{pmatrix} + \begin{pmatrix} -1/2 \\ 0 \end{pmatrix},$$

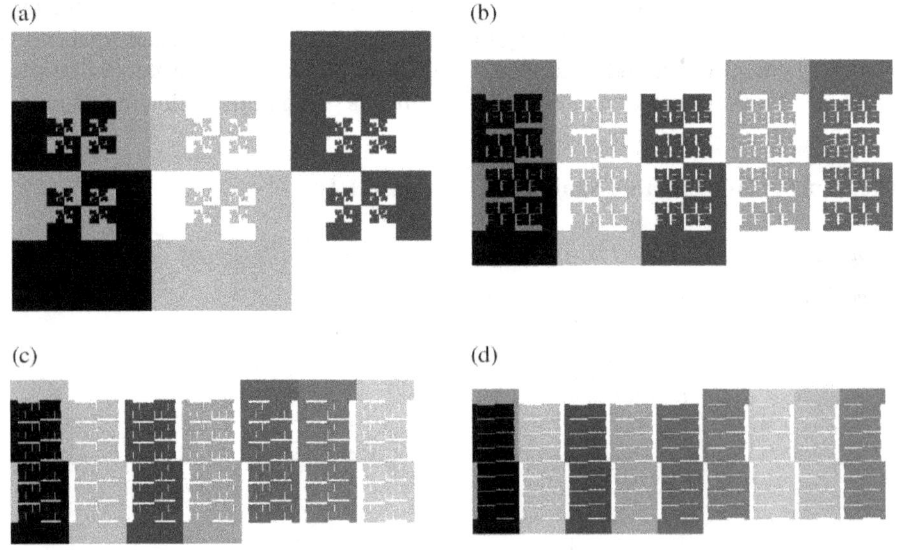

FIGURE 9.1 (a) $A_4(m=2)$, (b) $A_6(m=3)$, (c) $A_8(m=4)$, (d) $A_{10}(m=5)$.

$$w_0\begin{pmatrix} x \\ y \end{pmatrix} = \begin{pmatrix} 0 & 1/2 \\ -1/2 & 0 \end{pmatrix}\begin{pmatrix} x \\ y \end{pmatrix} + \begin{pmatrix} 1/2 \\ 0 \end{pmatrix},$$

$$f\begin{pmatrix} x \\ y \end{pmatrix} = \begin{pmatrix} 0 & -1/2 \\ -1/2 & 0 \end{pmatrix}\begin{pmatrix} x \\ y \end{pmatrix} + \begin{pmatrix} -3/2 \\ 0 \end{pmatrix}.$$

IFS for $m = 3$:

$$w_{-3}\begin{pmatrix} x \\ y \end{pmatrix} = \begin{pmatrix} 0 & -1/\sqrt{6} \\ 1/\sqrt{6} & 0 \end{pmatrix}\begin{pmatrix} x \\ y \end{pmatrix} + \begin{pmatrix} -5/2 \\ 0 \end{pmatrix},$$

$$w_{-2}\begin{pmatrix} x \\ y \end{pmatrix} = \begin{pmatrix} 0 & -1/\sqrt{6} \\ 1/\sqrt{6} & 0 \end{pmatrix}\begin{pmatrix} x \\ y \end{pmatrix} + \begin{pmatrix} -3/2 \\ 0 \end{pmatrix},$$

$$w_{-1}\begin{pmatrix} x \\ y \end{pmatrix} = \begin{pmatrix} 0 & -1/\sqrt{6} \\ 1/\sqrt{6} & 0 \end{pmatrix}\begin{pmatrix} x \\ y \end{pmatrix} + \begin{pmatrix} -1/2 \\ 0 \end{pmatrix},$$

$$w_0\begin{pmatrix} x \\ y \end{pmatrix} = \begin{pmatrix} 0 & 1/\sqrt{6} \\ -1/\sqrt{6} & 0 \end{pmatrix}\begin{pmatrix} x \\ y \end{pmatrix} + \begin{pmatrix} 1/2 \\ 0 \end{pmatrix},$$

$$w_1\begin{pmatrix} x \\ y \end{pmatrix} = \begin{pmatrix} 0 & 1/\sqrt{6} \\ -1/\sqrt{6} & 0 \end{pmatrix}\begin{pmatrix} x \\ y \end{pmatrix} + \begin{pmatrix} 3/2 \\ 0 \end{pmatrix},$$

$$f\begin{pmatrix} x \\ y \end{pmatrix} = \begin{pmatrix} 0 & -1/\sqrt{6} \\ -1/\sqrt{6} & 0 \end{pmatrix}\begin{pmatrix} x \\ y \end{pmatrix} + \begin{pmatrix} -5/2 \\ 0 \end{pmatrix}$$

and similar expressions for reptiles with $m = 4, 5$.

9.3 THE METHOD FOR CONSTRUCTING REPTILES WITH HOLES

We now introduce some notations, which will be helpful to describe the method for constructing reptiles and will be used throughout.

Let G be a finite group of orthogonal transformations (e.g., reflections and rotations) satisfying:

$$g(A_0) = A_0, \ \forall g \in G, \ G(A_1) = A_0 \tag{9.8}$$

Define a subset $\mathcal{F}_1 \subset \mathcal{F}$ induced by G as follows:

$$\mathcal{F}_1 = \{ f \in \mathcal{F} : f\,G \subseteq \mathcal{F} \}$$

Thus, $\mathcal{F}_1$ is the subset of $\mathcal{F}$ containing those mappings, which takes left cosets of G into $\mathcal{F}$.

Let $P = \{\mathcal{F}_1, \mathcal{F}_2\}$ be a partition of $\mathcal{F}$ into $\mathcal{F}_1$ and $\mathcal{F}_2 = \mathcal{F} \setminus \mathcal{F}_1$ such that

$$\mathcal{F}_1 \neq \varnothing, \ \{rf : f \in \mathcal{F}_2\} = \{fr : f \in \mathcal{F}_2\} \tag{9.9}$$

that is, $r\mathcal{F}_1 = \mathcal{F}_1 r$.

It follows from equations (9.8) and (9.9) that

$$\mathcal{F}_1 G = \mathcal{F}_1, \ G\mathcal{F}_2 = \mathcal{F}_2 G \tag{9.10}$$

Define

$$B_0 = \{f(A_0) : f \in \mathcal{F}_1\}, \ B_k = \{f(U) : U \in B_{k-1} \text{ and } f \in \mathcal{F}_2\}, \ k \geq 1$$

$$B = \bigcap_{k=0}^{\infty} B_k, \ V_k = \bigcap_{S \in B_k} S, \ \forall\, k \geq 0, \ V = \bigcap_{k=0}^{\infty} V_k. \tag{9.11}$$

Note that if $\mathcal{F}_2 \neq \varnothing$, then $B_k = \varnothing$ for all $k \geq 1$. The following condition by Jordan and Ngai (2005) ensures that A is connected.

Condition 9.1

A_0 is connected and some component of $\bigcap_{i=0}^{k} V_i$ has non-empty intersection with $f(A_0^\circ)$ for every $f \in \mathcal{F}$ for some $k \geq 0$.

We now recall the main result from Jordan and Ngai (2005) on constructions of fractal reptiles with holes.

Theorem 9.2 (Jordan and Ngai, 2005, Theorem 1.5)

Let $\mathcal{F} = \{w_i\}_{i=1}^{m}$ be an IFS with attractor A, let A_0 be a non-empty set invariant under $\mathcal{F}$ such that A_0 is equal to the closure of its interior and $\mu(A_0^\circ) = A_0$, and let $\{A_k\}_{k=0}^{\infty}$ be defined as

$$A_k = \mathcal{F}(A_{k-1}), \ A = \bigcap_{k=1}^{\infty} A_k$$

Suppose there exists a finite group of orthogonal transformations G on $\mathbb{R}^d$ that satisfies condition (9.8) and induces a partition $\{\mathcal{F}_1, \mathcal{F}_2\}$ of $\mathcal{F}$ satisfying condition (9.9). Then the following hold.

1. *The attractor A is an n-reptile.*
2. *In addition, if Condition 1 is satisfied, then A is connected.*

Theorem 9.2 provides an extensive method for constructing reptiles that cannot be obtained by the general method given by Bandt (1991, Theorem 2) where each similitude in the IFS necessarily requires expanding integer matrices to create reptiles.

Jordan and Ngai (2005) constructed fractal reptiles with various interesting topological properties using this result. They gave an example of a connected reptile in $\mathbb{R}^2$ whose interior consists of infinitely many components, with the closure of some of them having holes and some of them being topological disks (see Example 9.3 in Section 9.4). They also found a connected reptile in $\mathbb{R}^2$ whose interior consists of infinitely many components, with the closure of each component having finitely many holes (see Example 9.4 in Section 9.4).

9.4 EXAMPLES OF REPTILES WITH HOLES AND OTHER PROPERTIES

In this section, we use the geometric method introduced by Jordan and Ngai (2005) to construct reptiles with holes and various other topological properties. Most of these reptiles are obtained using IFS that involve both rotations and reflections, except the reptile in Example 9.2, which is constructed from an IFS without any reflection. The smallest n for which it is possible to create such reptile is 16. All other reptiles discussed in this section are obtained by IFS that involve at least two distinct orthogonal transformations.

We construct examples of n-reptiles with holes for both even and odd values of n. It was claimed (without proof) by Jordan and Ngai (2005) that when n equals an odd integer, the smallest n such that an n-reptile has a hole is 9, as described in the following example.

Example 9.1

Let $A_0 = [-3, 3]\times\left[-\frac{3}{2},\frac{3}{2}\right]$ and $r = \frac{1}{3}$. Define an IFS $\mathcal{F} = \{w_i\}_{i=1}^{9}$ as

$$w_i(x) = \begin{cases} r\rho_{\frac{\pi}{2}}(x) + d_i, & i = 1 \\ \rho_{\frac{\pi}{2}}(x)\sigma_y(x) + d_i, & i = 2 \\ r\rho_0(x) + d_i, & i = 3,4 \\ r\sigma_y(x) + d_i, & i = 5,6,7 \\ r\rho_\pi(x) + d_i, & i = 8,9 \end{cases} \tag{9.12}$$

$$d_1 = d_2 = \left(-\frac{5}{2},\frac{1}{2}\right),\ d_3 = d_5 = (-2,-1),\ d_4 = d_6 = (-1,1),$$

$$d_7 = (0,-1),\ d_8 = (-1,0),\ d_9 = (1,0)$$

Writing mappings in the IFS in the matrix form

$$w_i\begin{pmatrix} x \\ y \end{pmatrix}=\begin{pmatrix} a & b \\ c & d \end{pmatrix}\begin{pmatrix} x \\ y \end{pmatrix}+\begin{pmatrix} e \\ f \end{pmatrix}=A_i\begin{pmatrix} x \\ y \end{pmatrix}+\begin{pmatrix} e \\ f \end{pmatrix}$$

we get the corresponding IFS table for the IFS in equation (9.11) in Table 9.1.

The last column in Table 9.1 contains probabilities p_i associated with the mappings w_i, which are chosen so that

$$\sum_{i=1}^{9} p_i = 1, \quad p_i > 0 \text{ for } i = 1, 2, \ldots, 9.$$

These probabilities p_i play a key role in the computing attractor of an IFS using the random iteration algorithm, and their approximate values are given by

$$p_i \approx \frac{|\det A_i|}{\sum_{i=1}^{9} |\det A_i|} = \frac{|a_i d_i - b_i c_i|}{\sum_{i=1}^{9} |a_i d_i - b_i c_i|} \text{ for } i = 1, 2, \cdots, 9.$$

Such an IFS is known as *IFS with probabilities*. The IFS in equation (9.12) with probabilities can therefore be written as

$$\mathcal{F} = \{w_i; p_i\}_{i=1}^{9}$$

Explicit reference to the probabilities is usually omitted if there is no confusion likely to occur. The attractor is obtained as follows: Choose $x_0 \in X$ and recursively compute

$$x_n = \{w_1(x_{n-1}), w_2(x_{n-1}), \ldots, w_N(x_{n-1})\}, \quad \text{for } n = 1, 2, \ldots$$

This produces a sequence $\{x_n\} \subset X$, which converges to the attractor of the IFS under various conditions (see, for example, Barnsley, 1993).

TABLE 9.1
Table for the IFS of a Connected 9-Reptile with Holes

w_i	*a*	*b*	*c*	*d*	*e*	*f*	*p*
w_1	0.000	–0.333	0.333	0.000	–2.500	0.500	0.065
w_2	0.000	–0.333	–0.333	0.000	–2.500	0.500	0.065
w_3	0.333	0.000	0.000	0.333	–2.000	–1.000	0.065
w_4	0.333	0.000	0.000	0.333	–1.000	1.000	0.065
w_5	–0.333	0.000	0.000	0.333	–2.000	–1.000	0.148
w_6	–0.333	0.000	0.000	0.333	–1.000	1.000	0.148
w_7	–0.333	0.000	0.000	0.333	0.000	–1.000	0.148
w_8	–0.333	0.000	0.000	–0.333	–1.000	0.000	0.148
w_9	–0.333	0.000	0.000	–0.333	1.000	0.000	0.148

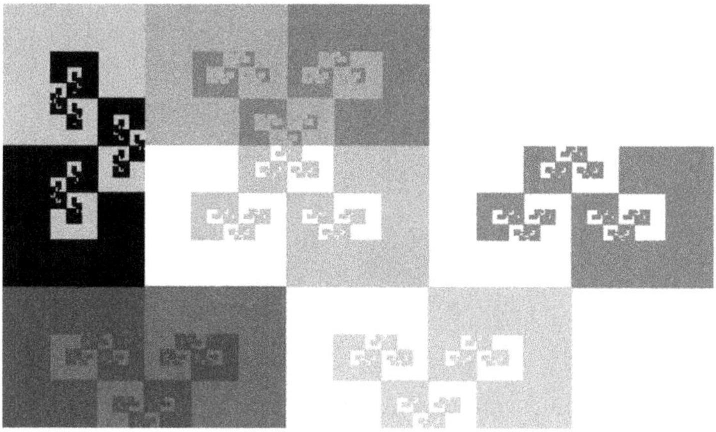

FIGURE 9.2 A connected 9-reptile with holes.

We add a brief note on the importance of these probabilities. In applying the random iteration algorithm, the rate at which an image of the attractor is generated depends on the probabilities. This suggests that in case of an IFS with probabilities, there is a unique *density* on the attractor of the IFS, which disappears with increasing number of iterations. This density defines *measure on fractals*! These measures can be used to describe intricate distributions of mass on fractals and reptiles.

The attractor A obtained from the IFS in equation (9.12) by applying the random iteration algorithm is shown in Figure 9.2.

We verify that this is a connected 9-reptile whose interior consists of infinitely many components, with the closure of each component consisting of infinitely many holes.

Proof: Clearly, A_0 satisfies conditions in equation (9.2) and it is a connected subset of $\mathbb{R}^2$. Let $G = \{id, \sigma_y\}$ be the group of orthogonal transformations. Then equation (9.8) holds and it follows that

$$\begin{aligned} w_1\sigma_y &= w_2, & w_2\sigma_y &= w_1, & w_3\sigma_y &= w_5, \\ w_5\sigma_y &= w_3, & w_4\sigma_y &= w_6, & w_6\sigma_y &= w_4, \\ w_7\sigma_y &= \sigma_y w_7, & w_8\sigma_y &= \sigma_y w_9, & w_9\sigma_y &= \sigma_y w_8. \end{aligned}$$

Thus, we can take $\mathcal{F}_1 = \{w_1, w_2, w_3, w_4, w_5, w_6\}$, $\mathcal{F}_2 = \{w_7, w_8, w_9\}$, and equation (9.9) holds. Thus, by Theorem 9.2(a), A is a 9-reptile. Also, it holds that for all $w \in \mathcal{F}$, $V_0 \cap w(A_0) \neq \varnothing$, where V_0 is defined as in equation (9.11). Thus, by Theorem 9.2(b), A is connected. To prove remaining assertions, we refer to Section 9.3 of Jordan and Ngai (2005) and omit details here. The image of the attractor generated by the IFS in 9.12 is illustrated in Figure 9.2, where each component is plotted in a separate color, and the closure of each component consists of infinitely many holes.

Example 9.2

In this example, we construct a reptile whose interior consists of finitely many components, with the closure of some component having a hole. It was claimed in [**3**] that the smallest n for which such an n-reptile exists is 16. To construct one such example, let $A_0 = [-8, 8]\times[-4,4]$ and $r = \frac{1}{4}$. Define an IFS $\mathcal{F} = \{w_i\}_{i=1}^{16}$ as

$$w_i(x) = \begin{cases} r\rho_{\frac{\pi}{2}}(x) + d_i, & 1 \le i \le 5 \\ r\rho_{-\frac{\pi}{2}}(x) + d_i, & 6 \le i \le 10 \\ r\rho_0(x) + d_i, & 11 \le i \le 13 \\ r\rho_{-\pi}(x) + d_i, & 14 \le i \le 16 \end{cases} \tag{9.13}$$

$$d_1 = d_6 = (-7,-2),\ d_2 = d_7 = (-7,2),\ d_3 = d_8 = (-5,0),$$

$$d_4 = d_9 = (-1,0),\ d_5 = d_{10} = (3,0),$$

$$d_{11} = d_{14} = (-4,-3),\ \ d_{12} = d_{15} = (-4,3),\ \ d_{13} = d_{16} = (0,3).$$

Note that reflections are not used in the IFS. Writing the IFS in the matrix form

$$w_i\begin{pmatrix} x \\ y \end{pmatrix} = \begin{pmatrix} a & b \\ c & d \end{pmatrix}\begin{pmatrix} x \\ y \end{pmatrix} + \begin{pmatrix} e \\ f \end{pmatrix} = A_i\begin{pmatrix} x \\ y \end{pmatrix} + \begin{pmatrix} e \\ f \end{pmatrix}$$

we get the corresponding IFS table for the IFS in equation (9.13) in Table 9.2. As in the previous example, the last column contains probabilities pi associated with the mappings w_i.

TABLE 9.2
Table for the IFS of a Connected 16-Reptile with Holes

w_i	a	b	c	d	e	f	p
w_1	0.000	–0.250	0.250	0.000	–7.000	–2.000	0.040
w_2	0.000	–0.250	0.250	0.000	–7.000	2.000	0.040
w_3	0.000	–0.250	0.250	0.000	–5.000	0.000	0.040
w_4	0.000	–0.250	0.250	0.000	–1.000	0.000	0.040
w_5	0.000	–0.250	0.250	0.000	3.000	0.000	0.040
w_6	0.000	0.250	–0.250	0.000	–7.000	–2.000	0.040
w_7	0.000	0.250	–0.250	0.000	–7.000	2.000	0.040
w_8	0.000	0.250	–0.250	0.000	–5.000	0.000	0.040
w_9	0.000	0.250	–0.250	0.000	–1.000	0.000	0.040
w_{10}	0.000	0.250	–0.250	0.000	3.000	0.000	0.091
w_{11}	0.250	0.000	0.000	–0.250	–4.000	–3.000	0.091
w_{12}	0.250	0.000	0.000	–0.250	–4.000	3.000	0.091
w_{13}	0.250	0.000	0.000	–0.250	0.000	3.000	0.091
w_{14}	–0.250	0.000	0.000	–0.250	–4.000	–3.000	0.091
w_{15}	–0.250	0.000	0.000	–0.250	–4.000	3.000	0.091
w_{16}	–0.250	0.000	0.000	–0.250	0.000	3.000	0.091

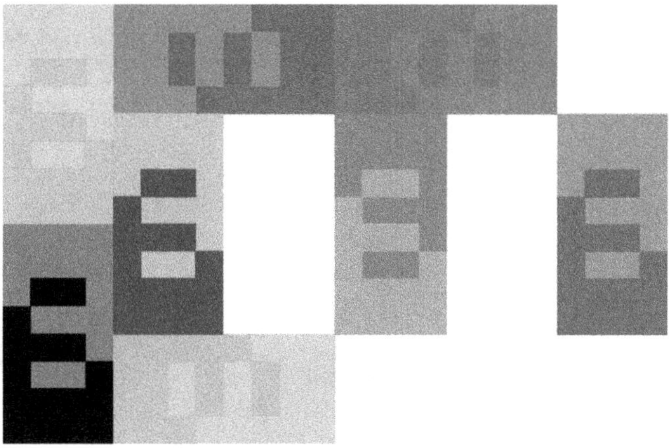

FIGURE 9.3 A connected 16-reptile having two components with the closure of one component having a hole and the closure of the other component a disk.

On applying the random iteration algorithm, we obtain the reptile shown in Figure 9.3. The attractor A is a connected 16-reptile, and its interior consists of two components, with the closure of exactly one of them having a hole.

Proof. Clearly, A_0 satisfies conditions in equation (9.2) and it is a connected subset of $\mathbb{R}^2$. Let $G = \{id, \rho_{-\pi}\}$ be the group of orthogonal transformations. Then conditions in equation 9.8 hold. Also, $w_i\rho_{-\pi} = w_{i+5}$, $w_{i+5}\rho_{-\pi} = w_i$ for $1 \le i \le 5$ and $w_i\rho_{-\pi} = w_{i+3}$, $w_{i+3}\rho_{-\pi} = w_i$ for $11 \le i \le 13$. Hence, $\mathcal{F}_1 = \mathcal{F}$, $\mathcal{F}_2 = \varnothing$ and the condition in equation (9.9) holds. Thus, by Theorem 9.2(a), A is a 16-reptile.

Moreover, $V_0 \cap w(A_0) \neq \varnothing$ for all $w \in \mathcal{F}$. Hence, A is connected. It is also clear from the figure that the interior of A consists of two components, with the closure of one component having a hole.

Our next example demonstrates a reptile whose interior consists of infinitely many components, with the closure of some components having holes and the closure of others being topological disks.

Example 9.3

Let $A_0 = [-4, 4] \times [-2, 2]$ and $r = \dfrac{1}{4}$. Define an IFS $\mathcal{F} = \{w_i\}_{i=1}^{16}$ as

$$w_i(x) = \begin{cases} r\rho_{\frac{\pi}{2}}(x) + d_i, & 1 \le i \le 5 \\ r\rho_{\frac{\pi}{2}}(x)\sigma_y(x) + d_i, & 6 \le i \le 10 \\ r\rho_0(x) + d_i, & 11 \le i \le 12 \\ r\sigma_y(x) + d_i, & 13 \le i \le 14 \\ r\rho_{\frac{\pi}{2}}(x) + d_i, & i = 15 \\ r\rho_{-\frac{\pi}{2}}(x) + d_i, & i = 16 \end{cases} \tag{9.14}$$

$$d_1 = d_6 = \left(-\frac{3}{2},-1\right),\ d_2 = d_7 = \left(-\frac{5}{2},1\right),\ d_3 = d_8 = \left(-\frac{7}{2},-1\right),\ d_4 = d_9 = \left(-\frac{7}{2},1\right),$$

$$d_5 = d_{10} = \left(-\frac{1}{2},1\right),\ d_{11} = d_{13} = \left(-2,\frac{3}{2}\right),\ d_{12} = d_{14} = \left(2,\frac{1}{2}\right),\ d_{15} = \left(-\frac{1}{2},-1\right),$$

$$d_{16} = \left(-\frac{1}{2},1\right)$$

The matrix form of the IFS

$$w_i\begin{pmatrix} x \\ y \end{pmatrix} = \begin{pmatrix} a & b \\ c & d \end{pmatrix}\begin{pmatrix} x \\ y \end{pmatrix} + \begin{pmatrix} e \\ f \end{pmatrix} = A_i\begin{pmatrix} x \\ y \end{pmatrix} + \begin{pmatrix} e \\ f \end{pmatrix}$$

is given in Table 9.3 along with the probabilities p_i associated with the mappings w_i.

The attractor A of the IFS $\mathcal{F}$ is shown in Figure 9.4, which is a connected 16-reptile with infinitely many components such that the closure of some of them contains holes and the closure of other components is topological disks.

Proof. Clearly, A_0 satisfies conditions in equation 9.2 and it is a connected subset of $\mathbb{R}^2$. Let $G = \{id, \sigma_y\}$ be the group of orthogonal transformations satisfying equation 9.8. With $\mathcal{F}_1 = \{w_i\}_{i=1}^{14}$ and $\mathcal{F}_2 = \{w_{15}, w_{16}\}$, it can be easily checked as before that the condition in equation (9.9) holds. Thus, by Theorem 9.2(a), A is a 16-reptile. The rest of the assertions are also similar to previous examples.

TABLE 9.3
Table for the IFS in Equation (9.14)

w_i	a	b	c	d	e	f	p
w_1	0.000	−0.250	0.250	0.000	−1.500	−1.000	0.040
w_2	0.000	−0.250	0.250	0.000	−2.500	−1.000	0.040
w_3	0.000	−0.250	0.250	0.000	−3.500	−1.000	0.040
w_4	0.000	−0.250	0.250	0.000	−3.500	1.000	0.040
w_5	0.000	−0.250	0.250	0.000	−0.500	1.000	0.040
w_6	0.000	−0.250	−0.250	0.000	−1.500	−1.000	0.040
w_7	0.000	−0.250	−0.250	0.000	−2.500	−1.000	0.040
w_8	0.000	−0.250	−0.250	0.000	−3.500	−1.000	0.040
w_9	0.000	−0.250	−0.250	0.000	−3.500	1.000	0.040
w_{10}	0.000	−0.250	−0.250	0.000	−0.500	1.000	0.091
w_{11}	0.250	0.000	0.000	0.250	−2.000	1.500	0.091
w_{12}	0.250	0.000	0.000	0.250	2.000	0.500	0.091
w_{13}	−0.250	0.000	0.000	0.250	−2.000	1.500	0.091
w_{14}	−0.250	0.000	0.000	0.250	2.000	0.500	0.091
w_{15}	0.000	−0.250	0.250	0.000	−0.500	−1.000	0.091
w_{16}	0.000	0.250	−0.250	0.000	0.500	−1.000	0.091

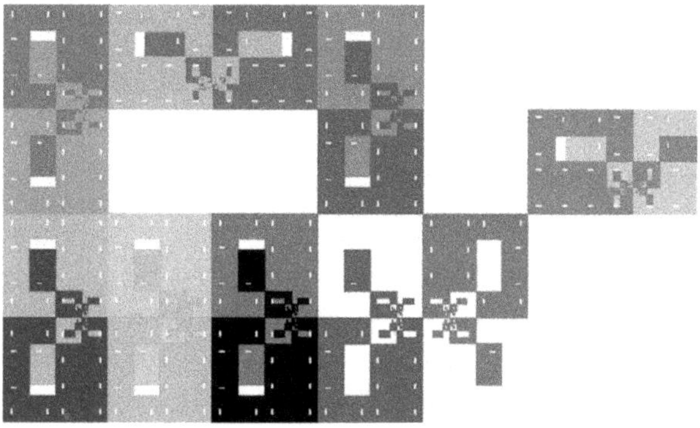

FIGURE 9.4 A connected 16-reptile with holes and topological disks.

A natural question arises now: Is it true that if the interior of a reptile consists of infinitely many components with the closure of some of them having holes, then the closure of these components must contain infinitely many holes? An interesting counterexample of a 36-reptile was given by Jordan and Ngai (2005) having infinitely many components with the closure of each component having only finitely many holes.

Example 9.4

Let $A_0 = [-6, 6]\times[-3,3]$ and $r = \frac{1}{6}$. Define an IFS $\mathcal{F} = \{w_i\}_{i=1}^{36}$ as

$$w_i(x) = \begin{cases} r\sigma_y(x)+d_i, & 1\le i\le 7 \\ r\rho_{\frac{\pi}{2}}(x)+d_i, & 8\le i\le 16 \\ r\rho_{\frac{\pi}{2}}(x)\sigma_y(x)+d_i, & 17\le i\le 25 \\ r\rho_0(x)+d_i, & 26\le i\le 34 \\ r\sigma_y(x)+d_i, & i=35 \\ r\rho_\pi(x)+d_i, & i=36 \end{cases} \tag{9.15}$$

$d_1 = d_{26} = \left(-4,-\frac{5}{2}\right)$, $d_2 = d_{27} = \left(-4,\frac{5}{2}\right)$, $d_3 = d_{28} = \left(-2,-\frac{5}{2}\right)$, $d_4 = d_{29} = \left(-2,-\frac{3}{2}\right)$,

$d_5 = d_{30} = \left(-2,\frac{5}{2}\right)$, $d_6 = d_{31} = \left(2,-\frac{1}{2}\right)$, $d_7 = d_{32} = \left(3,\frac{3}{2}\right)$, $d_8 = d_{17} = \left(-\frac{11}{2},-2\right)$,

$d_9 = d_{18} = \left(-\frac{11}{2},0\right)$, $d_{10} = d_{19} = \left(-\frac{11}{2},2\right)$, $d_{11} = d_{20} = \left(-\frac{9}{2},-1\right)$, $d_{12} = d_{21} = \left(-\frac{1}{2},0\right)$,

$d_{13} = d_{22} = \left(-\frac{1}{2},2\right)$, $d_{14} = d_{23} = \left(\frac{3}{2},1\right)$, $d_{15} = d_{24} = \left(\frac{7}{2},-1\right)$, $d_{16} = d_{25} = \left(\frac{9}{2},1\right)$,

$d_{33} = \left(-3,\frac{1}{2}\right)$, $d_{34} = \left(3,\frac{1}{2}\right)$, $d_{35} = \left(0,-\frac{3}{2}\right)$, $d_{36} = \left(0,-\frac{5}{2}\right)$

The matrix form of the IFS along with the probabilities p_i associated with the mappings w_i is given in Table 9.4.

TABLE 9.4
Table for the IFS in Equation (9.15)

w_i	a	b	c	d	e	f	p
w_1	–0.167	0.000	0.000	0.167	–4.000	–2.500	0.010
w_2	–0.167	0.000	0.000	0.167	–4.000	2.500	0.010
w_3	–0.167	0.000	0.000	0.167	–2.000	–2.500	0.010
w_4	–0.167	0.000	0.000	0.167	–2.000	–1.500	0.010
w_5	–0.167	0.000	0.000	0.167	–2.000	2.500	0.010
w_6	–0.167	0.000	0.000	0.167	2.000	–0.500	0.010
w_7	–0.167	0.000	0.000	0.167	3.000	1.500	0.010
w_8	0.000	–0.167	0.167	0.000	–5.500	–2.000	0.010
w_9	0.000	–0.167	0.167	0.000	–5.500	0.000	0.010
w_{10}	0.000	–0.167	0.167	0.000	–5.500	2.000	0.010
w_{11}	0.000	–0.167	0.167	0.000	–4.500	–1.000	0.010
w_{12}	0.000	–0.167	0.167	0.000	–0.500	0.000	0.010
w_{13}	0.000	–0.167	0.167	0.000	–0.500	2.000	0.010
w_{14}	0.000	–0.167	0.167	0.000	1.500	1.000	0.010
w_{15}	0.000	–0.167	0.167	0.000	3.500	–1.000	0.010
w_{16}	0.000	–0.167	0.167	0.000	4.500	1.000	0.042
w_{17}	0.000	–0.167	–0.167	0.000	–5.500	–2.000	0.042
w_{18}	0.000	–0.167	–0.167	0.000	–5.500	0.000	0.042
w_{19}	0.000	–0.167	–0.167	0.000	–5.500	2.000	0.042
w_{20}	0.000	–0.167	–0.167	0.000	–4.500	–1.000	0.042
w_{21}	0.000	–0.167	–0.167	0.000	–0.500	0.000	0.042
w_{22}	0.000	–0.167	–0.167	0.000	–0.500	2.000	0.042
w_{23}	0.000	–0.167	–0.167	0.000	1.500	1.000	0.042
w_{24}	0.000	–0.167	–0.167	0.000	3.500	–1.000	0.042
w_{25}	0.000	–0.167	–0.167	0.000	4.500	1.000	0.042
w_{26}	0.167	0.000	0.000	0.167	–4.000	–2.500	0.042
w_{27}	0.167	0.000	0.000	0.167	–4.000	2.500	0.042
w_{28}	0.167	0.000	0.000	0.167	–2.000	–2.500	0.042
w_{29}	0.167	0.000	0.000	0.167	–2.000	–1.500	0.042
w_{30}	0.167	0.000	0.000	0.167	–2.000	2.500	0.042
w_{31}	0.167	0.000	0.000	0.167	2.000	–0.500	0.042
w_{32}	0.167	0.000	0.000	0.167	3.000	1.500	0.042
w_{33}	0.167	0.000	0.000	0.167	–3.000	0.500	0.042
w_{34}	0.167	0.000	0.000	0.167	3.000	0.500	0.042
w_{35}	–0.167	0.000	0.000	0.167	0.000	–1.500	0.042
w_{36}	–0.167	0.000	0.000	–0.167	0.000	–2.500	0.042

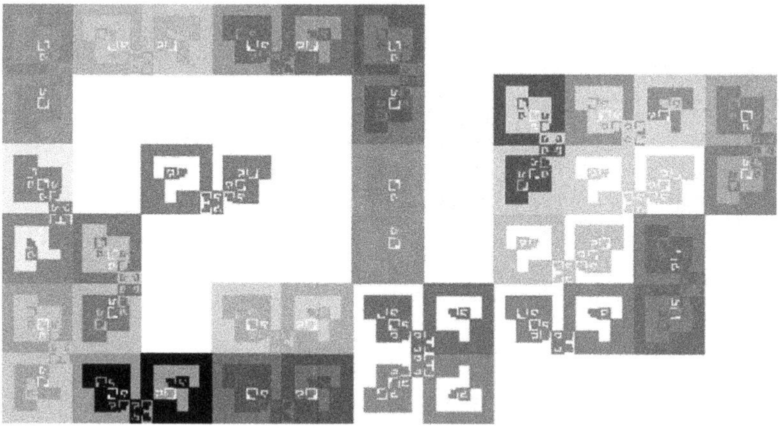

FIGURE 9.5 A connected 36-reptile with infinitely many components and the closure of each component having finitely many holes.

The attractor A of $\mathcal{F}$ is shown in Figure 9.5, which is a connected 36-reptile whose interior consists of infinitely many components, with the closure of each component having finitely many holes.

9.5 SOME VARIATIONS AND REPTILES WITH HOLES USING INTEGER MATRICES

In this section, we construct some variations of the reptiles described in Sections 9.2–9.4 by suitably modifying the transformations in the corresponding IFS. We also extend the method to construct an example of a 9-reptile with holes using expanding integer matrices. Matrices with integer entries have simple structure, integer determinants, algebraic eigenvalues and other interesting properties, which make them useful, and several authors have employed integer matrices in the construction of fractal reptiles (see, for example, Bandt (1991), Bandt and Wang (2001) and the recent paper by Husain et al. (2021)).

Example 9.5

Consider the IFS

$$\mathcal{F} = \{w_i\}_{-m}^{m-2} \cup \{f\},$$

where

$$w_i(x) = \frac{1}{\sqrt{n}} \rho_{\frac{\pi}{2}}(x) + \left(i + \frac{1}{2}, 0\right), \ -m \le i \le m-2$$

$$f(x) = \frac{1}{\sqrt{n}} \rho_{\frac{\pi}{2}} \sigma_y(x) + \left(-m + \frac{1}{2}, 0\right)$$

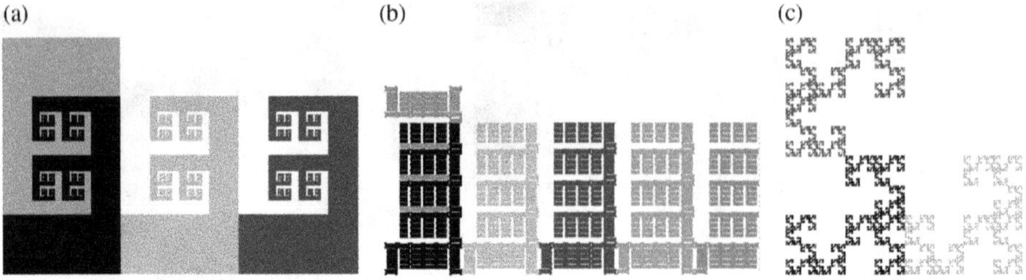

FIGURE 9.6 (a) A connected 4-reptile; (b) a connected 6-reptile; (c) a member of Sierpinski relatives.

Note that this IFS is obtained from the IFS in equations (9.5) and (9.6) by replacing the transformation $\rho_{-\frac{\pi}{2}}$ with $\rho_{\frac{\pi}{2}}$ in w_0. Thus, the IFS now contains a single rotation. The attractors of this IFS for $m = 2, 3$ are connected reptiles without holes, as shown in Figure 9.6a and b, respectively. Another fractal variation is shown in Figure 9.6c, which is obtained by removing the mapping w_0 from the IFS. This a self-similar fractal as expected and belong to the family of Sierpinski relatives. More reptiles and fractal variations can be obtained by deleting other mappings from the IFS.

We now consider an example of a connected 9-reptile with holes obtained via modification and extension of the IFS in equation (9.12) to the IFS involving expanding integer matrices.

Example 9.6

Let $A_0 = [-3,3]\times\left[-\frac{3}{2},\frac{3}{2}\right], r = \frac{1}{2}$ and choose the integer matrix M as

$$M = \begin{bmatrix} 1 & -1 \\ 1 & 1 \end{bmatrix}.$$

Clearly, M is expanding since the eigenvalues λ_i of M satisfy $|\lambda_i| > 1$ for $1 \le i \le 2$. Now, consider the IFS $\mathcal{F} = \{w_i\}_{i=1}^{9}$ defined by

$$w_i(x) = \begin{cases} r\rho_{\frac{\pi}{2}} M^{-1}(x) + d_i, & i = 1 \\ r\rho_{\frac{\pi}{2}} M^{-1}\sigma_y(x) + d_i, & i = 2 \\ r\rho_0 M^{-1}(x) + d_i, & i = 3,4 \\ r\sigma_y M^{-1}(x) + d_i, & i = 5,6,7 \\ r\rho_\pi M^{-1}(x) + d_i, & i = 8,9 \end{cases} \quad (9.16)$$

$$d_1 = d_2 = \left(-\frac{5}{2}, \frac{1}{2}\right), \; d_3 = d_5 = (-2,-1), \; d_4 = d_6 = (-1,1), \; d_7 = (0,-1), \; d_8 = (-1,0),$$

$$d_9 = (1,0).$$

The attractor is shown in Figure 9.7, which is a connected 9-reptile with infinitely many components in the interior and the closure of each component having infinitely many holes.

Proof: Clearly, A_0 satisfies conditions in equation (9.2) and it is a connected subset of $\mathbb{R}^2$. Let $G = \{id, \sigma_y\}$ be the group of orthogonal transformations. Then equation (9.8) holds and it follows that

$$\begin{array}{lll} w_1\sigma_y = w_2, & w_2\sigma_y = w_1, & \sigma_y w_3 = w_5, \\ \sigma_y w_5 = w_3, & \sigma_y w_4 = w_6, & \sigma_y w_6 = w_4, \\ \sigma_y w_7 = w_4, & w_8\sigma_y = w_9\sigma_y, & \sigma_y w_9 = \sigma_y w_8. \end{array}$$

Thus, we can take $\mathcal{F}_1 = \{w_1, w_2, w_3, w_4, w_5, w_6, w_7\}$, $\mathcal{F}_2 = \{w_8, w_9\}$, and equation (9.9) holds. By Theorem 9.2(a), A is a 9-reptile. Also, it holds that for all $w \in \mathcal{F}$, $V_0 \cap w(A_0) \neq \emptyset$, where V_0 is defined as in equation (9.11). Hence, by Theorem 9.2(b), A is connected. The closure of each component consists of infinitely many holes.

Notice that the dynamics of the points on the attractor in Figure 9.7 is more chaotic as compared to all other examples considered before. More examples of reptiles can be created by varying the integer matrix M.

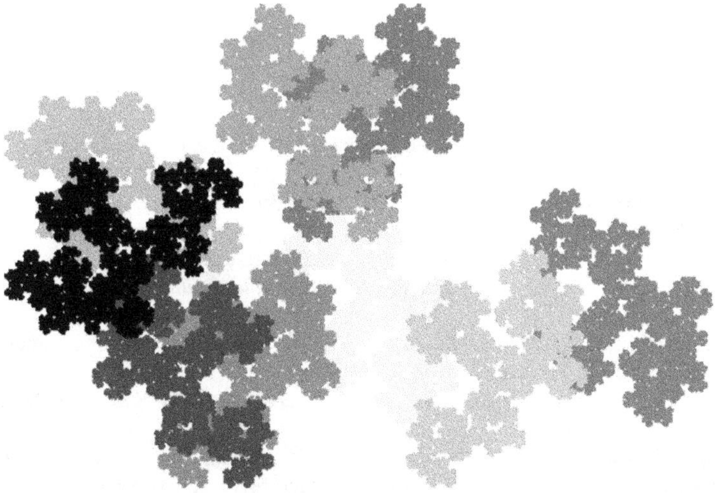

FIGURE 9.7 A connected 9-reptile with infinitely many components in the interior and the closure of each component having infinitely many holes.

9.6 SUMMARY

The study of tilings and patterns has a long history, and after the introduction of fractal reptiles, there have been continuous efforts from many researchers to characterize fractal tilings of the Euclidean plane and to understand their properties (both geometric and topological) as well as their connection in other fields. Reptiles with holes are the tiles such that the complement of the closure of some component of the interior of the tile has a bounded component and they form an important class of reptiles since they arise in connection with many open problems in geometry and topology. However, only limited work has been done so far on the structure and topology of 2D and 3D reptiles, and the work is still in progress.

In this chapter, a concise summary of the article by Jordan and Ngai (2005) on fractal reptiles of the plane with holes is presented using IFS mappings consisting of similitudes made up of rotations and reflections. Explicit tables for the reptiles are also given for a reader-friendly experience and construction purpose. All n-reptiles presented in the chapter make use of at least two distinct rotations, and it is not known whether there exist reptiles with holes which can be constructed by similitudes using a single rotation or reflection. The reptiles presented in this chapter provide answers to some interesting questions concerning geometry of reptiles with holes. In particular, the reptiles presented here reduce the original question of Croft, Falconer and Guy (1991) to: Are there any 2-reptiles or 3-reptiles in the plane with holes?

We have also discussed the extension of these reptiles to the fractal reptiles with holes using integer matrices and some other variations. Many more examples of reptiles with holes can be constructed by using the results presented here by changing various parameters. The IFS presented in this chapter can be modified suitably to produce examples of fractal reptiles with holes in 3D.

REFERENCES

Bandt, C. (1991). Self-similar sets, 5, Integer matrices and fractal tilings of $\mathbb{R}^n$, *Proceedings American Mathematics Society*, 112, 549–562.

Bandt, C., Wang, Y. (2001). Disk-like self-affine tiles in $\mathbb{R}^2$, *Discrete Computer Geometry*, 26, 591–601.

Barnsley, M. (1993). *Fractals Everywhere*, Academic Press, Elsevier.

Croft, H. T., Falconer, K. J., Guy, R. K. (1991). *Unsolved Problems in Geometry*, Springer.

Dekking, F. M. (1982), Recurrent Sets, *Advance Math*ematics, 44 (1), 78–104.

Falconer, K. J. (1990), *Fractal Geometry. Mathematical Foundations and Applications*, Wiley.

Husain, A., Karthik, G., Megham, M., Ashish S. (2021). Fractal rep tiles of cc and $\mathbb{R}^3$ using integer matrices, *Fractals*, 29, 2150027.

Jordan, F., Ngai, S. M. (2005). Reptiles with holes, *Proceedings of the Edinburgh Mathematical Society*, 48(3), 651–671.

Lagarias, J. C., Wang, Y. (1996). Self-affine tiles in $\mathbb{R}^n$, *Advance Mathematics*, 121, 21–49.

Luo, J., Rao, H., Tan, B. (2002). Topological structure of self-similar sets, *Fractals*, 10, 223–227.

Ngai, S. M., Tang, T. M. (2005). Topology of connected self-similar tiles in the plane with disconnected interiors, *Topological Applications*, 150, 139–155.

Schief, A. (1994). Separation properties for self-similar sets, *Proceedings American Mathematics Soc*, 122, 111–115.

10 Energy Approach for Free Vibration Analysis of Restrained Monoclinic Rectangular Plates of Varying Density and Thickness

Yajuvindra Kumar
Government Girls Degree College, Behat

CONTENTS

NOMENCLATURE

a, b, h	Length, breadth and thickness of the plate
$C_{11}, C_{12}, C_{22}, C_{33}$	Elastic constants
E	Young's modulus
w	Transverse displacement
$\bar{W}$	Maximum transverse displacement
V, V_{max}	Strain energy and maximum strain energy
T, T_{max}	Kinetic energy and maximum kinetic energy
U_{max}	Maximum strain energy due to rotational springs
r_1, r_2, r_3, r_4	Rotational spring constants
R_1, R_2, R_3, R_4	Flexibility parameters
N	Order of approximation
$\sigma_x, \sigma_y, \sigma_{xy}$	Stresses
$\varepsilon_x, \varepsilon_y, \varepsilon_{xy}$	Strains
ω	Circular frequency

DOI: 10.1201/9781003222255-10

Ω	Frequency parameter
υ	Poisson's ratio
ρ	Density of the plate material
δ_1, δ_2	Density parameters
ψ_1, ψ_2	Thickness parameters
$\phi_k, \hat{\phi}_k$	Orthogonal and orthonormal polynomials
δ_{jk}	The Kronecker delta

10.1 INTRODUCTION

Simply supported plates with elastically restrained edges against rotation find application in engineering problems. The initial work treating free transverse vibration of rectangular plates of uniform/non-uniform thickness with all possible classical boundary conditions was presented by Leissa [1–5]. As far as the work dealing with the vibration of rectangular plates of uniform/non-uniform thickness with elastically restrained edges is concerned, the following papers provide the relevant information: Grossi and Bhat [6], Zhou [7], Zhou [8], Cheung and Zhou [9], Zhou [10], Ashour [11], Li [12], Malekzadeh and Shahpari [13], Li et al. [14], Zhang and Li [15], Dal and Morgul [16], Lai and Xiang [17], Yang and Wang [18], Verma and Datta [19], Zhang and Zhang [20], Wang et al. [21], Zhang et al. [22] and Wattanasakulpong and Songsuwan [23]. Two-dimensional variation of thickness has been considered by many researchers and is reported as follows: Laura and Grossi [24], Singh and Saxena [25], Sakiyama and Huang [26], Cheung and Zhou [27], Zhou [28], Cheung and Zhou [29], Malekzadeh and Shahpari [13] and Huang et al. [30], Kumar and co-workers [31–34], Kumar and Lal [35], Bahmyari and Rahbar-Ranji [36], Semnani et al. [37], Xue et al. [38], Shufrin and Eisenberger [39] and Tran et al. [40].

There have been very few papers analyzing the vibrational characteristics of monoclinic plates and are presented as follows: Soldatos [41] developed a complex potential formulation for the bending of inhomogeneous monoclinic plates, including transverse shear deformation based on refined plate theories. A natural frequency analysis of thick monoclinic square plates was presented by Batra et al. [42]. Kumar and Tomar [43] studied free transverse vibration of monoclinic rectangular plates of varying density and thickness using the Chebyshev collocation method. A meshless method was used by Ferreira et al. [44] to study natural frequencies of thick monoclinic plates. Kumar [45] used differential transform method to study free transverse vibration of monoclinic rectangular plates resting on Winkler foundation. Bahrami et al. [46] presented a 3D static analysis of monoclinic plates using differential quadrature method.

The Rayleigh–Ritz method has extensively been used to analyze free and forced vibration of structural elements of various geometries. A comprehensive review of the work dealing with vibration analysis of plates using the Rayleigh–Ritz method was provided by Kumar [47]. Zhang et al. [48] used the Rayleigh–Ritz method to study free vibration of rectangular FGM plates with a cutout. Singh and Azam [49] investigated free vibration and buckling of transversely inhomogeneous FG nanoplate in thermal environment and resting on Pasternak foundation using the Rayleigh–Ritz method. Alanbay et al. [50] used the Ritz method to study free vibration of stiffened rectangular and quadrilateral plates employing orthogonal Jacobi polynomials based on first-order shear deformation theory. In another paper, Alanbay et al. [51] found up to 100 frequencies of isotropic rectangular plates using Jacobi polynomials in the Ritz method based on third-order shear and normal deformation plate theory. Free vibration analysis of porous FG rectangular plates was presented by Muc and Flis [52] using the Rayleigh–Ritz method. A free vibration analysis of laminated functionally graded carbon nanotube-reinforced composite plates using the Rayleigh–Ritz method was presented by Hung et al. [53].

To the author's knowledge, no work dealing with free transverse vibration of thin elastically restrained monoclinic rectangular plates with bilinearly varying thickness has been reported. To fill up this gap, this chapter deals with free transverse vibration of thin simply supported elastically restrained monoclinic rectangular plates of varying density and thickness based on classical plate theory. Two-dimensional boundary characteristic orthogonal polynomials (BCOPs) in the Rayleigh–Ritz method are used in the analysis. Orthogonal polynomials are generated using the Gram–Schmidt process, and the use of these orthogonal polynomials in the Rayleigh–Ritz method leads to the standard eigenvalue problem. The thickness and the density of the plate are dependent of both the in-plane variables x and y. The effects of density parameters, thickness parameters, flexibility parameters and aspect ratio on first three natural frequencies of rectangular plate have been studied. The first three 3D mode shapes for different combinations of parameters have been plotted. The results in special cases are also compared.

10.2 FORMULATION OF THE PROBLEM

The schematic diagram of a simply supported non-homogeneous monoclinic rectangular plate of length a, breadth b and varying thickness $h(x,y)$ with elastically restrained rotation along the edges is shown in Figure 10.1. The $x-$ and $y-$ axes are taken along the edges of the plate, and the $z-$ axis is perpendicular to the $xy-$ plane. The plane $z=0$ coincides with the middle surface of the plate, and the origin of the coordinate axes is O.

Based on Hook's law, the stress–strain relations for a thin plate made of monoclinic material are given by

$$\begin{bmatrix} \sigma_x \\ \sigma_y \\ \sigma_{xy} \end{bmatrix} = \begin{bmatrix} C_{11} & C_{12} & 0 \\ C_{12} & C_{22} & 0 \\ 0 & 0 & C_{33} \end{bmatrix} \begin{bmatrix} \varepsilon_x \\ \varepsilon_y \\ \varepsilon_{xy} \end{bmatrix} \tag{10.1}$$

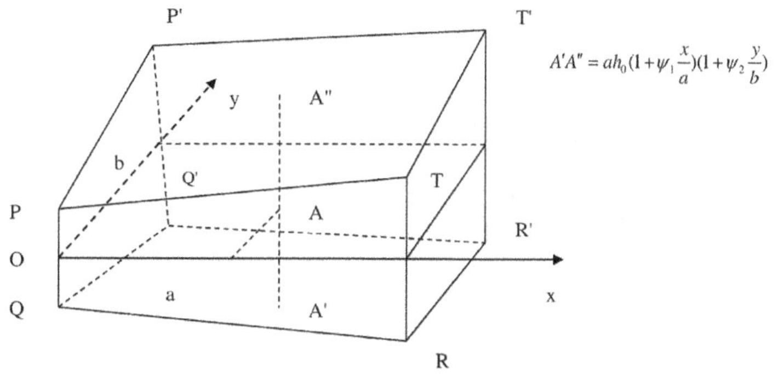

FIGURE 10.1 Schematic diagram of the plate.

The strains are related to displacements by

$$\varepsilon_x = -z\frac{\partial^2 w}{\partial x^2}, \varepsilon_y = -z\frac{\partial^2 w}{\partial y^2}, \varepsilon_{xy} = -2z\frac{\partial^2 w}{\partial x \partial y}. \tag{10.2}$$

The strain energy V of the plate is given by

$$V = \frac{1}{2}\int_0^a\int_0^b\int_{-h/2}^{h/2}\left(\sigma_x\varepsilon_x + \sigma_y\varepsilon_y + \sigma_{xy}\varepsilon_{xy}\right)dx\,dy\,dz. \tag{10.3}$$

By using (10.1) and (10.2), expression (10.3) becomes

$$V = \frac{1}{24}\int_0^a\int_0^b h^3\left[C_{11}{w_{xx}}^2 + 2C_{12}w_{xx}w_{yy} + 4C_{33}{w_{xy}}^2 + C_{22}{w_{yy}}^2\right]dx\,dy \tag{10.4}$$

and kinetic energy of the plate is given by

$$T = \frac{1}{2}\int_0^a\int_0^b \rho h\left(\frac{\partial w}{\partial t}\right)^2 dxdy, \tag{10.5}$$

where $\rho(x,y)$ is the density and t denotes the time.

For free vibration, the transverse deflection function $w(x,y,t)$ is taken as

$$w(x,y,t) = \bar{W}(x,y)\sin\omega t \tag{10.6}$$

where ω denotes the circular frequency and $\bar{W}(x,y)$ is the maximum transverse displacement.

Using relation (10.6) in expressions (10.4) and (10.5), the expressions for maximum strain energy and kinetic energy of the plate become

$$V_{\max} = \frac{1}{24}\int_0^a\int_0^b h^3\left[C_{11}{\bar{W}_{xx}}^2 + 2C_{12}\bar{W}_{xx}\bar{W}_{yy} + 4C_{33}{\bar{W}_{xy}}^2 + C_{22}{\bar{W}_{yy}}^2\right]dx\,dy \tag{10.7}$$

and

$$T_{\max} = \frac{\omega^2}{2}\int_0^a\int_0^b \rho h\bar{W}^2\,dx\,dy. \tag{10.8}$$

The maximum strain energy due to the rotational restraints is given by

$$U_{\max} = \frac{1}{2}\left[\begin{array}{l} r_1\int_0^a \bar{W}_y^2(x,0)dx + r_2\int_0^a \bar{W}_y^2(x,b)dx + r_3\int_0^b \bar{W}_x^2(0,y)dy \\ + r_4\int_0^b \bar{W}_x^2(a,y)dy \end{array}\right], \tag{10.9}$$

Here, $\bar{W}_i$ denotes the partial derivative with respect to the variable in the subscript and r_1, r_2, r_3, r_4 are the rotational spring constants.

The Rayleigh quotient is expressed as

$$\omega^2 = \frac{V_{\max} + U_{\max}}{\left(T_{\max}/\omega^2\right)} \tag{10.10}$$

We introduce the non-dimensional variables $X = x/a, Y = y/b, W = \bar{W}/a$ and assume that the density of the plate material and the thickness of the plate vary as follows:

$$\rho = \rho_0(1 + \delta_1 X + \delta_2 Y) \text{ and } h(X,Y) = ah_0(1 + \psi_1 X)(1 + \psi_2 Y) \tag{10.11}$$

Let us assume the deflection function

$$W(X,Y) = \sum_{m=1}^{N} d_m \hat{\phi}_m(X,Y), \tag{10.12}$$

where N is the order of approximation and d_m are unknowns. $\hat{\phi}_m$ are orthonormal polynomials. These orthonormal polynomials are generated by first generating orthogonal polynomials ϕ_m using the Gram–Schmidt process as reported by Singh and Chakraverty [54].

After some algebraic manipulations, the standard eigenvalue problem is obtained as follows:

$$\sum_{m=1}^{N} \left(a_{lm} - \Omega^2 \delta_{lm}\right) d_m = 0, l = 1,2,3,\ldots\ldots, N, \tag{10.13}$$

where

$$a_{lm} = \int_0^1 \int_0^1 F \begin{bmatrix} \hat{\phi}_l^{XX} \hat{\phi}_m^{XX} + \frac{C_{12}}{C_{11}} \left(\hat{\phi}_l^{XX} \hat{\phi}_m^{YY} + \hat{\phi}_l^{XX} \hat{\phi}_m^{YY} \right) + 4\left(\frac{C_{33}}{C_{11}}\right) \mu^2 \hat{\phi}_l^{XY} \hat{\phi}_m^{XY} \\ + \left(\frac{C_{22}}{C_{11}}\right) \mu^4 \hat{\phi}_l^{YY} \hat{\phi}_m^{YY} \end{bmatrix} dX\, dY$$

$$+ \mu R_1 \int_0^1 \left[\hat{\phi}_l^Y \hat{\phi}_m^Y\right]_{Y=0} dX + \mu R_2 \int_0^1 \left[\hat{\phi}_l^Y \hat{\phi}_m^Y\right]_{Y=1} dX + R_3 \int_0^1 \left[\hat{\phi}_l^X \hat{\phi}_m^X\right]_{X=0} dY$$

$$+ R_4 \int_0^1 \left[\hat{\phi}_l^X \hat{\phi}_m^X\right]_{X=1} dY, \tag{10.14}$$

$$R_1 = \frac{12r_1}{C_{11}h_0^3 b}, R_2 = \frac{12r_2}{C_{11}h_0^3 b}, R_3 = \frac{12r_3}{C_{11}h_0^3 a^2}, R_3 = \frac{12r_4}{C_{11}h_0^3 a^2}, \Omega^2 = \frac{12\rho_0 a^2 \omega^2}{C_{11}h_0^2},$$

$$\mu = \frac{a}{b}, \text{ and } \delta_{lm} = \begin{cases} 1, & if \quad l = m \\ 0, & if \quad l \neq m \end{cases}.$$

10.3 RESULTS AND DISCUSSION

The numerical values of the frequency parameter Ω have been obtained by solving equation (10.13). Appropriate value of N has been chosen by running a computer program developed in C++ to evaluate the frequency parameter Ω. In this study, the value of has been fixed as 36. The first three eigenvalues have been calculated for the plate made of rock gypsum (monoclinic material), and the values of the elastic coefficients for rock gypsum are taken from Haussuhl [55].

$$C_{11} = 7.859 \times 10^{11}, C_{12} = 4.101 \times 10^{11}, C_{22} = 6.274 \times 10^{11}, C_{33} = 1.044 \times 10^{11}$$

And the other plate parameters are as follows:

Density parameters: $\delta_1, \delta_2 = -0.5, -0.3, -0.1, 0.1, 0.3, 0.5.$
Thickness parameters: $\psi_1, \psi_2 = -0.5, -0.3, -0.1, 0.1, 0.3, 0.5.$
Aspect ratio: $a/b = 0.5, 1.0, 1.5, 2.0.$
Flexibility parameters: $R_1 = R_2 = R_3 = R_4 = R =$ 0, 10, 100, 1,000, 10,000, 1,000,000.

Table 10.1 shows the convergence of frequency parameter Ω with increasing value of N. A comparison of frequencies for simply supported homogeneous isotropic plate

TABLE 10.1
Convergence of Frequency Parameter Ω of Non-Homogeneous Simply Supported Monoclinic Plate with Elastically Restraint Edges Against Rotation for $R = 1{,}000{,}000$

			$\delta_1 = \delta_2 = \psi_1 = \psi_2 = -0.5$, $a/b = 1$		
			N		
Mode	**20**	**25**	**30**	**35**	**36**
I	26.6077	26.5276	26.3771	26.3557	26.3557
II	56.3323	55.2653	52.3754	52.1984	52.1984
III	61.9094	58.9763	57.0662	56.5278	56.5278
			$\delta_1 = \delta_2 = \psi_1 = \psi_2 = 0.5$, $a/b = 1$		
I	42.5109	42.4910	42.4289	42.4233	42.4233
II	87.7483	87.5532	84.2714	83.6881	83.6881
III	94.7286	91.7610	90.5527	89.7474	89.7474

with aspect ratio $a/b = 1$ and elastically restrained edges against rotation is shown in Table 10.2. The results are presented in Figures 10.2–10.9 and Tables 10.3 and 10.4. The effect of density parameter δ_1 on the frequencies for first two modes of vibration is shown in Figure 10.2 for $\delta_2 = \pm 0.5$, $R = 10{,}1000$, $\psi_1 = \psi_2 = 0.5$, $a/b = 1$. As the value of δ_1 increases, the frequency decreases. The trend is the same when we study the effect of density parameter δ_2 for $\delta_1 = \pm 0.5$, $R = 10{,}1000$, $\psi_1 = \psi_2 = 0.5$, $a/b = 1$ as shown in Figure 10.3. The graph of thickness parameter ψ_1 versus frequency is shown in Figure 10.4 for $\psi_2 = \pm 0.5$, $\delta_1 = \delta_2 = 0.5$, $R = 10{,}1000$, $a/b = 1$. It is observed that the thickness parameter ψ_1 increases the frequency of the plate. The same can be concluded from Figure 10.5, in which the effect of thickness parameter ψ_2 is shown on the frequency. Figure 10.6 shows the effect of $R_1 = R_2 = R_3 = R_4$ on frequency for $\delta_1 = \pm 0.5$, $\psi_1 = \pm 0.5$, $\delta_2 = \psi_2 = 0.5$, $a/b = 1$. It is concluded that frequency increases with the increasing values of R. It is also observed that frequency becomes constant in the neighborhood of $R = 100$ for $\psi_1 = -0.5$, while it becomes constant in the neighborhood of $R = 1000$ for $\psi_2 = 0.5$. The same trend can be seen from Figure 10.7, where the effect of R on frequency is depicted for $\delta_2 = \pm 0.5$, $\psi_2 = \pm 0.5$, $\delta_1 = \psi_1 = 0.5$, $a/b = 1$ The graph of aspect ratio a/b versus frequency parameter is shown in Figure 10.8 for $\delta_1 = \pm 0.5$, $R = 10{,}1000$, $\psi_1 = \psi_2 = 0.5$, $a/b = 1$. The frequency of the plate is more for larger values of aspect ratio. The rate of increase is more pronounced for $a/b > 1$. The same is true for $\psi_1 = \pm 0.5$, $R = 10{,}1000$, $\delta_1 = \psi_2 = 0.5$, $a/b = 1$ as evident

TABLE 10.2

Comparison of Frequency Parameter Ω of Homogeneous ($\delta_1 = \delta_2 = 0.0$) Simply Supported Isotropic ($C_{11} = E/(1-\upsilon^2)$, $C_{22} = C_{11}$, $C_{33}/C_{11} = (1-\upsilon)/2$, $C_{12}/C_{11} = \upsilon$, $\upsilon = 0.3$) Square ($a/b = 1$) Plate with Elastically Restraint Edges Against Rotation for $\psi_2 = 0$

Reference	ψ_1	R_1	R_2	R_3	R_4	Mode I	Mode II	Mode III
Grossi and Bhat [6]	0.2	1	0	0	0	22.27	-	-
Present						**22.1344**	-	-
Grossi and Bhat [6]	0.2	1000	0	0	0	26.14	-	-
Present						**25.8389**	-	-
Huang et al. [30]	0	1	1	0	0	20.639	49.721	50.830
Kobayashi and Sonoda [56]						20.639	49.721	50.830
Present						**20.6394**	**49.7207**	**50.8295**
Huang et al. [30]	0	100	100	0	0	28.165	54.109	67.133
Kobayashi and Sonoda [56]						28.165	54.109	67.133
Present						**28.1650**	**54.1090**	**67.1331**
Li et al. [14]	0	1	1	1	1	21.500	51.187	51.187
Present						**21.5019**	**51.1914**	**51.1923**
Li [12]	0	100	100	100	100	34.67	70.78	70.78
Present						**34.6777**	**70.7935**	**70.8030**
Li et al. [14]		1000	1000	1000	1000	35.842	73.103	73.103
Present						**35.8449**	**73.1309**	**73.1401**

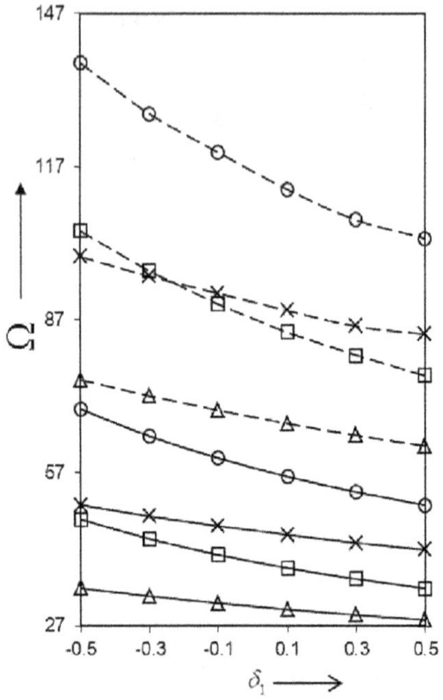

FIGURE 10.2 Frequency parameter Ω ——, mode I; ----, mode II; for $\psi_1 = \psi_2 = 0.5, a/b = 1$; □, $\delta_2 = -0.5, R = 10$; o, $\delta_2 = -0.5, R = 1000$; Δ, $\delta_2 = 0.5, R = 10$; ×, $\delta_2 = 0.5, R = 1000$.

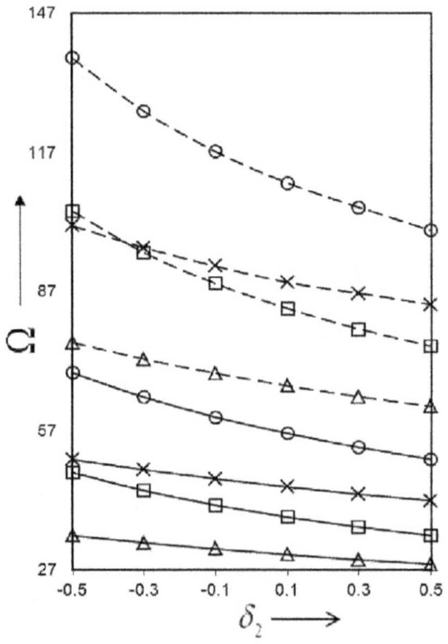

FIGURE 10.3 Frequency parameter Ω ——, mode I; ----, mode II; for $\psi_1 = \psi_2 = 0.5, a/b = 1$: □, $\delta_1 = -0.5, R = 10$; o, $\delta_1 = -0.5, R = 1000$; Δ, $\delta_1 = 0.5, R = 10$; ×, $\delta_1 = 0.5, R = 1000$.

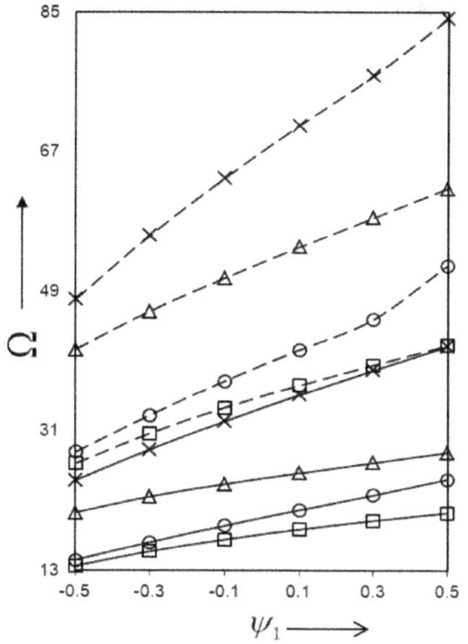

FIGURE 10.4 Frequency parameter Ω ——, mode I; ----› mode II; for $\delta_1 = \delta_2 = 0.5, a/b = 1$: □, $\psi_2 = -0.5, R = 10$; o, $\psi_2 = -0.5, R = 1000$; Δ, $\psi_2 = 0.5, R = 10$; ×, $\psi_2 = 0.5, R = 1000$.

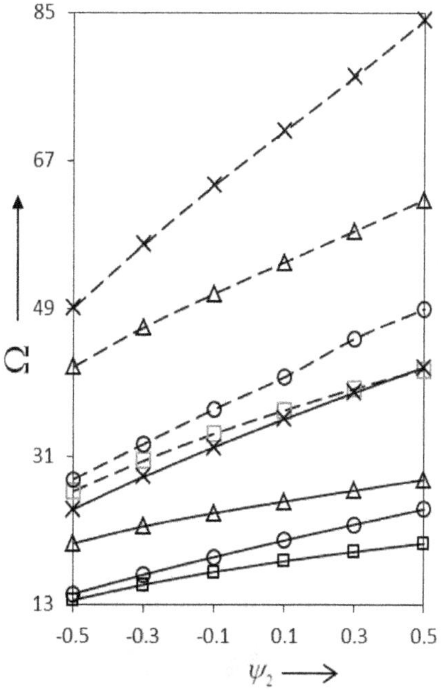

FIGURE 10.5 Frequency parameter Ω ——, mode I; ----› mode II; for $\delta_1 = \delta_2 = 0.5, a/b = 1$: □, $\psi_1 = -0.5, R = 10$; o, $\psi_1 = -0.5, R = 1000$; Δ, $\psi_1 = 0.5, R = 10$; ×, $\psi_1 = 0.5, R = 1000$.

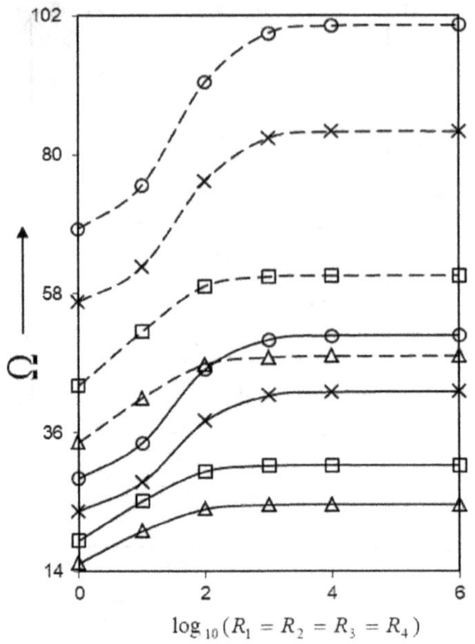

FIGURE 10.6 Frequency parameter Ω ——, mode I; ----, mode II; for $\delta_2 = \psi_2 = 0.5$, $a/b = 1$: □, $\delta_1 = -0.5, \psi_1 = -0.5$; o, $\delta_1 = -0.5, \psi_1 = 0.5$; Δ, $\delta_1 = 0.5, \psi_1 = -0.5$; ×, $\delta_1 = 0.5, \psi_1 = 0.5$.

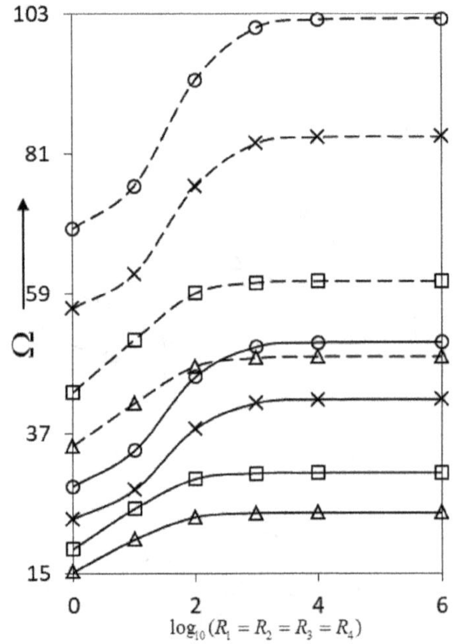

FIGURE 10.7 Frequency parameter Ω ——, mode I; ----, mode II; for $\delta_1 = \psi_1 = 0.5$, $a/b = 1$: □, $\delta_2 = -0.5, \psi_2 = -0.5$; o, $\delta_2 = -0.5, \psi_2 = 0.5$; Δ, $\delta_2 = 0.5, \psi_2 = -0.5$; ×, $\delta_2 = 0.5, \psi_2 = 0.5$.

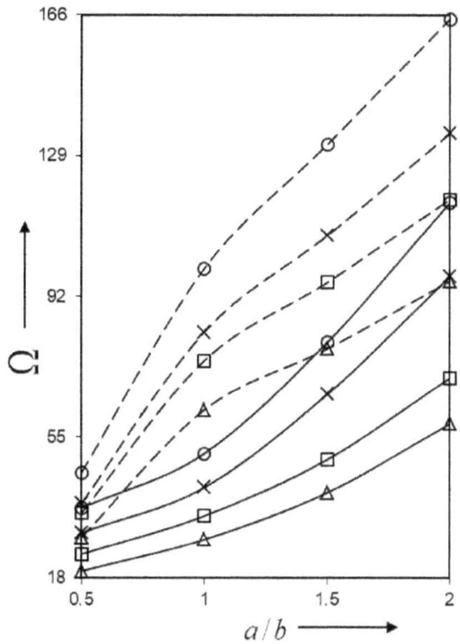

FIGURE 10.8 Frequency parameter Ω ——, mode I; ----, mode II; for $\delta_2 = \psi_1 = \psi_2 = 0.5$: □, $\delta_1 = -0.5$, $R = 10$; o, $\delta_1 = -0.5$, $R = 1000$; Δ, $\delta_1 = 0.5$, $R = 10$; ×, $\delta_1 = 0.5$, $R = 1000$.

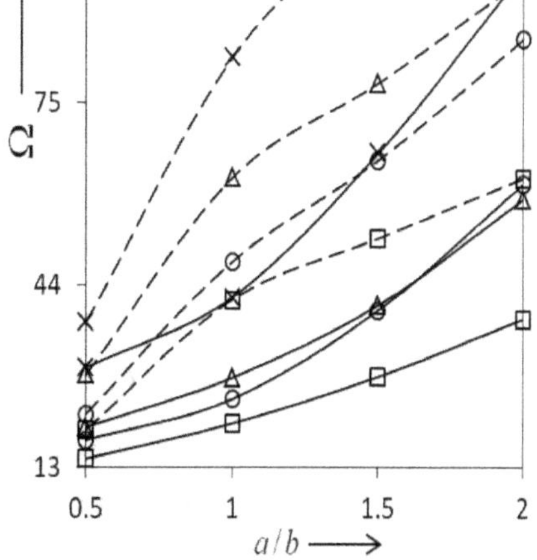

FIGURE 10.9 Frequency parameter Ω ——, mode I; ----, mode II; for $\delta_1 = \delta_2 = \psi_2 = 0.5$: □, $\psi_1 = -0.5$, $R = 10$; o, $\psi_1 = -0.5$, $R = 1000$; Δ, $\psi_1 = 0.5$, $R = 10$; ×, $\psi_1 = 0.5$, $R = 1000$.

TABLE 10.3
Frequency Parameter Ω of Simply Supported Monoclinic Plate for $\psi_1 = \psi_2 = 0$

			$R = 0$			$R = 1000000$		
			a/b			a/b		
δ_1	δ_2	Mode	0.5	1	2	0.5	1	2
0	0	I	11.8585	18.1271	44.2189	24.1387	33.6602	88.3883
		II	18.1274	44.2197	72.5087	30.1576	67.0579	116.9490
		III	28.9236	47.4341	121.0910	41.9500	71.2400	172.6370
	0.5	I	10.5861	16.2049	39.5408	21.5167	30.0882	79.0335
		II	16.2549	39.6344	64.8202	27.0415	60.0538	104.7290
		III	25.9440	42.3449	108.1940	37.7297	64.3371	154.4110
0.5	0	I	10.6041	16.2062	39.4901	21.5842	30.0914	78.8586
		II	16.2064	39.4909	65.0114	26.9627	59.8929	105.0010
		III	25.8484	42.5153	108.6070	37.5210	64.5368	155.1880
	0.5	I	9.6679	14.7908	36.0600	19.6584	27.4622	72.0279
		II	14.8231	36.1195	59.2817	24.6598	54.7458	95.7596
		III	23.8602	38.7344	98.9875	34.3835	58.8150	141.4500

TABLE 10.4
Frequency Parameter Ω of Simply Supported Monoclinic Plate for $\psi_1 = \psi_2 = 0.5$

			$R = 0$			$R = 1000000$		
			a/b			a/b		
δ_1	δ_2	Mode	0.5	1	2	0.5	1	2
0	0	I	18.0639	27.8231	67.5016	36.6011	51.4909	134.4330
		II	27.8984	67.6642	111.5370	46.5710	102.2740	180.3830
		III	44.4678	72.5067	185.6910	68.3314	109.4410	266.5120
	0.5	I	16.3072	24.9722	60.4797	33.1989	46.3117	120.7790
		II	25.0355	60.8618	100.1090	41.6385	91.9525	162.0230
		III	39.9421	65.4356	167.0910	60.3864	98.8174	250.6060
0.5	0	I	16.1828	24.9610	60.8487	32.8827	46.2951	121.6190
		II	25.0285	61.0150	100.1390	41.8225	92.0004	161.5680
		III	39.9761	65.1878	167.0950	61.0569	98.7108	248.1850
	0.5	I	14.8876	22.8438	55.5682	30.3631	42.4233	111.2710
		II	22.8950	55.8818	91.5986	38.1319	83.6881	147.9410
		III	36.7275	59.9179	153.1480	55.1352	89.7474	226.6040

from Figure 10.9. First, three 3D mode shapes for the plate have been plotted using MATLAB software and are shown in Figure 10.10.

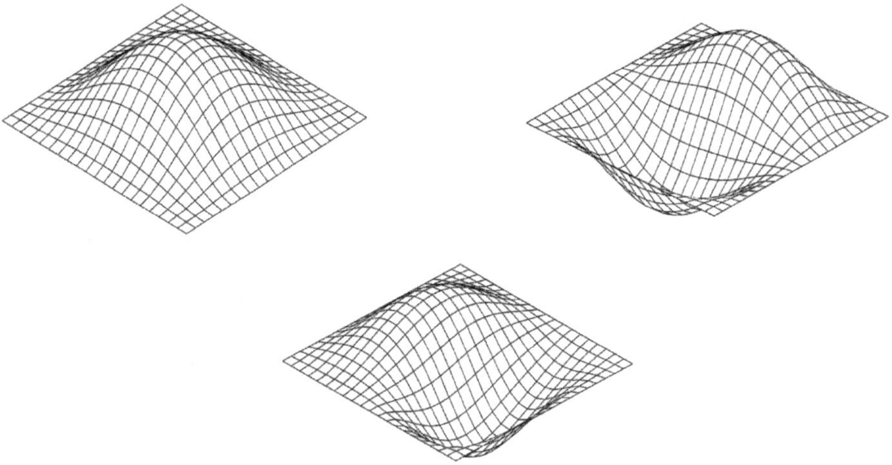

FIGURE 10.10 First three mode shapes of simply supported monoclinic square $(a/b = 1)$ plate for $\delta_1 = \delta_2 = \psi_1 = \psi_2 = 0.5$; $R = 100$.

10.4 CONCLUSIONS

This chapter presented the Rayleigh–Ritz solution to free transverse vibration problem of a thin simply supported monoclinic rectangular plate with elastically restrained edges against rotation. The Rayleigh–Ritz method used boundary characteristic orthogonal polynomials to obtain the standard eigenvalue problem. The density and the thickness of the plate depend on both the in-plane variables x and y. The effects of aspect ratio, density and thickness along with elastically restrained edges have been studied on first three natural frequencies of the plate. It is observed that

i. the frequency parameter Ω decreases as the plate becomes less stiff toward the edge $x = a$ and $y = b$ due to the increasing values of the density parameters δ_1 and δ_2.
ii. an increase in frequency is observed as the values of thickness parameters ψ_1 and ψ_2 increase.
iii. with the increase in the values of aspect ratio a/b and flexibility parameters $R_1 = R_2 = R_3 = R_4$, an increase in frequency is observed.
iv. the percentage variations in the value of frequency parameter Ω for the first three modes of vibration are 11.3 to −8.5, 10.6 to −8.2 and 11.9 to −8.6 for $R_1 = R_2 = R_3 = R_4 = 100$ and $\delta_2 = \psi_1 = \psi_2 = 0.5, a/b = 1$ when the density parameter δ_1 varies from −0.5 to 0.5.
v. the percentage variations in the value of frequency parameter Ω for the first three modes of vibration are −24.7 to 18.1, −25.8 to 19.2 and −24.8 to 18.9 for $R_1 = R_2 = R_3 = R_4 = 100$ and $\delta_1 = \delta_2 = \psi_2 = 0.5, a/b = 1$ when the density parameter ψ_1 varies from −0.5 to 0.5.

The frequencies of homogeneous ($\delta_1 = \delta_2 = 0.0$) simply supported isotropic square plates with elastically restraint edges against rotation have been obtained by considering $C_{11} = E/\left(1-\upsilon^2\right), C_{22} = C_{11}, C_{12} = \upsilon C_{11}, C_{33} = E/2(1+\upsilon), C_{33}/C_{11} = (1-\upsilon)/2, \upsilon = 0.3$ and are compared with available ones. A good agreement of results is observed.

REFERENCES

[1] Leissa, A.W. 1969. *Vibration of Plates.* NASA SP-160, Washington, D.C. U.S. Government Printing Office.

[2] Leissa, A.W. 1978. Recent research in plate vibrations 1973–1976: Complicating effects. *The Shock and Vibration Digest*, 10(12): 21–35.

[3] Leissa, A.W. 1981. Plate vibration research, 1976–1980: complicating effects. *The Shock and Vibration Digest*, 13(10): 19–36.

[4] Leissa, A.W. 1987. Recent studies in plate vibrations, 1981–1985 part I: classical theory. *The Shock and Vibration Digest*, 19(2): 11–18.

[5] Leissa, A.W. 1987. Recent studies in plate vibrations, 1981–1985 Part II: Complicating effects. *The Shock and Vibration Digest*, 19(3): 10–24.

[6] Grossi, R.O., and Bhat, R.B. 1995. Natural frequencies of edge restrained tapered rectangular plates. *Journal of Sound and Vibration*, 185:335–43.

[7] Zhou, D. 1995. Natural frequencies of elastically restrained rectangular plates using a set of static beam functions in the Rayleigh-Ritz method. *Computers and Structures*, 57:731–35.

[8] Zhou, D. 1996. An approximate solution of eigen-frequencies of transverse vibration of rectangular plates with elastical restraints. *Applied Mathematics and Mechanics* 17(5): 451–56.

[9] Cheung, Y.K., and Zhou, D. 2000. Vibrations of rectangular plates with elastic intermediate line-supports and edge constraints. *Thin-walled Structures* 37:305–31.

[10] Zhou, D. 2001. Vibrations of Mindlin rectangular plates with elastically restrained edges using static Timoshenko beam functions with the Rayleigh-Ritz method. *International Journal of Solids and Structures* 38:5565–580.

[11] Ashour, A.S. 2004. Vibration of variable thickness plates with edges elastically restrained against translation and rotation. *Thin-Walled Structures* 42:1–24.

[12] Li, W.L. 2004. Vibration analysis of rectangular plates with general elastic boundary supports. *Journal of Sound and Vibration* 273:619–35.

[13] Malekzadeh, P., and Shahpari, S.A. 2005. Free vibration analysis of variable thickness thin and moderately thick plates with elastically restrained edges by DQM. *Thin Walled Structures* 43(7):1037–050.

[14] Li, W.L., Xuefeng, Z., Jingtao, D., and Zhigang, L. 2009. An exact series solution for the transverse vibration of rectangular plates with general elastic boundary supports. *Journal of Sound and Vibration* 321:254–69.

[15] Zhang, X., and Li, W.L. 2009. Vibrations of rectangular plates with arbitrary non-uniform elastic edge restraints. *Journal of Sound and Vibration* 326:221–34.

[16] Hüseyin, D., and Ömer, K. M. 2011. Vibrations of elastically restrained rectangular plates. *Scientific Research and Essays* 6(34):6811–816.

[17] Lai, S.K., and Xiang, Y. 2012. Buckling and vibration of elastically restrained standing vertical plates. *Journal of Vibration and Acoustics* 134(1). doi.org/10.1115/1.4005007.

[18] Yang, B., and Wang, D. 2016. Buckling strength of rectangular plates with elastically restrained edges subjected to in-plane impact loading. *Proceedings of Institution of Mechanical Engineering Part C: J Mechanical Engineering Science* 231(20):3743–752.

[19] Verma, Y., and Datta, N. 2018. Comprehensive study of free vibration of rectangular Mindlin's plates with rotationally constrained edges using dynamic Timoshenko trial functions. *Engineering Transections* 66(2):129–60.

[20] Zhang, Y., and Zhang, S. 2019. Free transverse vibration of rectangular orthotropic plates with two opposite edges rotationally restrained and remaining others free. *Applied Science*, 22(9). doi: 10.3390/app9010022.

[21] Wang, G., Li, W., Feng, Z., and Ni, J. 2019. A unified approach for predicting the free vibration of an elastically restrained plate with arbitrary holes. *International Journal of Mechanical Sciences* 159:267–77.

[22] Zhang, J., Lu, J., Ullah, S., Gao, Y. and Zhao, D. 2020. Buckling analysis of rectangular thin plates with two opposite edges free and others rotationally restrained by finite Fourier integral transform method. *Journal of Applied Mathematics and Mechanics*. doi: 10.1002/zamm.202000153.

[23] Wattanasakulpong, N., and Songsuwan, W. 2020. Application of the Adomian modified decomposition method to free vibration analysis of thin plates with elastic supports. *Engineering Transactions* 23(2):115–25.

[24] Laura, P.A.A., and Grossi, R. 1979. Transverse vibrations of rectangular plates with thickness varying in two directions and with edges elastically restrained against rotation. *Journal of Sound and Vibration* 63(4):499–505.

[25] Singh, B., and Saxena, V. 1996. Transverse vibration of a rectangular plate with bidirectional thickness variation. *Journal of Sound and Vibration* 198(1):51–65.

[26] Sakiyama, T., and Huang, M. 1998. Free vibration analysis of rectangular plates with variable thickness. *Journal of Sound and Vibration* 216(3):379–97.

[27] Cheung, Y.K., and Zhou, D. 1999. The free vibrations of tapered rectangular plates using a new set of beam functions with the Rayleigh-Ritz method. *Journal of Sound and Vibration* 223(5):703–22.

[28] Zhou, D. 2002. Vibrations of point-supported rectangular plates with variable thickness using a set of static tapered beam functions. *International Journal of Mechanical Sciences* 44(1):149–64.

[29] Cheung, Y.K., and Zhou, D. 2003. Vibrations of tapered Mindlin plates in terms of static Timoshenko beam functions. *Journal of Sound and Vibration* 260:693–709.

[30] Huang, M., Ma, X.Q., Sakiyama, T., Matsuda, H., and Morita, C. 2007. Free vibration analysis of rectangular plates with variable thickness and point support. *Journal of Sound and Vibration* 300(3–5):435–52.

[31] Kumar, Y., and Lal, R. 2011. Buckling and vibration of orthotropic nonhomogeneous rectangular plates with bilinear thickness variation. *Journal of Applied Mechanics (ASME)* 78:1–11.

[32] Kumar, Y., and Lal, R. 2011. Vibrations of nonhomogeneous orthotropic rectangular plates with bilinear thickness variation resting on Winkler foundation. *Meccanica*. doi: 10.1007/s11012-011-9459–4.

[33] Lal, R., and Kumar, Y. 2011. Boundary characteristic orthogonal polynomials in the study of transverse vibrations of nonhomogeneous rectangular plates with bilinear thickness variation. *Shock and Vibration*. doi: 10.3233/SAV–2011–0635.

[34] Lal, R., and Kumar, Y. 2011. Characteristic orthogonal polynomials in the study of transverse vibrations of nonhomogeneous rectangular orthotropic plates of bilinearly varying thickness. *Meccanica*. doi: 10.1007/s11012-011-9430–4.

[35] Kumar, Y., and Lal, R. 2012. Vibrations of nonhomogeneous orthotropic rectangular plates with bilinear thickness variation resting on Winkler foundation. *Meccanica* 47:893–915.

[36] Bahmyari, E., and Rahbar-Ranji, A. 2012. Free vibration analysis of orthotropic plates with variable thickness resting on non-uniform elastic foundation by element free Galerkin method. *Journal of Mechanical Science and Technology* 26 (9):2685–694.

[37] Semnani, S.J., Attarnejad, R., and Firouzjaei, R.K. 2013. Free vibration analysis of variable thickness thin plates by two-dimensional differential transform method. *Acta Meccanica* 224:1643–658.

[38] Xue, K., Wang, J., Li, Q., and Wang, W. 2013. Free vibration analysis of rectangular plates with varying thickness in two directions. *Journal of Harbin Engineering University* 34(11):1456–459.

[39] Shufrina, I., and Eisenberger, M. 2016. Semi-analytical modeling of cutouts in rectangular plates with variable thickness-Free vibration analysis. *Applied Mathematical Modelling* 40:6983–7000.

[40] Tran, T.T., Quoc-Hoa, P., and Nguyen-Thoi, T. 2021. Static and free vibration analyses of functionally graded porous variable-thickness plates using an edge-based smoothed finite element method. *Defence Technology* 17:971–86.

[41] Soldatos, K.P. 2004. Complex potential formalisms for bending of inhomogeneous monoclinic plates including transverse shear deformation. *Journal of the Mechanics of Physics of Solids* 52:341–57.

[42] Batra, R.C., Qian, L.F., and Chen, L.M. 2004. Natural frequencies of thick square plates made of orthotropic, trigonal, monoclinic, hexagonal and triclinic materials. *Journal of Sound and Vibration* 270:1074–086.

[43] Kumar, Y., and Tomar, S.K. 2006. Free transverse vibrations of monoclinic rectangular plates with continuously varying thickness and density. *International Journal of Applied Mechanics and Engineering* 11(4):881–900.

[44] Ferreira, A.J.M., Fasshauer, G.E., and Batra, R.C. 2009. Natural frequencies of thick plates made of orthotropic, monoclinic, and hexagonal materials by a meshless method. *Journal of Sound and Vibration* 319:984–92.

[45] Kumar, Y. 2013. Differential transform method to study free transverse vibration of monoclinic rectangular plates resting on Winkler foundation. *Applied and Computational Mechanics* 7(2):145–54.

[46] Bahrami, K., Afsari, A., Janghorban, M., and Karami, B. 2019. Static analysis of monoclinic plates via a three-dimensional model using differential quadrature method. *Structural Engineering Mechanics* 72:131–39.

[47] Kumar, Y. 2018. The Rayleigh–Ritz method for linear dynamic, static and buckling behavior of beams, shells and plates: A literature review. *Journal of Vibration and Control* 24(7):1205–227.

[48] Zhang, J., Li, T., and Zhu, X. 2019. Free vibration analysis of rectangular fgm plates with a cutout. *IOP Conf. Series: Earth and Environmental Science* 283. doi: 10.1088/1755–1315/283/1/012037.

[49] Piyush, P.S., and Azam, M.S. 2020. Free vibration and buckling analysis of elastically supported transversely inhomogeneous functionally graded nanoplate in thermal environment using Rayleigh–Ritz method. *Journal of Vibration and Control.* doi: 10.1177/1077546320966932.

[50] *Alanbay*, B., Singh, K., and *Kapania*, R.K. 2020. Vibration of curvilinearly stiffened plates using Ritz method with orthogonal Jacobi polynomials. *Journal of Vibration and Acoustics* 142(1). doi.org/10.1115/1.4045098.

[51] Alanbay, B., Kapania, R.K., and Batra, R.C. 2020. Up to lowest 100 frequencies of rectangular plates using Jacobi polynomials and TSNDT. *Journal of Sound and Vibration* 480. doi.org/10.1016/j.jsv.2020.115352.

[52] Muc, A., and Flis, J. 2021. Flutter characteristics and free vibrations of rectangular functionally graded porous plates. *Composite Structures.* doi.org/10.1016/j.compstruct.2020.113301.

[53] Hung, D.X., Tu, T.M., and Hao, T.D. 2021. Free vibration analysis of laminated CNTRC plates using the pb2-Ritz method. *Journal of Mechanical Engineering* 18(1):213–32.

[54] Singh, B., and Chakraverty, S. 1994. Flexural vibration of skew plates using boundary characteristic orthogonal polynomials in two variables. *Journal of Sound and Vibration* 173(2):157–78.
[55] Haussuhl, V.S. 1965. Elastische und Thermoelastische Eigenschaften CaSO4.2H2O (Gips). *Zeitsehrift fur Kristallographite. Bd.* 122:311–14.
[56] Kobayashi, H., and Sonoda, K. 1991. Vibration and buckling of tapered rectangular plates with two opposite edges simply supported and the other two edges elastically restrained against rotation. *Journal of Sound and Vibration* 146:323–37.

11 A Numerical Note on Triple Diffusive Mixed Convection Flow over a Vertical Plate with Variable Physical Parameters of Water

Iyyappan G. and Govindaraj N.
Hindustan Institute of Technology and Science

A. K. Singh
VIT University, Chennai Campus

CONTENTS

LIST OF NOTATIONS

C	Species concentration
C_{1w}	Concentration at the wall
C_{2w}	Concentration at the wall
Cf_x	Skin friction coefficient
C_p	Constant pressure and specific heat
D_B	Brownian diffusion coefficient
D_T	Thermophoretic diffusion coefficient

DOI: 10.1201/9781003222255-11

F	Dimensionless velocity
f	Dimensionless stream function
g	Acceleration due to gravity
Gr	Grashof numbers due to temperature
Gr_c	Grashof numbers due to concentration
Le	Lewis numbers
H	Magnetic parameter
K*	Rosseland mean absorption coefficient
N	Viscosity ratio
Nb	Brownian motion
Nt	Thermophoresis
Nu_x	Nusselt number
Nc_1	Buoyancy ratio parameters for liquid C_1
Nc_2	Buoyancy ratio parameters for liquid C_2
Pr	Prandtl number
Re_L	Reynolds number
Rd	Thermal radiation
q_r	Heat radiation
T	Temperature of the nanofluid
T_w	Uniform temperature over the surface of the sheet
u	Velocity component in the x-direction
U_W	Moving plate velocity
U_∞	Free stream velocity
U	Reference velocity
v	Velocity component in the y-direction

GREEK SYMBOLS

φ	Stream function
λ	Mixed convection parameter
λ_c	Mixed convection concentration parameter
λ_{c2}	Mixed convection concentration parameter
μ	Dynamic viscosity
τ	Ratio between heat capacity of nanoparticle and base fluid
ρ	Fluid density
η	Kinematic viscosity
ρ_p	Heat capacity of the base fluid
ρ_f	Heat capacity of the nanoparticle material
σ^*	Stefan–Boltzmann constant
ε	The ratio between free stream velocity and the reference velocity

11.1 INTRODUCTION

Nanofluid is a type of nanotechnology based on heat and mass transfer by dissolving nanometer-sized fluid particles. Nanofluids are produced by scattering liquids with low thermal conductivity, such as water and ethylene, to nanometer-sized solid particles. Previous studies have shown that nanofluids improve thermal energy and control of mass transfer within the boundary layer. The current study presents an innovative approach to manipulating the chemical reactions of nanofluids with a variable viscosity and Prandtl number over a vertical plate. Recently, many researchers have focused on the studies of boundary layer flow on different surfaces by nanofluids, which are applicable in many industrial fields such as fiberglass production, food processing and aerodynamic emissions. The analysis of heat and mass transfer characteristics is applicable for many engineering processes such as polymer extrusion, glass blowing, wire drawing and paper production. Because of the significant effects of boundary layer, chemical reactions are considered in the water.

There has been a literary analysis focused on triple diffusion convection, and some literary discussion is included here. The authors of the study [1] looked into the chemical reaction of a nanofluid in a laminar boundary layer and the effects of different parameters, including buoyancy ratio and Lewis number. The authors in [2] technically described the boundary layer movement of a viscous fluid on a vertical flat plate traveling on the surface of the liquid fluid. The authors in study [3] investigated the porous nanofluid flow with two salts and discovered that increasing the volume friction and salt increases the heat transfer coefficient of the nanoparticles. The effects of nanoparticles and surface smoothness on fluid flow characteristics via MHD triple diffusion on rapidly elongated surface were explored in [4]. Also, they extended their work up to the stability of triple diffusive fluid on saturated porous layer using binary Maxwell methods. The effects of heat generation or absorption on the MHD composite convection porous medium of a nanofluid presented on an exponentially stretching sheet through a microscopic medium are discussed in [5] and [6].

The authors in study [7] presented the conjugate effects of heat and mass transfer on moving vertical plate with electrically conducting water-based Cu-nanofluid. Also, they used to compute on-heat transfer rate using Rosseland approximation for thermal radiation on boundary surface. The author of the study [8] discussed the effects of Brownian motion and thermophoresis on unsteady mixed convection flow over a hot vertical plate near the stagnation stage. The unsteady mixed convection flow over a vertical stretching sheet with the effects of suction/injection was studied in [9], and the governing partial differential equations are transformed to nonlinear ordinary differential equations that are solved using the Keller box method. In [10], the authors described the dual solution of nanofluid mixed convection fluid flow in a moving vertical plate. Three different types of nanoparticles are used to compute the magnitude of temperature and velocity profiles, namely aluminum oxide, titanium oxide and copper, which are considered by using water-based fluid. The authors in [11] studied the fluid flow of a fixed three-dimensional hybrid boundary layer past a vertical plate traveling in an upward direction parallel to a free stream. They also derived the results of a set of normal differential equations using finite differentiation methods with quasilinearization technique.

The Darcy–Brinkman model was used to explore the effect of mixed convection on a stagnation point flow of a thermo-micropolar hybrid nanofluid over a vertical surface in a saturated porous medium with inertial and microstructure properties in [12]. The moving plate and magnetic field effects are explored in a combined convection nanoliquid flow, including triple diffusive components, in [13]. In [14], the phenomena of triple diffusive nonlinear mixed convection flow across a vertical cone and the behavior of such flows were reported. Muhammad Khairul Anuar Mohamed et al. [15] investigated the mixed convection boundary layer flow and heat transfer on a moving vertical plate in a nanofluid with viscous dissipation and constant wall temperature. The heat transfer properties of a magnetohydrodynamic (MHD) hybrid nanofluid over a linear stretching and shrinking surface in the presence of suction and thermal radiation effects are reported by [16].

The aim of this study is to explain the triple diffusion mixed convection MHD nanoliquid flow and thermal radiation effects of nanoparticles with varying viscosity and Prandtl number. Thermal radiation has an effect on two molecules of different concentration profiles. In the current sense, the solutions to the water boundary layer with MHD nanoliquid have been found. Furthermore, the determination of Sherwood number and Nusselt number on wall skin friction has been carried out. The fluid flow pattern is generated by solving nonlinear partial differential equations with boundary conditions by using a finite difference scheme and the quasilinearization technique. The steady incompressible laminar boundary layer gets attention into a MHD flow for introducing the nanofluid concept, and water is considered as the base fluid. The thermal radiation on MHD fluid flow plays a very important role in many chemical industries and engineering and scientific fields.

11.2 FORMATION OF GOVERNING EQUATIONS

Consider a vertical plate with a steady laminar water-based nanoliquid mixed convection flow with variable viscosity and temperature. In the range of 0–45 degrees Celsius, Prandtl number (Pr) and dynamic viscosity (μ) are considered to differ as an inverse linear function of temperature. At constant strain, the variance of both specific heat (C_p) and density (ρ) is assumed to be less than 1%, so they are treated as constants. The values for thermophysical properties of water at various temperatures are provided in [17]. Buoyancy force rise to the fluid properties to relate the density changes to temperature, and the thermophysical properties of water with various temperature has been shown in [18,19]. In an incompressible fluid, the contribution of heat due to the contraction of the fluid is very small and it is ignored. The fluid at the surface is kept at a constant temperature T_w, and the boundary layer's edge is kept at the same temperature T_∞. The variable Prandtl number and variable viscosity are defined as:

$$Pr = \frac{1}{C_1 + C_2 T} \text{ and } \mu = \frac{1}{b_1 + b_2 T}$$

where b_1, b_2, C_1 and C_2 are the best approximation of water, and the numerical data are presented in [18,19]. The boundary layer equations are obtained in an analogical manner [18,19] (Figure 11.1).

$$\frac{\partial u}{\partial x} + \frac{\partial v}{\partial y} = 0 \tag{11.1}$$

$$u\frac{\partial u}{\partial x} + v\frac{\partial u}{\partial y} = \frac{1}{\rho}\frac{\partial}{\partial y}\left(\mu\frac{\partial u}{\partial y}\right) + \frac{\sigma B_0^2 u}{\rho} + g[\beta(T_w - T_\infty] + \beta c_1(C_{1w} - c_{1\infty}) + \beta c_2(C_{2w} - c_{2\infty}) \tag{11.2}$$

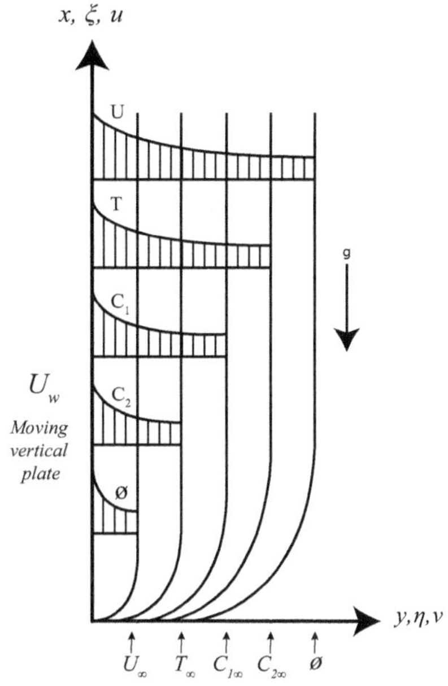

FIGURE 11.1 Physical model figure.

$$u\frac{\partial T}{\partial x}+v\frac{\partial T}{\partial y}=\frac{1}{\rho}\frac{\partial}{\partial y}\left(\frac{\mu}{Pr}\frac{\partial T}{\partial y}\right)+\frac{1}{(\rho c)_p}\frac{\partial q_r}{\partial y}+\tau\left[D_B\frac{\partial \varnothing}{\partial y}\frac{\partial T}{\partial y}+\frac{D_T}{T_\infty}\left(\frac{\partial T}{\partial y}\right)^2\right] \tag{11.3}$$

$$u\frac{\partial C_1}{\partial x}+v\frac{\partial C_1}{\partial y}=\frac{1}{\rho Sc_1}\frac{\partial}{\partial y}\left(\mu\frac{\partial C_1}{\partial y}\right) \tag{11.4}$$

$$u\frac{\partial C_2}{\partial x}+v\frac{\partial C_2}{\partial y}=\frac{1}{\rho Sc_2}\frac{\partial}{\partial y}\left(\mu\frac{\partial C_2}{\partial y}\right) \tag{11.5}$$

$$u\frac{\partial \phi}{\partial x}+v\frac{\partial \phi}{\partial y}=D_B\frac{\partial^2 \phi}{\partial y^2}+\frac{D_T}{T_\infty}\left(\mu\frac{\partial^2 T}{\partial y^2}\right) \tag{11.6}$$

Here, C_1, C_2 and $\varnothing$ represent the concentrations of the two species and nanoparticle volume fractions. Two water-based nanofluids with two different types of nanoparticles are considered, which are denoted as Nc_1 and Nc_2.

The initial velocity of the fluid and its components u, v and wall temperature are set as:

$$\begin{gathered} u(x,y)=U_w,\ v(x,y)=0,\ T(x,y)=T_w\ \ C_1(x,y)=C_{1w}, \\ C_2(x,y)=C_{2w},\ \varnothing(x,y)=\varnothing_w\ \text{ at } y=0,\ u(x,y)\to U_\infty,\ v(x,y)\to 0, \\ T(x,y)\to T_\infty,\ C_1(x,y)=C_{1\infty},\ C_2(x,y)=C_{2\infty},\ \varnothing(x,y)=\varnothing_\infty\ \text{ at } y\to\infty \end{gathered} \tag{11.7}$$

11.3 MATHEMATICAL ANALYSIS

To determine the velocity field and temperature profiles within the boundary layer, the problem is well posted with boundary condition equations (11.1–11.6) grouping into non-dimensional fluid parameters with the following similarity transformation:

$$\xi=\frac{x}{L},\eta=\left(\frac{U}{x\upsilon}\right)^{1/2},\Psi=(Ux\upsilon)^{1/2}f,u=\frac{\partial\Psi}{\partial x},v=-\frac{\partial\Psi}{\partial y},$$

$$u=UF,f_\eta=\mathrm{F},v=\frac{(Uvx)^{1/2}}{x}\frac{f}{2}+\xi f_\xi-\frac{\eta}{2}F,\frac{T\_T_\infty}{T_w-T_\infty}=\theta(\xi,\eta),$$

$$\frac{\varnothing\_\varnothing_\infty}{\varnothing_w-\varnothing_\infty}=S(\xi,\eta),\frac{C_1\_C_{1\infty}}{C_{1w}-C_{1\infty}}=Q(\xi,\eta),\frac{C_2\_C_{2\infty}}{C_{2w}-C_{2\infty}}=R(\xi,\eta)$$

Where continuity equation (11.1) is trivial, the momentum and energy equations (11.2–11.6) reduce to:

$$\left(NF_\eta\right)_\eta+F_\eta\frac{f}{2}-\xi HF+\xi\lambda\theta+Nc_1\xi\lambda Q+Nc_2\xi\lambda R=\xi\left(FF_\xi-f_\xi F_\eta\right)\quad(11.8)$$

$$\left(NPr^{-1}\theta_\eta\right)_\eta+\theta_{\eta\eta}Rd+\frac{f}{2}\theta_\eta+NbS_\eta\theta_\eta+Nt\theta_\eta^2=\xi\left(F\theta_\xi-f_\xi\theta_\eta\right)\quad(11.9)$$

$$\left(NQ_\eta\right)_\eta+Sc_1Q_\eta\frac{f}{2}=Sc_1\xi\left(FQ_\xi-f_\xi Q_\eta\right)\quad(11.10)$$

$$\left(NR_\eta\right)_\eta+Sc_2R_\eta\frac{f}{2}=Sc_2\xi\left(FR_\xi-f_\xi R_\eta\right)\quad(11.11)$$

$$\left(S_{\eta\eta}+\frac{Nt}{Nb}\right)\theta_{\eta\eta}+LeS_\eta\frac{f}{2}=Le\xi\left(FS_\xi-f_\xi S_\eta\right)\quad(11.12)$$

The transformed boundary conditions are as follows:

$$\begin{aligned}&F(\xi,\eta)=1-\varepsilon,\theta(\xi,\eta)=1,Q(\xi,\eta)=1,R(\xi,\eta)=1,S(\xi,\eta)=1\quad\text{at }\eta=0\\&F(\xi,\eta)=\varepsilon,\theta(\xi,\eta)=0,Q(\xi,\eta)=0,R(\xi,\eta)=0,S(\xi,\eta)=0\quad\text{at }\eta=\eta_\infty\end{aligned}\quad(11.13)$$

The physical parameter values of the skin friction, Nusselt number and Sherwood number related to the liquids Nc_1 and Nc_2 with nanoparticles are given below:

The Sherwood number of Nc_1 is $Sh_{x1}=x\left(\frac{\partial C}{\partial y}\right)_{y=0}(C_{1w}-C_{1\infty})^{-1}$

$$\text{i.e., }Sh_{x1}=-\left(Re_L\xi\right)^{1/2}Q_\eta(\xi,0)\quad(11.14)$$

The Sherwood number of Nc_2 is $Sh_{x2} = x\left(\frac{\partial C}{\partial y}\right)_{y=0}(C_{2w} - C_{2\infty})^{-1}$

i.e.,

$$Sh_{x2} = -(Re_L\xi)^{1/2} R_\eta(\xi,0) \tag{11.15}$$

The Sherwood number is $Sh_x = x\left(\frac{\partial C}{\partial y}\right)_{y=0}(C_w - C_\infty)^{-1}$

$$\text{i.e., } Sh_x = -(Re_L\xi)^{1/2} S_\eta(\xi,0) \tag{11.16}$$

The skin friction coefficient is defined by $Cf_x = \mu 2\left(\frac{\partial u}{\partial y}\right)(\rho U^2)^{-1}$

i.e.,

$$Cf_x = -(Re_L\xi)^{1/2} F_\eta(\xi,0) \tag{11.17}$$

The heat transfer coefficient is defined by $Nu_x = -x\left(\frac{\partial T}{\partial y}\right)_{y=0}(T_w - T_\infty)^{-1}$

i.e.,

$$Nu_x = -(Re_L\xi)^{1/2} \phi_\eta(\xi,0) \tag{11.18}$$

11.4 NUMERICAL COMPUTATION

In the process of numerical computation, the set of nonlinear governing equations play a vital role in view of their efficiency and accuracy in yielding the solutions. The nonlinear equations (11.8) to (11.12) with boundary conditions (11.13) have been linearized numerically by using quasilinearization technique, to obtain the following coupled PDEs.

$$X_1^k F_{\eta\eta}^{k+1} + X_2^k F_\eta^{k+1} + X_3^k F^{k+1} + X_4^k F_\xi^{k+1} + X_5^k \theta_\eta^{k+1} + X_6^k \theta^{k+1}$$
$$+X_7^k Q^{k+1} + X_8^k R^{k+1} = X_9^k \tag{11.19}$$

$$Y_1^k \theta_{\eta\eta}^{k+1} + Y_2^k \theta_\eta^{k+1} + Y_3^k \theta^{k+1} + Y_4^k \theta_\xi^{k+1} + X_5^k F^{k+1} + Y_6^k S_\eta^{k+1} = Y_7^k \tag{11.20}$$

$$Z_1^k Q_{\eta\eta}^{k+1} + Z_2^k Q_\eta^{k+1} + Z_3^k Q^{k+1} + Z_4^k \theta_\eta^{k+1} + Z_5^k \theta^{k+1} + Z_6^k F^{k+1} = Z_7^k \tag{11.21}$$

$$D_1^k R_{\eta\eta}^{k+1} + D_2^k R_\eta^{k+1} + D_3^k R_\xi^{k+1} + D_4^k \theta_\eta^{k+1} + D_5^k \theta^{k+1} + D_6^k F^{k+1} = D_7^k \tag{11.22}$$

$$E_1^k S_{\eta\eta}^{k+1} + E_2^k S_\eta^{k+1} + E_3^k S_\xi^{k+1} + E_4^k \theta_{\eta\eta}^{k+1} + E_5^k F^{k+1} = E_6^k \tag{11.23}$$

Here, the coefficient of the known iterative index k and that of unknown iterative index $k+1$ are determined. The boundary conditions in accordance with equations (11.19–11.23) are the following:

$$F^{k+1}(\xi,\eta)=1-\varepsilon,\ \theta^{k+1}(\xi,\eta)=1,\ \varnothing^{k+1}(\xi,\eta)=1,\ R^{k+1}(\xi,\eta)=1,$$

$$S^{k+1}(\xi,\eta)=1 \text{ at } \eta=0$$

$$F^{k+1}(\xi,\eta)=\varepsilon,\ \theta^{k+1}(\xi,\eta)=0,\ \varnothing^{k+1}(\xi,\eta)=0,\ R^{k+1}(\xi,\eta)=0,$$

$$S^{k+1}(\xi,\eta)=0 \text{ at } \eta=\eta_\infty \tag{11.24}$$

The coefficients in equations (11.19–11.23) are as follows:

$$D_1^k=N,\ D_2^k=-a_1\theta_\eta N^2+Sc_1\frac{f}{2}+Sc_1\xi f_\xi,\ D_3^k=-Sc_2\xi F,\ D_4^k=-a_1R_\eta N^2,$$

$$D_6^k=-2a_1^2\theta_\eta R_\eta N^3-a_1R_{\eta\eta}N^2,\ D_7^k$$

$$=-a_1\theta_\eta R_\eta N^2+2a_1^2\theta_\eta R_\eta N^3\theta-a_1\theta R_{\eta\eta}N^2-Sc_2\xi R_\xi F$$

$$E_1^k=Nb,\ E_2^k=NbLe\frac{f}{2}+NbLe\xi f_\xi,\ E_3^k=NbLe\xi F,\ E_4^k=Nt,\ E_5^k=NbLe\xi S_\xi$$

$$E_6^k=NbLe\xi FS_\xi$$

$$X_1^k=N,\ X_2^k=\xi f_\xi-a_1^2N^2\theta_\eta+\frac{f}{2},\ X_3^k=\xi F_\xi-\xi H,\ X_4^k=\xi F,\ X_5^k=-a_1N^2F_\eta,$$

$$X_6^k=\xi\lambda+2a_1^2N^3F_\eta\theta_\eta-a_1N^2F_{\eta\eta},\ X_7^k=\xi\lambda Nc_1,\ X_8^k=\xi\lambda Nc_2$$

$$X_9^k=-\ \xi FF_\xi+2a_1^2N^3F_\eta\theta_\eta\theta-a_1N^2F_{\eta\eta}\theta-a_1N^2F_\eta\theta_\eta$$

$$Y_1^k=Rd+NPr^{-1},\ Y_2^k=2Nt\theta_\eta+NbS_\eta+\xi f_\xi+\frac{f}{2}+2a_3\theta_\eta N-2a_1N^2Pr^{-1}\theta_\eta$$

$$Y_3^k=2a_1^2N^3Pr^{-1}\theta_\eta^2-2a_1a_3N^2\theta_\eta^2-a_1N^2Pr^{-1}\theta_{\eta\eta}+a_3N\theta_{\eta\eta},\ Y_4^k=-\xi F,\ Y_5^k=-\xi\theta_\xi$$

$$Y_6^k=-Nb\theta_\eta$$

$$Y_7^k=-Nb\theta_\eta S_\eta-\xi F\theta_\xi+Nt\theta_\eta^2+a_3\theta_\eta^2N-a_1\theta_\eta^2Pr^{-1}N^2+2Pr^{-1}a_1^2\theta_\eta^2\theta N^3$$

$$-2a_1a_3\theta N^2\theta_\eta^2+a_3\theta_{\eta\eta}\theta N-a_1\theta_{\eta\eta}\theta Pr^{-1}N^2$$

$$Z_1^k=N,\ Z_2^k=-a_1\theta_\eta N^2+Sc_1\frac{f}{2}+Sc_1\xi f_\xi,\ Z_3^k=-Sc_1\xi F,\ Z_4^k=-a_1Q_\eta N^2$$

$$Z_5^k=-2a_1\theta_\eta Q_\eta N^3-a_1Q_{\eta\eta}N^2,\ Z_6^k=-Sc_1\xi Q_\xi,$$

$$Z_7^k=-a_1Q_\eta\theta_\eta N^2+2a_1^2Q_\eta\theta_\eta\theta N^3-a_1\theta Q_{\eta\eta}N^2-Sc_1\xi Q_\xi F$$

11.5 RESULTS AND DISCUSSION

To obtain non-similar solutions, the numerical procedure in the above section is implemented. The triple diffusion mixed convection flow with variable viscous fluid and variable Prandtl number and consisting of two separate chemical components of a fluid mixture is considered. Numerical computation process has been carried out for different parameter values of Nc_1 ($0.1 \le Nc_1 \le 0.6$), Rd ($0.5 \le Rd \le 1.5$), H ($1 \le H \le 5$), Nb ($0.5 \le Nb \le 6$), Nt ($0.5 \le Nt \le 1.0$), λ ($0.5 \le \lambda \le 3$), N ($1 \le N \le 3$), ξ ($0 \le \xi \le 2.25$) and ε ($0.1 \le \varepsilon \le 1$). The effects of Prandtl number on skin friction, Nusselt number, concentration C_1, concentration C_2 and nanoparticle volume friction are presented here for $\lambda = R = H = Rd = \xi = \varepsilon = 0$, $Le = 10$, $Nb = Nt = 0.5$, $Sc_1 = 0.84441$ and $Sc_2 = 1.14540$. Table 11.1 shows the effects of Prandtl number on various non-dimensional parameters.

Table 11.2 shows the effects of various non-dimensional parameters on skin friction coefficient for ε, λ, $Nc_1 = 0.1$, $Nc_2 = 0.1$, Le, $Nb = 0.2$, Rd, H, $Nt = 0.2$ and $\xi = 1.0$.

Figure 11.2 shows how the buoyancy parameter (λ) and the velocity ratio (ε) affect the velocity profile for $\xi = 2.0$, $Nb = 0.1$, $Nc_1 = 0.1$, $Le = 2.0$, $Nc_2 = 0.1$, $Sc_1 = 0.84441$, $Sc_2 = 1.14540$, $Nt = 0.1$, $Rd = 1.0$ and $H = 1.0$. When $\lambda = 0.4$, the magnitude of the

TABLE 11.1
Effects of Prandtl Number on Various Non-Dimensional Parameters

Pr	Skin Friction	Nusselt Number	Concentration (C_1)	Concentration (C_2)	Mass Transfer
2	−0.8980	0.202465	0.4036	0.4855	1.6766
5	−0.8981	0.075141	0.4025	0.4851	1.7927
7	−0.8980	0.035574	0.4050	0.4847	1.8181
10	−0.8960	0.012007	0.4001	0.4842	1.8260

TABLE 11.2
Effects of Various Non-Dimensional Parameters on Skin Friction Coefficient

λ	H	Rd	ε	Le	Skin Friction Coefficient
0.5	1.0	1.0	0.1	2.0	−1.1734
1.0					−0.3545
2.0					1.1329
	1.0				1.1329
	3.0				−1.0081
	5.0			5	−2.4226
0.5	1.0		0.1		−0.6620
0.5	1.0		0.3		−0.1604
		1.5			−0.1277
	2.0	3.0			−0.7761

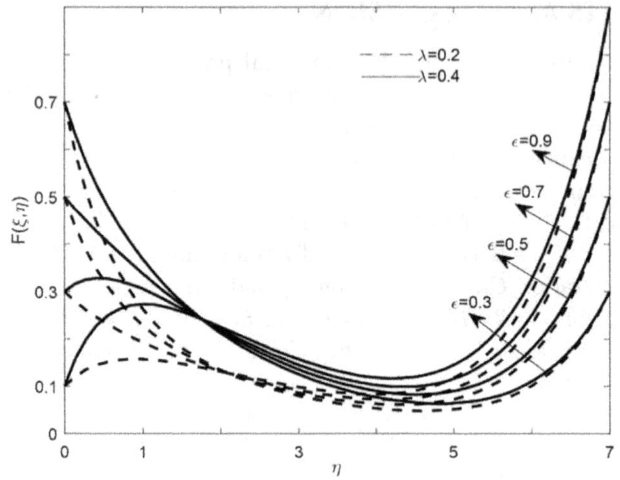

FIGURE 11.2 Velocity profile on F for $\xi = 2.0$, Nb = 0.1, $Nc_1 = 0.1$, Le = 2.0, $Nc_2 = 0.1$, $Sc_1 = 0.84441$, $Sc_2 = 1.14540$, Nt = 0.1, Rd = 1.0 and H = 1.0.

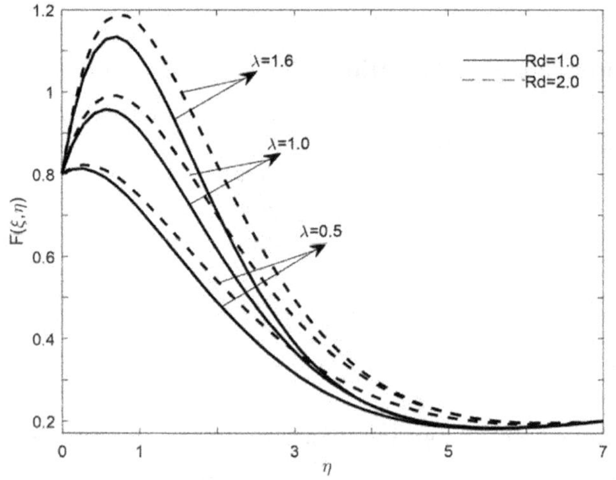

FIGURE 11.3 Velocity profile for $\xi = 1.0$, $\varepsilon = 0.2$, $Nc_1 = 0.4$, $Nc_2 = 0.3$, Nb = 0.5, $Sc_1 = 0.84441$, $Sc_2 = 1.14540$, Nt = 0.5, Le = 2.0 and H = 0.2.

velocity profile is found to be higher than when $\lambda = 0.2$. This is due to the sub-floating force centered on the upper λ, which acts as a gradient of the compressor increasing the magnitude of velocity profile. As ε increases, the velocity profile F also increases. It's important to note that, depending on the boundary conditions, velocity profiles adopt a monotonic pattern in the domain.

The effects of thermal radiation (Rd) and buoyancy parameter (λ) on the velocity profile for $\xi = 1.0$, $\varepsilon = 0.2$, $Nc_1 = 0.4$, $Nc_2 = 0.3$, Nb = 0.5, $Sc_1 = 0.84441$, $Sc_2 = 1.14540$, Nt = 0.5, Le = 2.0 and H = 0.2 are shown in Figure 11.3. As the buoyancy parameter

(λ) and Rd increase, the magnitude of velocity increases. Inside the boundary layer near the wall, the velocity profile overshoots with the increase in buoyancy parameter, while the velocity profile maintains its asymptotic nature. As the free stream velocity approaches zero, the fluid moves away from the plate.

Figure 11.4 presents the impact of Nc_1 and λ on velocity profile (F) for $\xi=2.0$, $\varepsilon=0.2$, $Nc_2=0.1$, Nb=0.5, $Sc_1=0.84441$, $Sc_2=1.14540$, Rd=1.0, Le=5.0, H=1.0 and Nt=0.1. It has been observed that increasing Nc_1 causes a sudden rise in fluid velocity immediately away from the wall. As the velocity reaches its peak, the thickness of the boundary layer exceeds the wall. A high overshoot is noticed for higher value of $\lambda=2.0$ for all values of Nc_1, but the velocity profile overshoot is completely absent for lower value of $\lambda=1.0$ and $Nc_1=0.1$.

Figure 11.5 represents the temperature distribution of different values of Rd and Nc_1 for $\xi=2.0$, $\lambda=1.0$, $\varepsilon=0.2$, $Nc_2=0.5$, Nb=0.2, $Sc_1=0.84441$, $Sc_2=1.14540$, H=1.0, Le=5.0 and Nt=0.2. The fluid temperature reaches a maximum at the surface and then gradually decreases. The magnitude of temperature profiles increases near the wall with a rise in Rd and decrease in Nc_1.

Figures 11.6 and 11.7 depict the chemical reaction H and N on concentration profile C_1. Figure 11.6 is plotted for the values of $\xi=1.0$, Nb=0.5, $Nc_1=0.1$, Le=2.0, $Nc_2=0.1$, $Sc_1=0.84441$, $Sc_2=1.14540$, Nt=0.1, Rd=1.0, $\varepsilon=0.1$ and $\lambda=3.0$, and Figure 7 is plotted for the values of $\xi=1.0$, H=1.0, $Nc_1=0.1$, Le=5.0, $Nc_2=0.1$, $Sc_1=0.84441$, $Sc_2=1.14540$, Nt=0.1, Rd=1.0, $\varepsilon=0.2$ and $\lambda=1.0$. The boundary layer thickness monotonically increases significantly when H and N increase at $\lambda=3.0$. In comparison with the stretching velocity, an increase in the value of H and N implies a decrease in free stream velocity. On the other hand, as ∇T_w and Nb

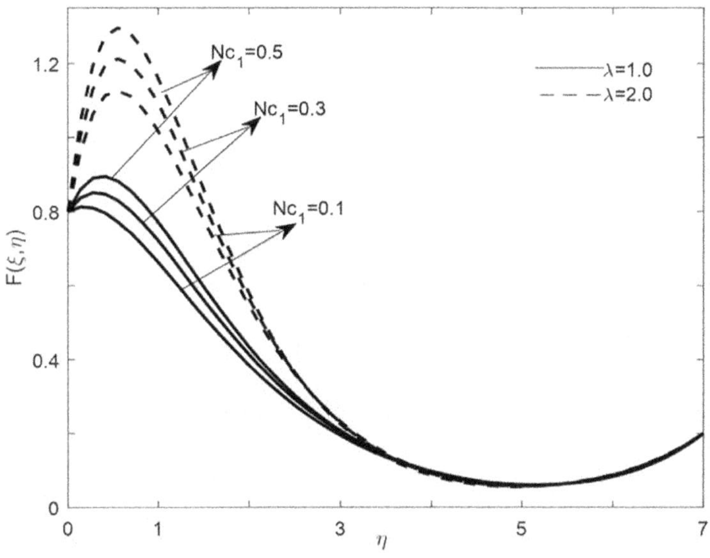

FIGURE 11.4 Velocity profile for $\xi=2.0$, $\varepsilon=0.2$, $Nc_2=0.1$, Nb=0.5, $Sc_1=0.84441$, $Sc_2=1.14540$, Rd=1.0, Le=5.0, H=1.0 and Nt=0.1.

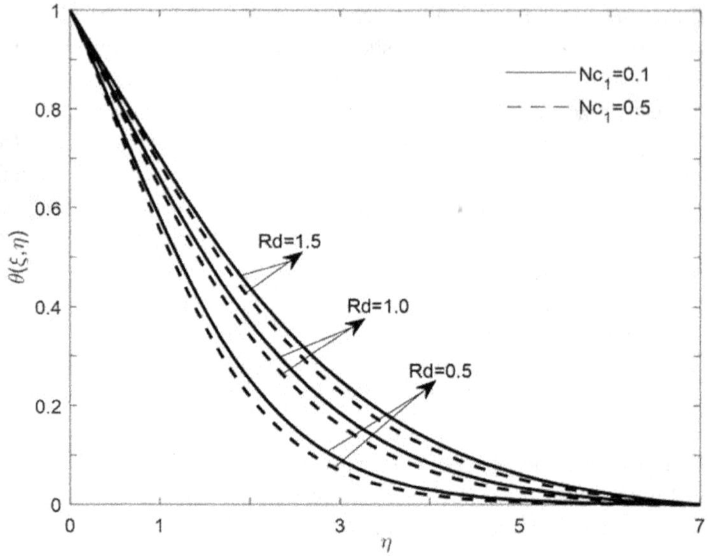

FIGURE 11.5 Temperature profile for $\xi=2.0$, $\lambda=1.0$, $\varepsilon=0.2$, $Nc_2=0.5$, Nb=0.2, $Sc_1=0.84441$, $Sc_2=1.14540$, H=1.0, Le=5.0 and Nt=0.2.

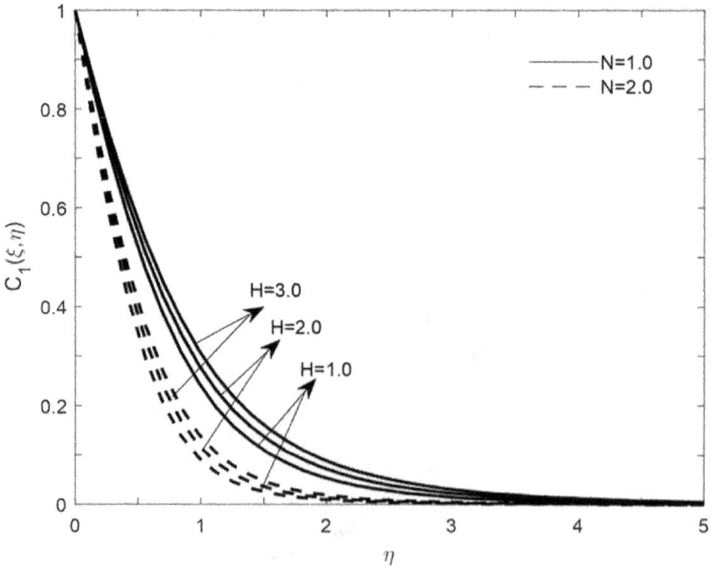

FIGURE 11.6 Concentration profile on C_1 for $\xi=1.0$, Nb=0.5, $Nc_1=0.1$, Le=2.0, $Nc_2=0.1$, $Sc_1=0.84441$, $Sc_2=1.14540$, Nt=0.1, Rd=1.0, $\varepsilon=0.1$ and $\lambda=3.0$.

decrease, the thickness of the concentration boundary layer decreases, as shown in Figure 11.7.

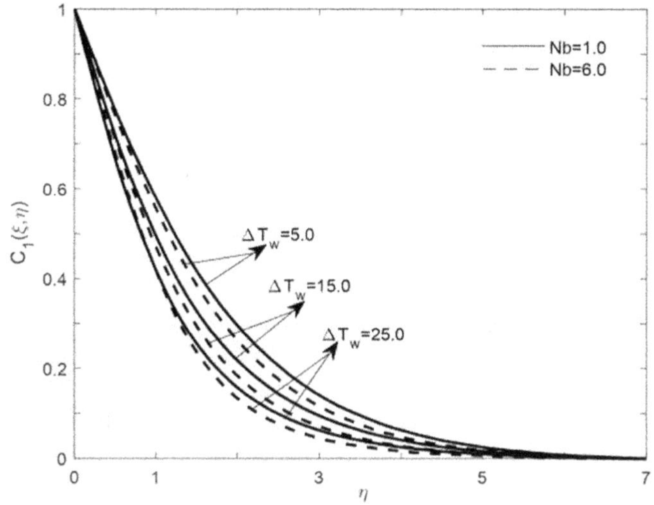

FIGURE 11.7 Effects of Nb and ∇T_w on concentration profile C_1 for $\xi = 1.0$, H = 1.0, $Nc_1 = 0.1$, Le = 5.0, $Nc_2 = 0.1$, $Sc_1 = 0.84441$, $Sc_2 = 1.14540$, Nt = 0.1, Rd = 1.0, $\varepsilon = 0.1$ and $\lambda = 3.0$.

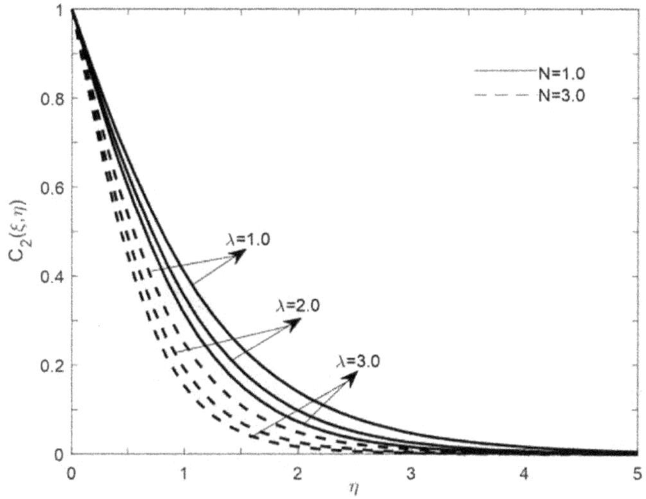

FIGURE 11.8 Concentration profile on C_2 for $\xi = 1.0$, Nb = 0.5, $Nc_1 = 0.1$, Le = 2.0, $Nc_2 = 0.1$, $Sc_1 = 0.84441$, Sc_2 1.14540, Nt = 0.1, Rd = 1.0, $\varepsilon = 0.1$ and H = 1.0.

Figure 11.8 displays the chemical reaction H and N on concentration profile C_2 for $\xi = 1.0$, Nb = 0.5, $Nc_1 = 0.1$, Le = 2.0, $Nc_2 = 0.1$, $Sc_1 = 0.84441$, $Sc_2 = 1.14540$, Nt = 0.1, Rd = 1.0, $\varepsilon = 0.1$, H = 1.0, $\lambda(\lambda = 1.0$, 2.0 and 3.0) and N (N = 1 and 3). As expected, the decreasing trend of C_2 for irrespective values of all the other parameters. It is noteworthy that the size of the concentration profile decreases as λ and N increase.

Therefore, for a lower value of N, the steepness of the boundary layer thickness is reduced.

The effects of Nb and Nt on heat transfer rate for $\varepsilon=0.2$, $Nc_1=2.0$, Le=2.0, $Nc_2=2.0$, $Sc_1=0.84441$, $Sc_2=1.14540$, Le=5.0, Rd=0.5, H=0.2 and $\lambda=0.5$ are displayed in Figure 11.10. The variation of Nusselt number with ξ is almost constant for Nb=1.5 and Nb=1.0, but an increasing trend is observed for Nb=0.5 for all values of λ. The magnitude of Nusselt number decreases due to the increase in Nb and Nt.

Figure 11.9 depicts the effects of N and λ on skin friction coefficient for $\varepsilon=0.2$, H=0.5, $Nc_1=0.1$, $Nc_2=0.1$, $Sc_1=0.84441$, $Sc_2=1.14540$, Nt=0.3, Nb=0.5, Rd=2.0 and Le=2.0. As N increases from N=1.0 to N=3.0, skin friction coefficient increases and attains maximum magnitude at $\xi=2.25$. This is because the buoyancy force raises the boundary force, which contributes to the desirable pressure gradient, thereby accelerating the fluid. As a consequence, thin hydrodynamic boundary layers form. The analytical behavior of the magnitude of the skin friction coefficient increases as λ and N increase.

Figure 11.11 illustrates the effects of N and λ on mass transfer rate $Sh_1\,(\xi\ Re_L)^{-1/2}$ for $\varepsilon=0.2$, H=0.5, $Nc_1=0.1$, $Nc_2=0.1$, $Sc_1=0.84441$, $Sc_2=1.14540$, Nt=0.3, Nb=0.5, Rd=2.0, Le=2.0, H=0.5, $Nc_1=Nc_2=0.1$, $Sc_1=0.84$ and $Sc_2=1.14$. It is noted that as N and λ increase, the rate of mass transfer enhanced the boundary layer thickness. Also, the Sherwood number increases with an increase in N and λ simultaneously. In fact, the combined viscous ratio and buoyancy force act on favorable pressure gradient of fluid flow moving on the momentum boundary layer.

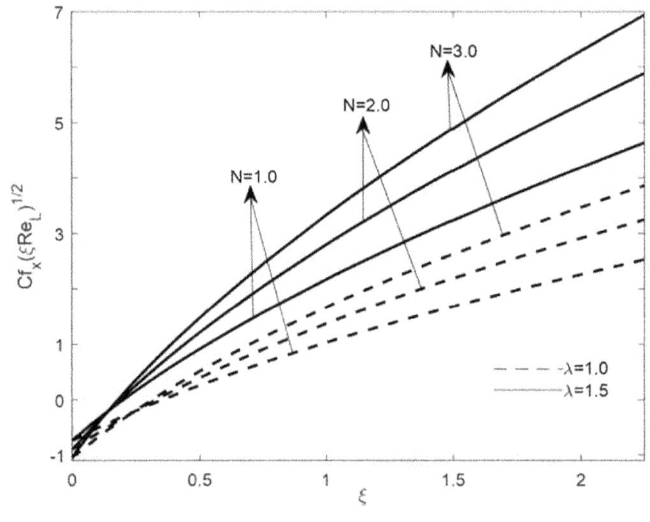

FIGURE 11.9 Effects of N and λ on skin friction coefficient for $\varepsilon=0.2$, H=0.5, $Nc_1=0.1$, $Nc_2=0.1$, $Sc_1=0.84441$, $Sc_2=1.14540$, Nt=0.3, Nb=0.5, Rd=2.0 and Le=2.0.

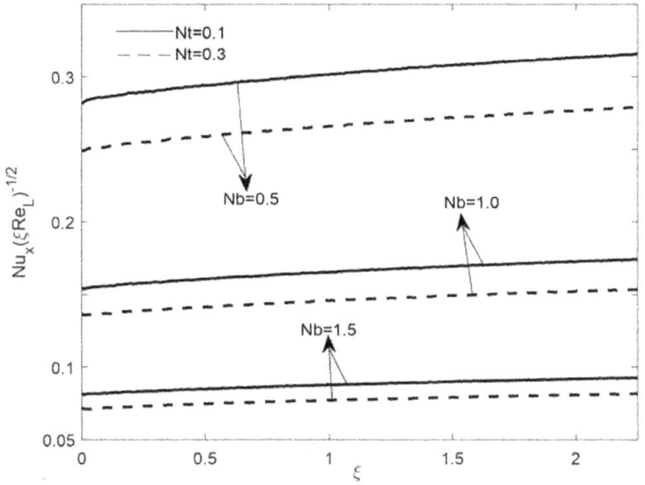

FIGURE 11.10 Nusselt number profile for $\varepsilon=0.2$, $Nc_1=2.0$, Le=2.0, $Nc_2=2.0$, $Sc_1=0.84441$, $Sc_2=1.14540$, Le=5.0, Rd=0.5, H=0.2 and $\lambda=0.5$.

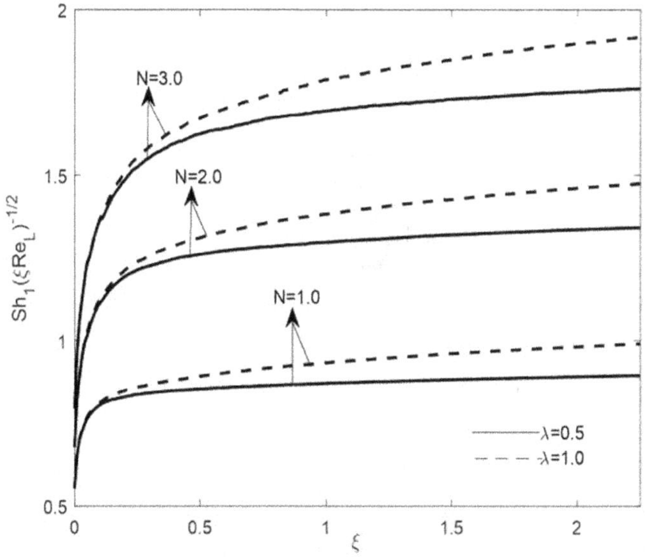

FIGURE 11.11 Mass transfer Sh_1 for $\varepsilon=0.2$, H=0.5, $Nc_1=0.1$, $Nc_2=0.1$, $Sc_1=0.84441$, $Sc_2=1.14540$, Nt=0.3, Nb=0.5, Rd=2.0 and Le=2.0.

11.6 CONCLUSIONS

The non-similar solution to steady laminar incompressible flow on MHD triple diffusion convection with effects of thermal radiation and chemical reaction of the nanofluid over a vertical plate with suitable boundary condition is analyzed. As a result of buoyancy (λ), composite velocity (ε) and buoyancy ratio parameter with concentration (Nc_1) enhanced the velocity profile. The influence of buoyancy ratio parameter, concentration (Nc_1) and thermal radiation (Rd) acts in the same direction on temperature profiles. Chemical reaction of the concentration profiles (C_1) and (C_2) decreases with the increase in viscosity (N), Brownian motion (Nb) and (∇T_w), but the reverse trend observed when there is a decrease in magnetic parameter (H). Also, the presence of viscosity ratio (N) and buoyancy force (λ) enhanced the skin friction coefficient.

REFERENCES

1. Khan Z. H., Khan W. A. and Pop I. (2013) Triple diffusive free convection along a horizontal plate in porous media saturated by nanofluid with convective boundary condition, *International Journal Heat and Mass Transfer*, 66, 603–612.
2. Bachok N., Ishak A., Pop I. (2013) Mixed convection boundary layer flow over a moving vertical flat plate in an external fluid flow with viscous dissipation effect, *PLoS ONE*, 2013, 8(4).
3. Khan Z. H., Culham J. R., Khan W. A., Pop I. (2014) Triple convective-diffusion boundary layer along a vertical flat plate in a porous medium saturated by a water-based nanofluid, *International Journal Thermal Science*, 99, 53–61.
4. Zhao M., Wang S., Zhanga, Q. (2014) Onset of triply diffusive convection in a Maxwell fluid saturated porous layer, *Applied Mathematical Modelling*, 38, 2345–2352.
5. Eid M.R. (2016) Chemical reaction effect on MHD boundary layer flow of two-phase nanofluid model over an exponentially stretching sheet with a heat generation, *Journal MolnLiquids*, 220, 718–725.
6. Ganesh Kumar K., Gireesha B. J., Manjunatha S., Rudraswam N. G. (2017) Effect of nonlinear thermal radiation on double diffusive mixed convection boundary layer flow of viscoelastic nanofluid over a stretching sheet, *International Journal Mechanical Materials Engineering*, 2017, 12–18.
7. Mahanthesh B, Gireesha B J, Rama Subba Reddy G. (2016) Heat and mass transfer effects on the mixed convective flow of chemically reacting nanofluid past a moving stationary vertical plate, *Alexandria Engineering Journal*, 55(1), 569–581.
8. Abdullaha A., Abdullha, F., Ibrahim, S. and Chamkha, A.J. (2018) Non similar solution of unsteady mixed convection flow near the stagnation point of a heated vertical plate in a porous medium saturated with nanofluid, *International Journal Physics Science*, 2018, 21, 363–388.
9. Daba, M., Devaraj, P. (2017) Unsteady double diffusive mixed convection flow over a vertically stretching sheet in the presence of suction/injection, *Journal of Applications Mechanical Technology Physics*, 58, 232–243.
10. Subhashini, S.V. and Sumathi, R. (2014) Dual solutions of a mixed convection flow of nanofluids over a moving vertical plate. *International Journal Heat and Mass Transfer*, 71, 117–124.
11. Patila, P.M., Shashikanta, A., Hiremath, P.S. (2018) Influence of liquid hydrogen and nitrogen on MHD triple diffusive mixed convection nanoliquid flow in presence of surface roughness, *International Journal Hydrogen Energy*, 2018, 43, 20101–20117.
12. Khan, U., Zaib, A., Bakar, S.A., Roy, N.C. and Ishak, A. (2021) Buoyancy effect on the stagnation point flow of a hybrid nanofluid toward a vertical plate in a saturated porous medium, *Case Studies in Thermal Engineering*, 2021, 27, 101342.
13. Patil, P.M., Shashikant, A., Momoniat, E. (2021) Analysis of sodium chloride and sucrose diffusions in mixed convective nanoliquid flow, *Ain Shams Engineering Journal*, 12, 2117–2124

14. Patil, P.M., Shashikant, A., Hiremath, P.S. (2019) Diffusion of liquid hydrogen and oxygen in nonlinear mixed convection nanofluid flow over vertical cone, *International Journal of Hydrogen Energy*, 44.
15. Muhammad Khairul Anuar, M., Mohd Zuki, S., Anuar, S. (2020) Effects of viscous dissipation on mixed convection boundary layer flow past a vertical moving plate in a nanofluid, *Journal of Advanced Research in Fluid Mechanics and Thermal Sciences* 69, Issue 2, 1–18.
16. Yashkun, U., Zaimi, K., Abu Bakar, N.A., Ishak, A. and Pop, I. (2021) MHD hybrid nanofluid flow over a permeable stretching/shrinking sheet with thermal adiation effect, *International Journal of Numerical Methods for Heat & Fluid Flow*, Vol. 31 No. 3, pp. 1014–1031.
17. D.R. Lide (Ed.) (1990) *CRC Handbook of Chemistry and Physics*, 71st ed., CRC Press, BocaRaton, FL.
18. Govindhasamy I., Singh A.K. (2020) Boundary layer flow and stability analysis of forced convection over a diverging channel with variable properties of fluids, *Heat Transfer*, 49(8), 5050–5065.
19. Govindaraj N., Singh A. K., Shukla P. (2020) MHD nanofluid flow with variable physical parameters via thermal radiation: A numerical study, *Heat Transfer*, 49(8), 202.

12 Differential Equations of Thermodynamics

Achintya Kumar Pramanick
National Institute of Technology, Durgapur

CONTENTS

12.1 MOTIVATION

The search for a generalized formulation is eternal. A theory, a formulation, or a concept that is too general is frequently too vague. The concern that corners around us is how to encapsulate the whole body of information across antagonistic avenues of facets, in a nutshell. Admittedly, it is not as easy as it sounds. But the assimilation of a large body of facts is always possible out of a single formalism, at least in principle. It is proposed that if a single postulate is valid in one field of application, the same is going to be valid in other fields of applications too, at least qualitatively. This proposition can rightly be termed as "comparing apples with oranges" or more succinctly "A to O" principle (Pramanick, 2014a). This "A to O" principle is inculcated to cultivate some formalism of thermodynamics, in general. This is at any rate by the opinion of the author; economy of words supports the process of understanding: inception, conception, and perception. In contrast to presenting large number figures or computer-generated images, it is preferable to establish the central idea in simple

DOI: 10.1201/9781003222255-12

terms (Alembert, 1777; Truesdell, 1966). It may be emphasized that the lack or even absence of figures accompanying the analysis or description actually stimulates the abstract thinking processes, which is eventually the key to the aspect of problem solving (Pramanick, 2014a). Frequently, we cannot solve problems because we fail to apprehend the situation as a chain of related solved issues and hence the purview and purpose of the following deliberations is to widen the opening of the mind (Hilbert, 1901).

12.2 INTRODUCTION

Every mathematical formulation has an underlying physical principle, and every physical principle may have a mathematical formulation. We cannot solve mathematical problems more often than not because we fail to recognize the physical principles behind a mathematical modeling. Recently, two laws of nature (thermodynamics) have been discovered: One is the "Constructal Law" by Adrian Bejan (1996) (Bejan, 1996a), and the other is the "Law of Motive Force" by Achintya Kumar Pramanick (2014) (Pramanick, 2014a). These two laws have greatly enabled physical solutions of a large variety of practical problems of interest. By virtue of the "Constructal Law," the problems of fluid mechanics, heat transfer, and thermodynamics can be brought under the single umbrella that is none other than thermodynamic optimization. On the other hand, the "Law of Motive Force" facilitates the understanding of true physical nature of problems and consequently the formulation of corresponding governing equations. This is at any rate by the opinion of the author; a plethora of problems of physical reality can be solved by the applications of Reynolds transport theorem, momentum theorem, first and second laws of thermodynamics, Gouy–Stodola theorem, Patankar–Spalding-type partial differential equations, continuity equation, Navier–Stokes equations, energy conservation equation, and $k - \varepsilon$ turbulence model, employing elementary mathematics alone.

The approximate description of nature as well as engineered system is possible through the introduction of differential equations: ordinary and partial. It is more often than not the general nature of differential equations, which is quite obscure. Without an underlying physical principle, it is impossible to accept the solution to a differential equation verbatim.

The present chapter encompasses some differential equations encountered by the sophomores of engineering, mathematical physics, and applied mathematics. With the advent of celebrated works of Adrian Bejan, the branches of thermodynamics, heat transfer, and fluid mechanics are brought under the aegis of a single umbrella of thermodynamic optimization (Bejan, 1982a).

Normally, the beginners of applied mathematics and mathematical physics learn the fundamental laws from the perspective of a systems approach, for which the inventory of mass remains fixed. Contrarily, many physical problems of practical interest demand a control volume approach, for which the entity of mass varies. For some reasons, it is advisable to reformulate all the pertinent laws of physics from the standpoint of a control volume. This in turn demands a general formulation that can pave the way for a smooth transition from a systems approach to the control volume approach and vice versa. The Reynolds transport theorem (RTT) (Reynolds,

1903; Granger, 2020; Shames, 1992) is such a mathematical apparatus that serves the purpose. The RTT is also known as Leibnitz–Reynolds transport theorem. It is a three-dimensional generalization of Leibniz integral rule or simply differentiation under the integral sign. It is to be remarked here that in the fluid mechanics literature, the systems approach is recognized as material or Lagrangian description and the control volume approach as spatial or Eulerian description although historians argue that both descriptions should be attributed to Euler (Yih, 1979).

Generally, a common person tends to get rid of mathematical intricacies of the formulation of a physical situation. On the other hand, in devoid of mathematics, quantification of either of exact or approximate nature is next to impossible. Instead of suppressing the underlying mathematics or referring them from a treatise with no immediate physical example makes the subject obscure to the beginner. The scope of the present discussion is to employ as little mathematics as possible with physical interpretations. Gradient and divergence operators are of paramount importance to describe many natural phenomena and as such will be presented.

It goes without saying that only a limited class of physical problems confers upon the so-called conservation principle. The present attempt is to fit, at least, all classes of physical problems, under the aegis of a conservation principle. It turns out to be a fact that our understanding of the conservation principle is poor, vague, and scanty (Mach, 1911; Tolhoek et al., 1952). The RTT can easily be admitted purely as a conservation principle.

Riding on the vehicle of RTT, we shall arrive at momentum theorem (Bejan, 2013), from the momentum theorem to the first law of thermodynamics for open system (control volume) (Bejan, 2016a), and from the first law of thermodynamics for control volume to the second law of thermodynamics (Bejan, 2016b). The Gouy–Stodola theorem (Gouy, 1889; Stodola, 1905) as central proposition of classical thermodynamics is deduced combining together the first and second laws of thermodynamics for control volume. It is generally to be credited the discovery of Bejan's Constructal Law (Bejan, 2016c), which accounts for the evolution of the system and not necessarily as an entropy generation minimization (EGM) principle (Bejan, 1995).

The conservation principle is also exploited to Patankar–Spalding (Patankar, 2011a) type of partial differential equation. It is to be recognized that the Law of Motive Force (Pramanick, 2014b) as proposed by the author of this chapter greatly simplifies the physical understanding of a natural process or engineered system. From the Patankar–Spalding equation alone, the conservation of chemical species, continuity equation toward the conservation of mass, energy conservation equation, and Navier–Stokes equations for the conservation of momentum are seamlessly derived.

Finally, in this present discourse, turbulence phenomenon (Frisch, 1999) is also regarded as conservation of some entity. Accordingly, the $k - \varepsilon$ turbulence model (Wilcox, 2006) is presented in line with the conservation principle.

12.3 BEJAN'S CONSTRUCTAL LAW

Generally, when we design a system with anything: animate and inanimate (fluid, heat, stress, people, money, etc.) that flows, employing laws of physics, we do not consider the time frame associated with it. But a system either natural or engineered has to evolve with time in order to exist (Bejan, 2014). The underlying law of physics was formulated by Adrian Bejan in 1996 (Bejan, 1996a; Bejan, 1996b; Bejan, 1997) and phrased as (Bejan, 2016c):

"For a flow system to persist in time (to survive) it must evolve in such a way that it provides easier and easier access to the currents that flow through it."

In view of the above law, a boundary value problem can be treated almost by ordinary algebraic calculations. As we know, an initial value problem can further be transformed to an appropriate boundary value problem by employing Laplace transform (Bellman and Roth, 1984).

Another important aspect of the application of Constructal Law is that it can solve the problems of branching networks (Rozanov, 1969) in both animate and inanimate systems almost effortlessly. It is a standard practice in the branches of applied mathematics to view branching processes as a probabilistic event (Chung, 1960; Harris, 1963), whereas contrastingly and remarkably in Constructal Law branching events are very much deterministic.

The Constructal Law is much more than the Steiner problem (Hwang et al., 1992) alone. Physical solutions to Steiner problems exist by exercising the Constructal Law considering the entire flow paths as combinations of slower paths (diffusion) and faster paths (convection) and balancing them. Three fundamental types of partial differential equations (Courant and Hilbert, 2008), namely parabolic, hyperbolic, and elliptic, can be obtained from the random walk model (Zauderer, 2006). Random walk (Pearson, 1905; Einstein and Furth, 1956; Révész, 2013) problems are in turn amenable to the formulation offered by the Constructal Law.

12.4 LAW OF MOTIVE FORCE

Another fundamental law of nature (thermodynamics) that has a self-standing is named as Law of Motive Force. The terminology "motive force" has been coined in all branches of exact sciences, non-exact sciences, philosophy, psychology, and what not. But the true meaning and implications of "motive force" have not been realized until recently. The author Achintya Kumar Pramanick started working in this direction since 1989 when he was an undergraduate student (Pramanick, 1990–1991), and the work appeared in full bloom in 2014 through the publication of a book (Pramanick, 2014a). The Law of Motive Force as phrased by the author Achintya Kumar Pramanick is as follows (Pramanick, 2014b):

"Every motive force is self-contradictory in its existence."

It is now important to quantify the "commodity" named "motive force." From a mechanistic standpoint, "motive force" can be recognized as the motion that has a changed form (Ellis, 1962). In a simple situation, it turns out to be a fact of mutual transformation between potential energy and kinetic energy. In a greater sense, the essence of the Law of Motive Force is that it represents a competition between two opposing tendencies: one favoring and the other antagonizing an event. A cause that is favoring an occurrence is termed as "forward motivation," and the element that hinders the event is described as "backward motivation" (Pramanick, 1990–91). The compelling competition between two antagonizing avenues brings about a resultant, which frequently fluctuates around a mean value. The above intuitive description can be mathematically represented as: $M_f + M_b = \overline{F}(a,b,c\ldots)$, where the term M_f stands for the forward motivation, M_b represents the backward motivation, $\overline{F}$ is the mean value of the resultant outcome, and a, b, and c are some influencing factors. The above mathematical expressions may be manifested and interpreted in a number of ways.

It can be suggested to view the Law of Motive Force as a conflict between the cause(s) and effect(s) (Bunge, 1980). Being contextual, the approximate quantitative description of a phenomenon is possible through differential equations. A differential equation in turn can only be formulated if and only if we properly identify the cause(s), which are the so-called independent variables, and effect(s), which are commonly termed as dependent variables in mathematics. In the light of above discussions, the readers may reformulate a plethora of differential equations listed by Zwillinger (1998).

12.5 REYNOLDS TRANSPORT THEOREM

Let us think of an example that can illustrate the physical essence of the RTT. Suppose we have prepared some tea in a saucepan and pour it into a cup. It is stated that the size of the saucepan is relatively large and that of the cup is comparatively small. It is quite likely that while pouring the tea from the saucepan to the cup some amount of tea will be poured inside the cup and rest of it will be spilled outside the cup. The RTT accounts for the entirety of the tea and thus emphatically belongs to a conservation principle.

Generally, in thermodynamics it is useful to distinguish between the intensive quantity, which is not dependent on the mass of the system, and extensive quantity, which depends on the mass of the system. Consider Ψ to be any extensive quantity, ψ to be the corresponding intensive quantity, and ρ to be the uniform density in a local thermodynamic equilibrium model (Glansdorff and Prigogine, 1971). For an elemental volume dv in a differential sense, the relation between Ψ and ψ can be established through a distributive measure of the type $\Psi = \int d\Psi = \iiint \psi\rho \, dv$. The rate of change of Ψ pertaining to a system is known as system derivative and denoted as $\left(\frac{d\Psi}{dt}\right)_{system} = \frac{D\Psi}{Dt}$. The job is to find out a relation of the system derivative to that pertaining to the control volume.

Next, consider a flow of fluid from left to right and into this flowing fluid, let us drop a droplet of liquid ink. This droplet of ink must have an instantaneous mass and volume. The instantaneous mass pertains to a system, and the instantaneous volume belongs to the description of control volume. It is quite obvious that to start with at the initial instant of time t the system and control volume overlap each other. Only after a small time lag of Δt, which means at the time step $t + \Delta t$, the spatial position of the system and control volume will differ. Thus, at the time step $t + \Delta t$ the control surfaces of the system and the control volume will make a crossover.

This translates into a geometric situation that comprises of three regions, namely I, II, and III. Region II is common to both the system and control volume. Region I is bounded by a convex curve through which the fluid enters the control volume with a uniform horizontal velocity $\vec{V}$. On the other hand, Region III is bounded by a concave curve through which the fluid leaves the control volume with a uniform horizontal velocity $\vec{V}$.

We can now calculate the rate of change of Ψ for the system as a limiting process as follows:

$$\left(\frac{d\Psi}{dt}\right)_{system} = \frac{D\Psi}{Dt}$$
$$= \underset{\Delta t \to 0}{Lt}\left[\frac{\left(\iiint\limits_{II} \psi\rho\, dv - \iiint\limits_{III} \psi\rho\, dv\right)_{t+\Delta t} - \left(\iiint\limits_{I} \psi\rho\, dv + \iiint\limits_{II} \psi\rho\, dv\right)_{t}}{\Delta t}\right]. \quad (12.1)$$

Applying now the rule that the sum of limits equals the limits of the sums, equation (12.1) assumes the following form:

$$\frac{D\Psi}{Dt} = \underset{\Delta t \to 0}{Lt}\left[\frac{\left(\iiint\limits_{II} \psi\rho\, dv\right)_{t+\Delta t} - \left(\iiint\limits_{II} \psi\rho\, dv\right)}{\Delta t}\right]$$
$$+ \underset{\Delta t \to 0}{Lt}\left[\frac{\left(\iiint\limits_{III} \psi\rho\, dv\right)_{t+\Delta t}}{\Delta t}\right] - \underset{\Delta t \to 0}{Lt}\left[\frac{\left(\iiint\limits_{I} \psi\rho\, dv\right)_{t}}{\Delta t}\right]. \quad (12.2)$$

The first term on the right-hand side of equation (12.2) is an event that occurs completely inside the control volume and it admits an immediate differentiation given as below:

$$\frac{\partial}{\partial t}\iiint\limits_{cv} \psi\rho\, dv = \underset{\Delta t \to 0}{Lt}\left[\frac{\left(\iiint\limits_{II} \psi\rho\, dv\right)_{t+\Delta t} - \left(\iiint\limits_{II} \psi\rho\, dv\right)_{t}}{\Delta t}\right]. \quad (12.3)$$

where the subscript "cv" under the sign of triple integral refers to the control volume. The second term on the right-hand side indicates an event commencing across the convex control surface in the direction of the flow from left to right. Such an occurrence is known as efflux, meaning some amount of fluid is exiting the control volume through the control surface. Similarly, the third and last term in equation (12.2) is related to the fact that some amount of fluid is entering the control volume through the control surface that is concave. This event is recognized as influx, indicating some amount of fluid enters the control volume through the control surface. The difference in the second and the third terms of equation (12.2) is labeled as the net rate of efflux of the extensive quantity Ψ out of the control volume as written below:

The net rate of efflux of Ψ from the control volume

$$= \underset{\Delta t \to 0}{Lt}\left[\frac{\left(\iiint\limits_{III} \psi\rho\, dv\right)_{t+\Delta t}}{\Delta t}\right] - \underset{\Delta t \to 0}{Lt}\left[\frac{\left(\iiint\limits_{I} \psi\rho\, dv\right)_{t}}{\Delta t}\right]. \tag{12.4}$$

Here, the terminologies "influx" and "efflux" are far-reaching. It generally means all types of physical quantities, for example electric and magnetic fluxes (Maxwell, 1954).

Since the second and the third terms of equation (12.2) are related to some boundary phenomena, it would be mathematically useful to convert the volume integrals into surface integrals. It is to be noted that the elemental volume dv of fluid that occupies the region swept by the elemental area dA in small time step dt forms an elemental stream tube. Let the direction of horizontal flow velocity $\vec{V}$ from the left to right be positive. Then for the right-hand side control surface of Region III we can write elemental volume dv considering the vector dot products between velocity and area vector as:

$$(dv)_{III} = \left(\vec{V}.\vec{d}A\, dt\right)._{III} \tag{12.5}$$

It is to be noted that the surface for Region III is concave in the direction of flow. In view of equation (12.5) the first term of equation (12.4) assumes the form:

$$\iiint\limits_{III} \psi\rho\, dv = \iint\limits_{III} \psi\left(\rho\vec{V}.d\vec{A}\right). \tag{12.6}$$

Similarly, for the left-hand side control surface of Region I, we can write elemental volume dv considering the vector dot products between velocity and area vector as:

$$(dv)_{I} = -\left(\vec{V}.\vec{d}A\, dt\right)._{I} \tag{12.7}$$

It is to be remarked here that the surface for Region I is convex in the direction of flow. In view of equation (12.7), the second and last term of equation (12.4) assumes the form:

$$\iiint\limits_{I} \psi\rho\, dv = -\iint\limits_{I} \psi\left(\rho\vec{V}.d\vec{A}\right). \tag{12.8}$$

Collecting the terms from equations (12.6) and (12.8), we have:

$$\underset{\Delta t\to 0}{Lt}\left[\frac{\left(\iiint\limits_{III}\psi\rho\, dv\right)_{t+\Delta t}}{\Delta t}\right]-\underset{\Delta t\to 0}{Lt}\left[\frac{\left(\iiint\limits_{I}\psi\rho\, dv\right)_{t}}{\Delta t}\right]$$
$$=\iint\limits_{III}\psi\left(\rho\vec{V}.d\vec{A}\right)-\left[-\iint\limits_{I}\psi\left(\rho\vec{V}.d\vec{A}\right)\right]=\oiint\limits_{cs}\psi\left(\rho\vec{V}.d\vec{A}\right) \quad (12.9)$$

where the subscript "cs" under the double integral refers to the control surface.

Providing information from equations (12.3) and (12.9) into (12.2), we finally arrive at:

$$\left(\frac{d\psi}{dt}\right)_{system}=\frac{D\Psi}{Dt}=\frac{\partial}{\partial t}\iiint\limits_{cv}\psi\rho\, dv+\oiint\limits_{cs}\psi\left(\rho\vec{V}.d\vec{A}\right). \quad (12.10)$$

Equation (12.10) is recognized as RTT and enables us to carry out the formulation from the systems approach to the control volume approach. A variety of practical applications of equation (12.10) can be found in the engineering literature (Granger, 2020; Shames, 1992).

12.6 MOMENTUM THEOREM

A class of problems in classical mechanics (Dugas, 1988; Sommerfeld, 1964; Goldstein et al., 2002) can be solved by Newton's second law of motion, and it is stated as (Chandrasekhar, 2020):

> **"The change of motion is proportional to the motive force impressed; and is made in the direction of the right line in which that force is impressed."**

Let M be the total mass of the system and V_x the component in the horizontal direction due to the sum of impressed forces $\sum F_x$ resolved along the same direction. Then according to Newton's second law of motion, we arrive at the mathematical expression:

$$\frac{\partial}{\partial t}(MV_x)=\sum F_x. \quad (12.11)$$

For an alternative as well as philosophical discussion on force, readers can refer to the treatises by Hertz (1899) and Lützen (2010).

Equation (12.11) is of limited application as is only concerned with a system. But, in principle, equation (12.11) should be equally applicable to a flow system as well. In turn, it demands the extension of the formulation presented in equation (12.11) via RTT.

Consider now a control volume that has a velocity component V_x in the horizontal direction and an instantaneous mass entrapped M, and the sum of all forces resolved along the horizontal direction is $\sum F_x$. The control volume may have a number of openings through which some amount of fluid may be inducted, and such an opening on the control surface is termed as "inlet port" or simply "inlet." Similarly, through a number of outlets on the control surface, some amount of fluid may be exhausted, and such an opening is labeled as "outlet port" or more succinctly "outlet." In view of this, intuitively, equation (12.11) for the system has an easy extension to that of control volume as:

$$\frac{\partial}{\partial t}(MV_x)_{cv} = \sum_{in} \dot{m}v_x - \sum_{out} \dot{m}v_x + \sum F_x \tag{12.12}$$

where $\dot{m} = \frac{dm}{dt}$ is the average fluid flow rate through the inlets and outlets and v_x is the velocity component of fluid flows through the inlets and outlets resolved along any horizontal direction.

It is easy to observe that the first two terms on the right-hand side bear a one-to-one correspondence with the term on the left-hand side. It is to be remarked that equation (12.12) is a momentum balance equation and hence belongs to a conservation principle. In what follows, qualitatively, equation (12.12) translates into a fact that a change term balances a transfer term, or in other words, the change term is in competition with the transfer term. Thus, without loss of any generality, we can write, at least qualitatively, for all classes of transport problems pertaining to some conservation principles as:

$$\textit{Rate of Change Term} = \textit{Rate of Transfer Terms}. \tag{12.13}$$

Further, equation (12.12) talks about the transport of momentum and hence a transport equation. The currency of transport here is momentum. The quantitative equivalence between the two apparently non-similar entities, namely the "force" and the "momentum," is made possible by virtue of Newton's second law of motion. Equation (12.12) is a clear-cut exposition of RTT and is called momentum theorem (Bejan, 2013). For all engineering calculations, equation (12.12) provides a fairly accurate account.

12.7 FIRST LAW OF THERMODYNAMICS

Equation (12.2) not only casts merely the momentum theorem, but also represents all classes of transport equations. Let us have a second look at equation (12.12), and it is easy to recognize that the "currency" of transport in the momentum theorem is momentum. The momentum is an extensive (total) quantity, and the corresponding intensive (specific) quantity is velocity. In view of this argument, equation (12.12) can be qualitatively rewritten as:

$$\frac{\partial}{\partial t}\left(Total\ Currency\right)_{cv} = \sum_{in} \dot{m}(Specific\ Currency) - \sum_{out} \dot{m}(Specific\ Currency) + \sum Nonsimilar\ Currencies. \tag{12.14}$$

The first law of thermodynamics is historically explored (Bejan, 2016e; Epstein, 1937) in connection with power generation. A heat engine is fed with some amount of total heat input rate $\dot{Q}$ that yields $\dot{W}$ rate of work output. In heat engine notion (Truesdell and Bharatha, 1977), the heat provided is taken as positive. As generally uppercase letters are taken for extensive quantities and the corresponding lowercase letters for intensive quantities, we consider here E to be the total energy and e the corresponding specific energy. Recognizing the first law of thermodynamics as energy transport equation, we have for equation (12.14) as:

$$\frac{\partial}{\partial t}(E)_{cv} = \sum_{in} \dot{m}e - \sum_{out} \dot{m}e + \sum_{in} \dot{Q} - \sum_{out} \dot{W}. \tag{12.15}$$

For most of the engineering applications, the specific energy is a shorthand and can be written as:

$$e = h + \frac{1}{2}V^2 + gz \tag{12.16}$$

where h represents the specific enthalpy, V is the component of velocity in the direction considered, g is the gravitational acceleration, and z is the average height of the appliance above datum level. Traditionally, we consider a single heat input source and single work output; equations (12.15) and (12.16) can be combined to cast as:

$$\frac{\partial}{\partial t}(E)_{cv} = \sum_{in} \dot{m}\left(h + \frac{1}{2}V^2 + gz\right) - \sum_{out} \dot{m}\left(h + \frac{1}{2}V^2 + gz\right) + \dot{Q} - \dot{W}. \tag{12.17}$$

When the inlet and outlet ports are closed, equation (12.17) reduces to the mathematical apparatus of a system instead of control volume and can be represented as:

$$\frac{\partial}{\partial t}(E)_{cv} = \dot{Q} - \dot{W}. \tag{12.18}$$

It is to be remarked that the generalization of the first law of thermodynamics from the system to the control volume was also presented by Prigogine (1947) and de Groot (1951), but on different footings. A large number of applications as well as conceptual issues of equation (12.17) can be found from Bejan (2016f). The first law of thermodynamics from the perspective of differential equations is nicely treated by Sychev (1983).

12.8 SECOND LAW OF THERMODYNAMICS

From a large number of experiments (Wien et al., 1929), it is known that the entity "work" can be converted completely into "heat," but the contrary is not true.

Generally, in engineering community, the experiments in this regard due to Rumford and Joule are much popular (Saha and Srivastava, 1931). Then, obviously, equation (12.13) does not hold good, and instead, it can be accommodated via an inequality of the form:

$$\textit{Rate of Change Term} > \textit{Rate of Transfer Terms}. \tag{12.19}$$

Admittedly, equation (12.19) does not fit into the framework of a conservation principle. Equations (12.13) and (12.19) can be combined to yield:

$$\textit{Rate of Change Term} \geq \textit{Rate of Transfer Terms}. \tag{12.20}$$

By introducing some non-negative accommodating factor ε, equation (12.20) can be fitted to look like a conservation principle as:

$$\textit{Rate of Change Term} = \textit{Rate of Transfer Terms} + \varepsilon \tag{12.21}$$

where, by definition, $\varepsilon \geq 0$. Thus, while transforming heat into work, some entity is being created.

In analogy with transport equation (12.17), the second law of thermodynamics can be treated as the transport of entropy. Until recently, there has been no rigorous mathematical interpretation of entropy in macroscopic formalism, and this was possible due to Leib and Yngvason (1999). An affordable discussion on the Leib–Yngvason principle to the sophomore can be availed from the lucid work of Thess (2011). The terminology entropy has been coined in all branches of knowledge (Landsberg, 1961). In layman's terms, the entropy can be regarded as a competition between the quantity and quality of an entity. In particular, in the field of classical thermodynamics, entropy can be treated as the competition between the quantity of heat Q and quality of heat, that is temperature T. This competition is represented in ratio form as: $dS = \frac{\delta Q}{T}$, where the symbol δ stands for inexact differential, S is the total entropy, and T is the average (arithmetic mean) temperature at which the heat flows. For most of the engineering applications, the ultimate heat flow reservoir (sink) is the environment whose temperature being fairly constant is denoted as T_0. It can be seen from the definition of entropy alone that for a definite amount of heat (δQ) to flow from high to low temperature, entropy has to increase. In what follows, for an amount of heat transfer rate $\dot{Q}$ at the environmental sink with temperature T_0, the corresponding entropy transfer rate is $\dot{S} = \frac{\dot{Q}}{T_0}$ Owing to the fact that complete conversion of heat is not possible, there is accumulation of an entity termed as entropy while converting heat into work. Contrastingly, the entropy transfer associated with the rate of transfer of work $\dot{W}$ is zero as complete transformation of work into heat is possible, and hence, there is no accumulation or depletion of any entity. From this standpoint of entropy production alone work is treated as the highest grade of energy.

Combining these embedded arguments, we gather from equations (12.17) and (12.20) as:

$$\frac{\partial}{\partial t}(S)_{cv} \geq \sum_{in} \dot{m}s - \sum_{out} \dot{m}s + \frac{\dot{Q}}{T_0} - 0,$$

where dropping the non-contributing last term on the right-hand side, we write:

$$\frac{\partial}{\partial t}(S)_{cv} \geq \sum_{in} \dot{m}s - \sum_{out} \dot{m}s + \frac{\dot{Q}}{T_0}. \tag{12.22}$$

Conforming equation (12.22) into the framework of equation (12.21), we arrive at:

$$\frac{\partial}{\partial t}(S)_{cv} = \sum_{in} \dot{m}s - \sum_{out} \dot{m}s + \frac{\dot{Q}}{T_0} + \dot{S}_{gen} \tag{12.23}$$

where $\dot{S}_{gen}$ is recognized as entropy generation rate in the literature (Bejan, 1982b) and can be recast as:

$$\dot{S}_{gen} = \frac{\partial}{\partial t}(S)_{cv} - \sum_{in} \dot{m}s + \sum_{out} \dot{m}s - \frac{\dot{Q}}{T_0}. \tag{12.24}$$

Physically, the term entropy generation implies the degree of irreversibility and it is not unbounded (Planck, 1998). It may require a passing mention that the equality sign holds good for reversible processes and inequality sign holds good for irreversible processes. Equations (12.22) and (12.24) can be reformulated for the system by closing the ports and assume the respective forms as:

$$\frac{\partial}{\partial t}(S)_{cv} \geq \frac{\dot{Q}}{T_0} \tag{12.25}$$

and

$$\dot{S}_{gen} = \frac{\partial}{\partial t}(S)_{cv} - \frac{\dot{Q}}{T_0}. \tag{12.26}$$

Interesting as well as applied discussions on the second law of thermodynamics can be found from Bejan (2016g). An alternative formulation of the second law of thermodynamics that results in some partial differential equations is due to Carathéodory (1909) and subsequent criticism due to Born (1921). An account of these developments can also be found in Landé (1926). However, for a lucid discussion at the level of sophomore, the reader is referred to the treatise of Sneddon (2006).

12.9 GOUY–STODOLA THEOREM

There are many misunderstandings in the scientific community (Bejan, 1996c; Salamon et al., 1980) since the first and second laws of thermodynamics are not placed side by side. This communication gap was adequately abridged by Bejan (1996c). Concerning the conversion of heat into work, two opposite events take place: one is the production of work that is the quality upgradation of heat energy $\left(Q_{upgradation}\right)$, and the other is the generation of entropy that is the quality degradation of heat energy $(Q_{\text{deg}radation})$. The first law of thermodynamics dictates that the total quantity of heat as energy is conserved:

$$Q_{upgradation} + Q_{\text{deg}radation} = W + T\left(\frac{Q}{T}\right) = W + TS = \text{Constant}.$$

Thus, it fits in the framework of the Law of Motive Force (Pramanick, 2014b). The entities heat and entropy are two ever competing forces in nature (Alekseev, 1986; Müller and Weiss, 2005).

From the perspective of heat engine, the quantity of heat transfer rate $\dot{Q}$ is generally treated as a floating quantity. From the standpoint of mathematics, it implies an elimination of $\dot{Q}$ between equations (12.17) and (12.22) that leads to the following result:

$$\dot{W} \leq \sum_{in} \dot{m}\left(h + \frac{1}{2}V^2 + gz - T_0 s\right) - \sum_{out} \dot{m}\left(h + \frac{1}{2}V^2 + gz - T_0 s\right) - \frac{\partial}{\partial t}(E - T_0 S)_{cv}. \quad (12.27)$$

Equation (12.27) admits the natural optimum that is maxima, and the event of maximum work production rate $\dot{W}$ is:

$$\dot{W}_{\max} = \sum_{in} \dot{m}\left(h + \frac{1}{2}V^2 + gz - T_0 s\right) - \sum_{out} \dot{m}\left(h + \frac{1}{2}V^2 + gz - T_0 s\right) - \frac{\partial}{\partial t}(E - T_0 S)_{cv}. \quad (12.28)$$

Equation (12.27) is applicable to both reversible and irreversible processes, whereas equation (12.28) is exclusively relevant to reversible processes. The difference of level of work production rates provided by equations (12.27) and (12.28) is known as the lost available work (LAW) and is given by:

$$\dot{W}_{lost} = \dot{W}_{\max} - \dot{W}. \quad (12.29)$$

Providing expression for $\dot{W}_{\max}$ from equation (12.28) and another expression for $\dot{W}$ from equation (12.17) into equation (12.29), we arrive at:

$$\dot{W}_{lost} = T_0\left[\frac{\partial}{\partial t}(S)_{cv} - \sum_{in} \dot{m}s + \sum_{out} \dot{m}s - \frac{\dot{Q}}{T_0}\right]. \quad (12.30)$$

In view of equation (12.24), equation (12.30) takes the form:

$$\dot{W}_{lost} = T_0 \dot{S}_{gen}. \quad (12.31)$$

Recognizing T_0 is fairly constant being environmental temperature, equation (12.31) turns out to be:

$$\dot{W}_{lost} \propto \dot{S}_{gen}. \quad (12.32)$$

In the pertinent literature (Bejan, 2016h), equation (12.32) is recognized as the Gouy–Stodola theorem. The proportionality constant in equation (12.32) is not always the environmental constant; it depends on the constitution and makeup of the system. In contrast to the old and prevalent literature (Bejan, 2016h; Chambadal, 1963; Szargut and Petela, 1965), an extensive as well as modern treatment is due to Bejan (2016h).

12.10 PATANKAR–SPALDING-TYPE DIFFERENTIAL EQUATION

In principle, equation (12.20) accommodates all classes of physical problems in the form of both a conservation and non-conservation principle. The paramount importance of the Law of Motive Force (Pramanick, 2014b) is that it admits a competition even between non-similar entities such as space and time (Borel, 2003). It was also set forth in the treatise (Pramanick, 2014b) that slower motion in nature originates first compared to the faster motion. In fact, the slower motion gets converted into the faster motion, and thus, they are in competition with each other. In view of this argument, equation (12.20) can be expounded in somewhat more explicit form of a conservation principle as:

$$\begin{aligned}&\text{Time-Dependent Processes} + \text{Faster Processes}\\&= \text{Slower Processes} + \text{Time-Independent Processes.}\end{aligned} \tag{12.33}$$

In the light of the Law of Motive Force (Pramanick, 2014b), it is easy to regard equation (12.33) purely as conservation formalism. In reality, each term of equation (12.33) competes with each other, rendering a very complex situation. In other words, equation (12.33) serves as a transport equation. This qualitative relation (12.33) demands now quantification in order to serve the purpose of exact sciences.

Without loss of any generality, let us consider an elemental control volume in an orthogonal rectangular Cartesian coordinate system $X - Y - Z$ with elementary lengths dx, dy, and dz along X-, Y-, and Z-directions, respectively. A flux-like vector quantity $\vec{F}$ can be represented as $\vec{F} = \vec{i}F_x + \vec{j}F_j + \vec{k}F_z$, where F_x, F_y, and F_z are the scalar components of $\vec{F}$ along the respective directions X, Y, and Z with unit vectors $\vec{i}$, $\vec{j}$, and $\vec{k}$. Thus, along the $Y - Z$ face, two opposite events take place: one is the influx of any physical quantity, and the other is the efflux of the same. It is easy to visualize that along the $Y - Z$ face, the total influx is $F_x dydz$ Admitting a Taylor's series variation of the flux inside along the X-direction of the control volume and retaining only the first–order terms, we have for the total efflux along the same $Y - Z$ face as $\left(F_x + \frac{\partial F_x}{\partial x}dx\right)dydz$. The net efflux (efflux minus influx) along the X-direction through the $Y - Z$ face is $\left(F_x + \frac{\partial F_x}{\partial x}dx\right)dydz - F_x dydz = \frac{\partial F_x}{\partial x}dxdydz$. Repeating the discussions for $Z - X$ and $X - Y$ planes of the control volume, the analogous quantities are $\frac{\partial F_y}{\partial y}dxdydz$ and $\frac{\partial F_z}{\partial z}dxdydz$, respectively. So the net efflux per unit control volume is the sum total of $\frac{\partial F_y}{\partial y}dxdydz$, $\frac{\partial F_y}{\partial y}dxdydz$, and $\frac{\partial F_y}{\partial z}dxdydz$. Thus, the net efflux per unit control volume is $\frac{\partial F_x}{\partial x}+\frac{\partial F_y}{\partial y}+\frac{\partial F_z}{\partial z}$ The resulting expression is thus a scalar quantity that emerged out. It is possible to cast this scalar quantity as a dot product of two vector quantities as:

$$\frac{\partial F_x}{\partial x}+\frac{\partial F_y}{\partial y}+\frac{\partial F_z}{\partial z}=\left[\vec{i}\frac{\partial}{\partial x}(\ )+\vec{j}\frac{\partial}{\partial y}(\ )+\vec{k}\frac{\partial}{\partial z}(\ )\right]\cdot\left[\vec{i}F_x+\vec{j}F_y+\vec{k}F_z\right]=div\left(\vec{F}\right).$$

This vector dot product is known as the "divergence" operation and denoted by "div" or "$\vec{\nabla}$". Physically, it represents net efflux per unit control volume. The analogous counterpart operation of the "del operator" $\left(\vec{\nabla}\right)$ applied to any scalar quantity, say F_x on any direction, looks like $\left[\vec{i}\frac{\partial}{\partial x}(\ )\right][F_x]=\vec{i}\frac{\partial}{\partial x}(F_x)$, which is clearly a flux-like quantity. This vectorial operation is termed as "gradient" and denoted as "grad" or "$\vec{\nabla}$". In general, the gradient operation on a scalar quantity S in three-dimensional Cartesian coordinate system assumes the form: $\left[\vec{i}\frac{\partial}{\partial x}(\ )+\vec{j}\frac{\partial}{\partial y}(\ )+\vec{k}\frac{\partial}{\partial z}(\ )\right]S=\vec{i}\frac{\partial}{\partial x}(S)+\vec{j}\frac{\partial}{\partial y}(S)+\vec{k}\frac{\partial}{\partial z}(S)=grad(S)$. Thus, the gradient operation simply indicates a tendency (on the verge of a flow of any entity) of a system. For an insightful discussion on "divergence" and "gradient" operations, the reader may lay hands on the treatment of Milne-Thomson (1974). Contextually, readers are also referred to the derivation of Ostrogradski's formula alternatively also known as Green's lemma, or Gauss's theorem by Sobolev (1989). Although the physical meaning and implications of divergence and gradient are furnished with reference to rectangular Cartesian coordinate system, it can easily be conceived in any other orthogonal coordinate systems as well (Mikhailov and Özişik, 1994; Pipes and Harvill, 2014).

The slower flow process is always a gradient-dependent process and usually proportional to the gradient, for example the diffusion of heat that is the propagation of heat in the interior of solids (Fourier, 1955a). The diffusion process in turn demands a non-zero diffusion coefficient. On the other hand, the faster processes are always velocity-dependent processes, such as the heat transfer due to convection (Poisson, 1855).

Now, we are in a position to supplement the mathematical expression for each of the four qualitative terms in equation (12.33). For the intensive quantities ψ and ρ to be of the uniform density in a local thermodynamics model (Glansdorff and Prigogine, 1971), we immediately have: $Time-Dependent\,\mathrm{Pr}ocesses=\frac{\partial}{\partial t}(\rho\psi)$. In terms of the intensive quantity ψ, local uniform density ρ, and average velocity $\vec{v}$, we can write: $Faster\,\mathrm{Pr}ocesses=div(\rho\psi\,\vec{v})$. With the introduction of non-vanishing diffusion coefficient C_ψ for the intensive transport quantity ψ, we can furnish: $Slower\,\mathrm{Pr}ocesses=div(C_\psi\,grad\,\psi)$. At this stage of development, let the fourth and remaining term be: $Time-Independent\,\mathrm{Pr}ocesses=S_\psi$. Coining mathematical expressions for all the four terms and substituting in equation (12.33), we finally arrive at the following differential equation:

$$\frac{\partial}{\partial t}(\rho\psi)+div(\rho\psi\vec{v})=div\left(C_\psi grad\psi\right)+S_\psi. \tag{12.34}$$

Equation (12.34) may be appropriately named as Patankar–Spalding equation (Patankar, 2011a) with due credit to Suhas V. Patankar and Dudley Brian Spalding. A lucid numerical study of equation (12.34) is available in the open literature (Patankar, 2011a).

In principle, equation (12.34) is applicable at least qualitatively to a very large class of physical problems. It is only important to recognize the appropriate meaning of intensive quantity ψ, identification of the law of diffusion $\left(C_{\psi}\, grad\psi\right)$, and the pertinent nature of time-independent processes $\left(S_{\psi}\right)$ in a case-specific manner.

12.11 CONSERVATION OF CHEMICAL SPECIES

A class of problems of practical interest pertaining to chemical equations can be studied fundamentally (Le Chatelier, 1884, 1887; Braun, 1887). Such chemical equations are amenable to general differential equation as portrayed in equation (12.34). Appropriately, we first recognize the definition of ψ in this case. It can be the mass fraction denoted by f_m for a chemically reactive mixture. The law of diffusion of chemical species can be attributed to Fick's law (Fick, 1855) of diffusion and is treated as a gradient-dependent phenomenon: $\vec{F} = C_m \left|gradf_m\right|$. The time-independent processes are due to the resultant production, destruction, or neither of these, and this can be labeled as a scalar term S_m. Gathering these information for equation (12.34), we arrive at:

$$\frac{\partial}{\partial t}\left(\rho f_m\right) + div\left(\rho f_m \vec{v}\right) = div\left(C_m gradf_m\right) + S_m. \tag{12.35}$$

12.12 CONTINUITY EQUATION

In order to arrive at some consolidated account of comprehension of a large class of physical problems, it is advisable to consider simpler situations, at the outset. Such a simplification can begin with equation (12.35). Accordingly, in order to serve our purpose, we choose a single component and chemically non-reactive system that renders the values $f_m = 1$ and $S_m = 0$. Thus, equation (12.35) assumes a simplified form as:

$$\frac{\partial}{\partial t}(\rho) + div\left(\rho \vec{v}\right) = 0. \tag{12.36}$$

Equation (12.36) in the pertinent literature is recognized as the continuity equation. Equation (12.36) precisely describes the conservation principle for mass. Equation (12.36) was derived by Sommerfeld (1950) through physical consideration of a deformable body. Equation (12.36) presented here is in differential form. An account of equation (12.36) in integral form is furnished by Bateman (1932) by the application of Green's theorem.

12.13 NAVIER–STOKES EQUATIONS

Equation (12.34) can also be explored to seek the meaning and implications of the conservation of momentum. In accordance with the meaning of ψ in this context is velocity (specific momentum) $\vec{v}$ with its scalar components u, v, and w along X-, Y-, and Z-directions, respectively. Physically, the momentum is actually diffused (Schlichting and Gersten, 2000) through shear stress, and for a large class of fluid, it can be represented by Newton's law of viscosity (Newton, 1999) μ of the form: $F_x = \mu \frac{\partial u}{\partial x}$, where F_x is the shear stress along the $X - Y$ plane, μ is the dynamic coefficient of viscosity, and u is the velocity in X-direction. The origin of fluid motion primarily may be due to shear-driven Couette flow (Stokes, 1856; Rayleigh, 1911), pressure-driven Poiseuille flow (Poiseuille, 1840, 1841), or a combination of these two. Further, for a fluid element to be in a flow, there is a continuous drop in pressure gradient. Generally speaking, for many practical problems, the fluid element is also exposed to the gravitational field. Thus, it turns out to be a fact that the time-independent term assumes the form $-\frac{\partial P}{\partial x} + F_{b,x} + F_{s,x}$ along the X-direction considering all possible generalities, where P stands for pressure, $F_{b,x}$ is the body forces resolved along the X-direction, and $F_{s,x}$ is the surface forces resolved along the X-direction. Thus, the conservation of momentum along the X-direction appears as:

$$\frac{\partial}{\partial t}(\rho u) + div(\rho u \vec{v}) = div(\mu grad u) - \frac{\partial P}{\partial x} + F_{b,x} + F_{s,x}. \tag{12.37}$$

Analogously, the conservation of momentum along the Y-direction stands as:

$$\frac{\partial}{\partial t}(\rho v) + div(\rho v \vec{v}) = div(\mu grad v) - \frac{\partial P}{\partial z} + F_{b,y} + F_{s,y}. \tag{12.38}$$

Similarly, the conservation of linear momentum along the Z-direction is furnished as:

$$\frac{\partial}{\partial t}(\rho w) + div(\rho w \vec{v}) = div(\mu grad w) - \frac{\partial P}{\partial z} + F_{b,z} + F_{s,z}. \tag{12.38}$$

In the relevant literature, equations (12.37), (12.38), and (12.39) are recognized as Navier–Stokes equations (Navier, 1827; Stokes, 1845). A compact account of Navier–Stokes equation is available in Batchelor (2009), and Landau and Lifshitz (2005), whereas a critical account of it is due to Birkhoff (1960). It is worthy to mention that the so-called Navier–Stokes equations are primarily due to Navier alone (Navier, 1827; Poission, 1831; Fourier, 1833) and were derived by employing assumptions about intermolecular forces. The same equations were arrived macroscopically by St. Venant (De Saint-Venant, 1843) in 1843 and then by Stokes (1845) in 1845.

12.14 ENERGY CONSERVATION EQUATION

Equation (12.34) can further be inculcated for the conservation of energy. At the outset, it is to be remarked that heat transmission primarily takes place in the form of heat conduction that is the slower motion, which originates first (Pramanick, 2014b). So the most natural variable that accounts for the energy transport in the mode of heat can be most suitably chosen as enthalpy. Therefore, in this situation, we can write: $\psi \equiv h$, where h is the specific enthalpy of the heat transporting fluid. The law of diffusion of heat can be attributed to Fourier's law of conduction (Fourier, 1955b) and is treated as a gradient-dependent phenomenon: $\vec{F} = C_h|gradh| = C_h|gradc_pT| = \frac{C_h}{c_p}|gradT| = k|gradT|$, where C_h is some constant, c_p is the specific heat at constant pressure of the heat-carrying fluid media, and T is the temperature of the fluid. In obtaining the above expression, we have assumed a temperature-independent specific heat at constant pressure (Kondepudi and Prigogine, 2015) and as such specific enthalpy h scales with temperature T that is $h \propto T$. Accordingly, equation (12.34) assumes the form:

$$\frac{\partial}{\partial t}(\rho h) + div(\rho h\vec{v}) = div(kgradT) + S_h. \tag{12.39}$$

In the pertinent literature, equation (12.39) is known as energy conservation equation for a moving medium. Alternative and comprehensive derivations of equation (12.39) are available in the published literature and are due to Kutatelazde (1963), Luikov (1968), Eckert and Drake (1972), Schneider (1955), Özisik (1989), and Bird et al. (1960).

12.15 THE $k - \varepsilon$ TURBULENCE MODEL

Turbulence is a complex fluid motion, and a descriptive account due to Leonardo da Vinci is available in Tokaty (1994). Generally, such an obscure motion of fluid is studied employing statistical tools (Monin and Yaglom, 1975; Lin, 1961). Still riding on the "A to O" principle, a macroscopic description of turbulence is quite possible. First, following the Law of Motive Force (Pramanick, 2014b), we identify two opposing tendencies (competitive mechanisms) that are present in turbulence production. Intuitively, one mechanism is favoring the generation and the other is antagonizing the turbulence. The concept introduced by Prandtl (1945) is a valuable means to understand turbulence from the physical perspective. The mechanism of growth of turbulence can be inferred from the work of Chou (1945). On the other hand, the mechanism of decay of turbulence can be reasoned out from the contribution of Harlow and Nakayama (1968).

In what follows, equation (12.34) can be cast in the following form to account for the phenomenological aspect of turbulent motion:

$$\frac{\partial}{\partial t}(\rho k) + div(\rho k\vec{v}) = div(C_k gradk) + G - \rho\varepsilon. \tag{12.40}$$

In equation (12.40), k is the specific kinetic energy of the turbulent motion and synonymous to ψ, C_k is the diffusion coefficient for the specific kinetic energy of the turbulent motion, G is the rate of generation of turbulent kinetic energy, and ε is the kinematic rate of dissipation of turbulence energy. In particular, the respective terms in equation (12.34) assume the form: $\frac{\partial}{\partial t}(\rho\psi) = \frac{\partial}{\partial t}(\rho k)$, $div(\rho\psi\vec{v}) \equiv div(\rho k\vec{v})$, $div(C_\psi grad\psi) \equiv div(C_k gradk)$, and $S_\psi \equiv G - \rho\varepsilon$. In order to complement equation (12.40), an accompanying equation for ε is needed (Patankar, 2011b). The equation for ε is analogous to equation (12.40): it is merely a conservation equation in ε. Two equations: one for the conservation of k and the other for the conservation of ε, constitute the complete model and are known as two-equation $k - \varepsilon$ model.

A good agreement of the $k-\varepsilon$ model is reported by Jones and Launder (1972). The two-equation $k-\varepsilon$ turbulence was made popular by Launder and Spalding (1972). A comprehensive account of the two-equation $k-\varepsilon$ model as well as other two-equation turbulence models is given by Reynolds (1976), Patel et al. (1985), Speziale et al. (1992), and Wilcox (2006). From the perspective of energy balance, $k-\varepsilon$ can be well understood from the celebrated lectures of Chandrasekhar (2010). A good physical insight into the mechanism of turbulence affordable to the beginners is due to Arpaci (1997).

12.16 CONCLUSIONS

True advancement in research is possible, if and only if each and every new discovery has its fitting pedagogical counterpart. The "A to O" principle as outlined by the author (Pramanick, 2014a) is a step in this direction. Purely from the pedagogical perspective alone, it is reasoned out here that the so-called conservation principle (Mach, 1911; Tolhoek and De Groot, 1952) is all pervasive. David Hilbert rightly pointed out: "A mathematical theory is not to be considered complete until you have made it so clear that you can explain it to the first man you meet on the street." The difficulty for most of the beginners mostly lies not on the physical concepts, but on the symbolism of mathematics. For reasons here, the scalar notation is followed as portrayed by Love (2003). The basic mathematics required to outline a physical process must be expounded with contextual physical examples as suggested by Lumley (2007). In order to assimilate the foregoing discussions, a mere knowledge of vector operations such as "divergence" and "gradient" is needed and these are embedded here with physical insights.

Generally speaking, physicists seek a theoretically well-founded formulation and subsequent elegant solution, mathematicians search the beauty and well-posedness of the problem and its solution, but the engineer's quest is the amicable practical solution to a problem. Surely, addressing this diverse population of approaches is not as difficult as it sounds, if we were to work in a body. In 1994, Bejan appealed that working as a team is the need of the hour (Bejan, 1994). The present effort is another step in this direction.

Last but not the least, about the "forget-first-the-engineer" syndrome, Rankine (1888) wrote as early as 1859: "The improvers of mechanical arts were neglected by biographers and historians, from a mistaken prejudice against practice, as being inferior in dignity to contemplation; and even in the case of men such as Archytas (an ancient Greek philosopher) and Archimedes, who combined practical skill with scientific knowledge, the records of their labours that have reached our time give but vague and imperfect accounts of their mechanical inventions, which are treated as matters of trifling importance in comparison with their philosophical speculations. The same prejudice, prevailing with increased strength during the middle ages, and aided by the prevalence of the belief in sorcery, rendered the records of the progress of practical machines, until the end of fifteenth century, almost a blank. These remarks apply, with peculiar force, to the history of those machines called PRIME MOVERS." Bejan (2016i) echoed: "Which is why Rankine—the engineer and cofounder of classical thermodynamics (next to Clausius and Kelvin)—is almost never mentioned by the philosophers." The previous comments apply as well at least to Bejan (2000)—a mechanical engineer and the discoverer of the Constructal Law (Bejan, 2016c). Thus, the ambit of the present effort is to alter the mindset of our contemporary thinkers.

Further, from the listed references, it can be traced that how nationalism and forgetfulness affected the spreading of science in general (Bejan, 2021). According to Bejan (2021): "The effect of the big language is obscure, as it is connected with other tendencies such as nationalism, chauvinism, jingoism, group think, grandiosity, and the belief that the bigger size and power originated the better the idea." Galileo Galilei described it best: "In questions of science, the authority of a thousand is not worth the humble reasoning of a single individual." The same is echoed by Pramanick (2014a).

REFERENCES

Alekseev, G.N. (1986). Energy and entropy. (trans: Taube, U.M.). Mir, Moscow.
Alembert, J.R.D'Le (1777). Nouvelles experiences sur la résistance des fluids, Lambert, Paris.
Arpaci, V.S. (1997). Microscales of turbulence: heat and mass transfer correlations. Gordon and Breach, Amsterdam.
Batchelor, G.K. (2009). An introduction to fluid dynamics. Cambridge University Press, Cambridge, 147–148.
Bateman, H. (1932). Partial differential equations of mathematical physics. Cambridge University Press, Cambridge, 115–118.
Bejan, A. (1982a). Entropy generation through heat and fluid flow. Wiley, New York, ix.
Bejan, A. (1982b). Entropy generation through heat and fluid flow. Wiley, New York, 22.
Bejan, A. (1994). Engineering advances on finite-time thermodynamics. American Journal of Physics, 62, 11–12.
Bejan, A. (1995). Entropy generation minimization: the method of thermodynamic optimization of finite-size systems and finite-size processes. CRC, Boca Raton.
Bejan, A. (1996a). Street network theory of organization in nature. Journal of Advanced Transportation, 30(2), 85–107.
Bejan, A. (1996b). Constructal-theory network for conducting paths for cooling a heat generating volume. International Journal of Heat Mass Transfer, 40, 799–816.
Bejan, A. (1996c). Models of power plants that generate minimum entropy while operating at maximum power. American Journal of Physics, 64(8), 1054–1059.
Bejan, A. (1997). Advanced engineering thermodynamics. Wiley, New York, 807.
Bejan, A. (2000). Shape and structure, from engineering to nature. Cambridge University Press, Cambridge, xv-xix.
Bejan, A. (2013). Convection heat transfer. Wiley, Hoboken, 4–8.
Bejan, A. (2014). Maxwell's demons everywhere: evolving design and the arrow of time. Scientific Reports, 4(4017), 1–4.
Bejan, A. (2016a). Advanced engineering thermodynamics. Wiley, Hoboken, 1–38.
Bejan, A. (2016b). Advanced engineering thermodynamics. Wiley, Hoboken, 39–94.
Bejan, A. (2016c). Advanced engineering thermodynamics. Wiley, Hoboken, 646–724.
Bejan, A. (2016d). Advanced engineering thermodynamics. Wiley, Hoboken, 646.
Bejan, A. (2016e). Advanced engineering thermodynamics. Wiley, Hoboken, 23–31.
Bejan, A. (2016f). Advanced engineering thermodynamics. Wiley, Hoboken, 1–38.
Bejan, A. (2016g). Advanced engineering thermodynamics. Wiley, Hoboken, 39–94.
Bejan, A. (2016h). Advanced engineering thermodynamics. Wiley, Hoboken, 95–139.
Bejan, A. (2016i). Advanced engineering thermodynamics. Wiley, Hoboken, 21.

Bejan, A. (2021). Nationalism and forgetfulness in the spreading of thermal sciences. International Journal of Thermal Sciences, 163, 1–9.
Bellman, R.E., & Roth, R.S. (1984). The Laplace transform. World Scientific, Singapore.
Bird, R.B., Stewart, W.E., & Lightfoot, E.N. (1960). Transport phenomena. Wiley, New York, 310–317.
Birkhoff, G. (1960). Hydrodynamics. Princeton University Press, Princeton, 29–30.
Borel, E. (2003). Space and time. Dover, New York.
Born, M. (1921). Kritische betrachtungen zur tradionellen danstellung der thermodymik. Physikalische Zeitschrift, 22, 218–224, 249–254, 282–286.
Braun, F. (1887). Untersuchungen über und die den vorgang der lösung begleiterden. Volum- und energieänderungen. Zeitschrift für Physikalische Chemie, 1, 259–269.
Bunge, M. (1980). Causality and modern science. Dover, New York.
Carathéodory, C. (1909). Untersuchungen uber die grundlagen der thermodynamic. Mathathematische Annalen, 67, 355–386.
Chambadal, P. (1963). Evolution et applications du concept d'entropie. Dunod, Paris, Sec. 30.
Chandrasekhar, S. (2010). The theory of turbulence (ed: Spiegel, E.A.). Springer, New York.
Chandrasekhar, S. (2020). Newton's principia for the common reader. Clarendon Press, Oxford, 23.
Le Chatelier, H. (1884, 1887). Sur un énoncé general des lois des équilibres chimiques. Competes Rendus Hebdomadaires des Séances de l'Académie des Sciences, 99, 786–789, 104, 679.
Chou, P.Y. (1945). On the velocity correlations and the solution of the equations of turbulent fluctuations. Quarterly of Applied Mathematics, 3, 38–54.
Chung, K.L. (1960). Markov chains with stationary transition probabilities. Springer, Berlin.
Courant, R., & Hilbert, D. (2008). Methods of mathematical physics, Vol. II. Wiley-VCH, New York.
Dugas, R. (1988). A history of mechanics (trans: Maddox, J.R.). Dover, New York.
Eckert, E.R.G., & Drake, Jr., R.M. (1972). Analysis of heat and mass transfer. McGraw-Hill, Tokyo, 9–12.
Einstein, A., & Furth, R.V. (1956). Investigations on the theory of the Brownian motion. Dover, New York.
Ellis, B.D. (1962). Newton's concept of motive force. Journal of the History of Ideas, 23(2), 273–278.
Epstein, P.S. (1937). Textbook of thermodynamics. Wiley, New York, 27–34.
Fick, A. (1855). Ueber diffusion. Annalen der Physik, 94, 59–86.
Fourier, M. (1833). Mémoire d'analyse sur le movement de la chaleur dans les fluides. Mémoires des l'Academie Royale des Sciences, 12, 507–530.
Fourier, J. (1955a). The analytical theory heat (trans: Freeman, A.). Dover, New York, 104–115.
Fourier, J. (1955b). The analytical theory of heat (trans: Freeman, A.). Dover, New York.
Frisch, U. (1999). Turbulence: the legacy of A. N. Kolmogorov. Cambridge University Press, Cambridge.
Glansdorff, P., & Prigogine, I. (1971). Thermodynamic theory of structure, stability and fluctuations. Wiley-Interscience, London.
Goldstein, H., Poole, C.P., & Safko, J. (2002). Classical mechanics. Pearson, London.
Gouy, G. (1889). Sur l'énergie utilisable. Journal of Physics, 501–518.
Granger, R.A. (2020). Fluid mechanics. Dover, New York, 154–159.
De Groot, S.R. (1951). Thermodynamics of irreversible processes. North Holland. Amsterdam.
Harris, T.E. (1963). The theory of branching processes. Prentice-Hall, New Jersey.
Harlow, F.H., & Nakayama, P.I. (1968). Transport of turbulence energy decay rate. Los Alamos Scientific Laboratory, University of California, Report LA–3854.
Hertz, H. (1899). The principles of mechanics. Macmillan, London.
Hilbert, D. (1901). Mathematical problems. Archiv der Mathematik, 3(1), 44–63, 213–237.

Hwang, F.K., Richards, D.S., & Winter, P. (1992). The Steiner problem. North-Holland, Amsterdam.
Jones, W.P., & Launder, B.E. (1972). The prediction of laminarization with a two-equation model of turbulence. International Journal of Heat Mass Transfer, 15, 301–314.
Kondepudi, D., & Prigogine, I. (2015). Modern thermodynamics: from heat engines to dissipative structures. Wiley, New York, 65–66.
Kutateladze, S.S. (1963). Fundamentals of heat transfer (trans: Cess, R.D.). Edward Arnold, London, 10–14.
Landau, L.D., & Lifshitz, E.M. (2005). Fluid mechanics (trans: Sykes, J.B., & Reid, W.H.). Butterworth-Heinemann, Oxford, 44–51.
Landé, A. (1926). Axiomatische begründung der thermodynamik durch Carathéodory, In: Bennewitz, K. et al. (eds.) Theorien der wärme. Handbuch der Physik, 9, Springer, Heidelberg, 281–300.
Landsberg, P.T. (1961). The entropy and the unity of knowledge. University of Wales Press, Cardiff.
Launder, B.E., & Spalding, D.B. (1972). Mathematical models of turbulence. Academic, New York.
Leib, E.H., & Yngvason, J. (1999). The physics and mathematics of the second law of thermodynamics. Physics Reports, 310, 1–96.
Lin, C.C. (1961). Statistical theories of turbulence. Princeton University Press, Princeton.
Love, A.E.H. (2003). A treatise on the mathematical theory of elasticity. Dover, New York.
Luikov, A.V. (1968). Analytical heat diffusion theory (ed: Hartnett, J.P.). Academic Press, New York, 15–20.
Lumley, J.L. (2007). Stochastic tools in turbulence. Dover, New York, x.
Lützen, J. (2010). Mechanistic images in geometric form: Heinrich Hertz's principles of mechanics. Oxford University Press, Oxford.
Mach, E. (1911). History and root of the principle of conservation of energy (trans: Jourdain, P.E.B.). Open Court, Chicago.
Maxwell, J.C. (1954). A treatise on electricity and magnetism, Vol. I, & Vol. II. Dover, New York.
Mikhailov, M.D., & Özişik, M.N. (1994). Unified analysis and solutions of heat and mass diffusion. Dover, New York, 14–17.
Milne-Thomson, L.M. (1974). Theoretical hydrodynamics. Dover, New York, 29–69.
Monin, A.S., & Yaglom, A.M. (1975). Statistical fluid mechanics: mechanics of turbulence (ed: Lumley, J.L.), Vol. I, & Vol. II. Dover, New York.
Müller, I., & Weiss, W. (2005). Entropy and energy: a universal competition, Springer, Heidelberg.
Navier, M. (1827). Mémoire sur les lois du mouvement des fluids. Mémoires de l'Académie Royale des Sciences, 6, 389–440.
Newton, I. (1999). The circular motion of fluids. In: The principia: mathematical principles of natural philosophy (trans: Cohen, I.B., Whitman, A., & Budenz, J.), Book II, Section IX. University of California Press, Berkley, 779–790.
Özisik, M.N. (1989). Boundary value problems of heat conduction. Dover, New York, 5–7.
Patankar, S.V. (2011a). Numerical heat transfer and fluid flow. Taylor & Francis, New York, xi-xiii.
Patankar, S.V. (2011b). Numerical heat transfer and fluid flow. Taylor & Francis, New York, 15.
Patel, V.C., Rodi, W., & Scheuerer, G. (1985). Turbulence models for near-wall and low-Reynolds number flows: a review. American Institute of Aeronautics and Astronautics Journal, 23, 1308–1319.
Pearson, K. (1905). The problem of the random walk. Nature, 72(1865), 294.

Pipes, L.A., & Harvill, L.R. (2014). Applied mathematics for engineers and physicists. Dover, New York, 918–923.

Planck, M. (1998). Eight lectures on theoretical physics (trans: Wills, A.P.). Dover, New York, 1–20.

Poiseuille, J.L.M. (1840, 1841). Reserches expéperimentelles sur le movement des liquids dans les tubes de très petis diameters. Comptes Rendus, 11, 961–967, 1041–1048, 12, 112–115.

Poission, S.-D. (1831). Mémoire sur les equations generals de l'equilibre et du movement des corps solides élastiques et des fluids. Journal de l'Ecole Royale Polytechnique, 13, 1–174.

Poisson, S.-D. (1855). Théorie mathématique dé la chaleur. Bachelier, Imprimeur-Libraire, Paris.

Pramanick, A.K. (2014a). The nature of motive force, Springer, Heidelberg, xi-xv.

Pramanick, A.K. (2014b). The nature of motive force. Springer, Heidelberg, 7–11.

Pramanick, A.K. (1990–91). Philosophy of nature. Reflection Magazine, Regional Engineering College Durgapur, 2–5.

Prandtl, L. (1945). Über ein neues formelsystem für die ausgebildete turbulenz. Nachrichten der Akademie der Wissenschaften in Göttingen, Mathematisch-Physikalische Klasse, 6–9.

Prigogine, I. (1947). Étude thermodynamique des phénoméne irréversibles. Thesis, Brussels.

Rankine, W.J.M. (1888). A manual of the steam engine and other prime movers (revised: Millar, W.J.). Charles Griffin & Company, London, xv.

Rayleigh, L. (1911). On the motion of solid bodies through viscous liquids. Philosophical Magazine, 21, 697–711.

Révész, P. (2013). Random walk in random and non-random environments. World Scientific, Singapore.

Reynolds, O. (1903). Papers on mechanical and physical subjects, Vol. 3: Sub-mechanics of the universe. Cambridge University Press, Cambridge.

Reynolds, W.C. (1976). Computation of turbulent flows. Annual Review of Fluid Mechanics, 8, 183–208.

Rozanov, Y.A. (1969). Probability theory: a concise course (trans: Silverman, R.A.). Dover, New York, 127–135.

De Saint-Venant, M. (1843). Dynamique–note a joindre an mémoire sur la dynamique des fluids. Comptes Rendus, 17, 1240–1247.

Saha, M.N., & Srivastava, B.N. (1931). A treatise on heat. The Indian Press, Allahabad, 89–97.

Salamon, P., Nitzan, A., Andresen, B., & Berry, R.S. (1980). Minimum entropy production and the optimization of heat engines. Physical Review A, 21, 2115–2129.

Schlichting, H., & Gersten, K. (2000). Boundary layer theory (trans: Mayes, K.). Springer, Heidelberg, 4–5.

Schneider, P.J. (1955). Conduction heat transfer, Addison-Wesley, Massachusetts, 3–5.

Shames, I.H. (1992). Mechanics of fluids. McGraw-Hill, New York, 132–140.

Sneddon, I.N. (2006). Elements of partial differential equations. Dover, New York, 39–42.

Sobolev, S.L. (1989). Partial differential equations of mathematical physics (trans: Dawson, E.R., trans. ed.: Broadbent, T.A.A.). Dover, New York, 1–3.

Sommerfeld, A. (1950). Mechanics of deformable bodies. Academic Press, New York, 37–41.

Sommerfeld, A. (1964). Mechanics. Academic Press, New York.

Speziale, C.G., Abid, R., & Anderson, E.C. (1992). A critical evaluation of two-equation models for near-wall turbulence. American Institute of Aeronautics and Astronautics Journal, 30, 311–319.

Stodola, A. (1905). Steam turbines (with an appendix on gas turbines and the future of heat engines) (trans: Loewenstein, L.C.). Van Nostrand, New York, 402.

Stokes, G.G. (1845). On the theories of the internal friction of fluids in motion, and of the equilibrium and motion of elastic solids. Transactions of the Cambridge Philosophical Society, 8 (part 3), 75–129.

Stokes, G.G. (1856). On the effect of internal friction of fluids on the motion of pendulums. Transactions of the Cambridge Philosophical Society, 9, Part II, 8–106.

Sychev, V.V. (1983). The differential equations of thermodynamics (trans: Yankovsky, E.). Mir, Moscow.

Szargut, J., Petela, R. (1965). Egzergia. Wydawnictwa Naukowo-Techniczne, Warsaw.

Thess, A. (2011). The entropy principle: thermodynamics for the unsatisfied. Springer, Heidelberg.

Tokaty, G.A. (1994). A history and philosophy of fluid mechanics. Dover, New York, 33–48.

Tolhoek, H.A., & De Groot, S.R. (1952). A discussion of the first law of thermodynamics for open systems. Physica, XVIII, 780–790.

Truesdell, C.A. (1966). Six lectures on modern natural philosophy. Springer, New York, 100–101.

Truesdell, C.A., & Bharatha, S. (1977). The concepts and logic of classical thermodynamics as a theory of heat engines. Springer, New York.

Wien, W., Harms, F., Wien, M., & Joos, G. (1929). Handbuch der experimentalphysik, Part 1. Academische Verlagsgesellschaft, Leipzig, 30–32.

Wilcox, D.C. (2006). Turbulence modeling for CFD. DCW Industries, California, 107–237.

Yih, C.-S. (1979). Fluid mechanics: a concise introduction to the theory. West River Press, Ann Arbor, 4–6.

Zauderer, E. (2006). Partial differential equations of applied mathematics. Wiley-Interscience, New York.

Zwillinger, D. (1998). Handbook of differential equations. Academic Press, San Diego.

13 Nonlinear Evolution of Weak Discontinuity Waves in Darcy-Type Porous Media

Mithilesh Singh
Rajkiya Engineering College, Sonbhadra, U. P.

CONTENTS

13.1 INTRODUCTION

The propagation of nonlinear traveling waves in porous media has received considerable attention, during the past few decades, due to its applications in geophysics, mining, oil exploration, etc. The problems of linear wave propagation in porous media have been analyzed in Morse and Ingard (1968) and Pascal (1986). In gas dynamics, the system of Euler's equations representing the conservation of mass, momentum, and energy for perfect gas in Darcy-type porous media are used to analyze the problem of nonlinear wave propagation (Nield and Bejan, 1999). The growth and decay of acceleration waves in porous media were studied by Jordon (2005). The properties of compressible gas flow in a porous medium were examined by Ville (1996). Hsiao (1999) analyzed the initial value problem for the system of compressible adiabatic flows through porous media in one-dimensional space with fixed boundary condition. Pan (2006) conjectured that Darcy's law governs the motion of compressible porous media flow in large time adiabatic flow. Jeffrey (1976) studied the quasilinear hyperbolic systems and waves. Boillat and Ruggeri (1965) developed the interaction between shock and acceleration waves in the ideal gas. The evolution of weak discontinuity waves in self-similar flow and secondary shock formation with point explosion model was analyzed by Virgopia and Ferraioli (1982). Ram (1978) studied the growth and decay of acceleration waves in radiating gas. Pandey and Sharma (2007) analyzed the characteristic shock with weak discontinuity in non-ideal gas.

DOI: 10.1201/9781003222255-13

Mentrelli et al. (2008) examined the shock and an acceleration wave in a perfect gas for increasing shock strength. Singh et al. (2010) analyzed the propagation of nonlinear traveling waves in Darcy-type porous media by wavefront expansion technique. Further, Jordan (2019, 2020) discussed the finite amplitude acoustics under the classical theory of particle-laden flows and poroacoustic traveling waves in porous media, respectively. Recently, Berjamin (2021) has analyzed the nonlinear waves in saturated porous media.

In this chapter, the characteristic coordinates are used to analyze the nonlinear wave propagation in a fluid that saturates a fixed, rigid, homogeneous, and isotropic porous medium. The effect of pores in the flow region has been incorporated using the Darcy law given as follows (Nield and Bejan, 1999):

$$\nabla P = -(\mu\chi/K)u, \tag{13.1}$$

where P is the intrinsic pressure, u is the velocity vector, and K, μ, and $\chi(<1)$ are the positive constants that denote the permeability, dynamic viscosity, and porosity, respectively. Also, the filtration velocity vector υ is related to the intrinsic velocity vector u by the Dupuit–Forchheimer relationship $\upsilon = \chi u$. Darcy's law is equivalent to a body force term.

To the best of our knowledge, such an analytical description of a complete history of acceleration waves has not been studied previously in Darcy-type porous media. Ram (1978) and Boillat and Ruggeri (1965) developed a theory for one-dimensional flow satisfying the Euler equations, to achieve a detailed comprehension of the consequences of the compression waves steepen into shock and expansion waves relax. This chapter brings out some interesting features of the evolutionary behavior of acceleration waves in Darcy-type porous media.

13.2 GOVERNING EQUATIONS

The governing equations for a viscous, compressible, and homogeneous fluid flowing in a fixed, rigid, non-thermally conducting, and isotropic porous medium of porosity χ and permeability K can be written as

$$\rho_t + u\rho_r + \rho u_r + \frac{m\rho u}{r} = 0, \tag{13.2}$$

$$\rho(u_t + uu_r) + p_r + \frac{\mu\chi}{K}u = 0, \tag{13.3}$$

$$e_t + ue_x - p\rho^{-2}\left(\rho_t + u\rho_x\right) = -\frac{\mu\chi}{K\rho}u^2, \tag{13.4}$$

where ρ denotes the gas density, u is the velocity, p represents the pressure, e is the internal energy, γ denotes the specific heat ratio, t is the time, and x is the spatial coordinate. Here, m takes values 0, 1, or 2 for planar, cylindrical, and spherical symmetry, respectively. The subscripts denote partial differentiation unless stated otherwise.

The internal energy of an ideal gas is given by

$$e = p/\rho(\gamma - 1).$$

Using the above relation, equations (13.2)–(13.4) may be written as

$$p_t + up_r - \frac{\gamma p}{\rho}(\rho_t + u\rho_r) + (\gamma - 1)\frac{\mu\chi}{K}u^2 = 0. \tag{13.5}$$

The above system of equations (13.2–13.5) are supplemented with an equation of state $p = \rho RT$, where R is gas constant and T represents the temperature.

The governing system of equations (13.2), (13.4), and (13.5) can be written in the following matrix form

$$V_t + AV_r + B = 0, \tag{13.6}$$

where

$$V = \begin{bmatrix} \rho \\ u \\ p \end{bmatrix} \quad A = \begin{bmatrix} u & \rho & 0 \\ 0 & u & 1/\rho \\ 0 & \gamma p & u \end{bmatrix} \text{and } B = \begin{bmatrix} m\rho u/r \\ \mu\chi u/\rho K \\ \gamma mpu/r + (\gamma - 1)\mu\chi u^2/K \end{bmatrix}. \tag{13.7}$$

An acceleration wave is a propagating singular surface across the motion, and the velocity and deformation gradient are continuous, but second- and higher-order derivatives of motion suffer finite jump discontinuity. The jump $[\ddot{r}]$ is taken to be an amplitude vector of the acceleration wave. The function $V(r,t)$ is continuous everywhere except at the characteristic curve $\Sigma(t)$, but V_r and V_t have finite jumps and it is said to be a weak discontinuity at this curve.

Furthermore, the matrix A has three families of characteristic curves C_0, C_-, and C_+, respectively:

$$\frac{dr}{dt} = u, \tag{13.8}$$

$$\frac{dr}{dt} = u - a, \tag{13.9}$$

$$\text{and } \frac{dr}{dt} = u + a, \tag{13.10}$$

where $a = (\gamma p/\rho)^{1/2}$ is the sound speed. For isentropic flow, the speed of sound is given by "Courant and Friedrichs (1948)."

$$a^2 = (\partial p/\partial \rho)_{s=const}, \tag{13.11}$$

where s is the entropy, which is constant along the particle's path.

13.3 EVALUATION OF THE WEAK DISCONTINUITY

To study the nonlinear effects on the wave pattern, we introduce a new coordinate system (α,β), where α is constant along with the outgoing wave $dr/dt = u + a$ in the (r,t) plane, if it originates at time $\alpha = t'$. β is a particle tag so that β is constant along the particle path $dr/dt = u$ in the (r,t) plane. If the wavefront transverses a particle at an instant t^*, these particles and their path are labeled by $\beta = t^*$.

For each pair of values (α,β), there is a corresponding pair (r,t) so that $r = r(\alpha,\beta)$, $t = t(\alpha,\beta)$.

The view of the nature of α and β will be satisfied by the following PDEs:

$$r_\alpha = ut_\alpha, \tag{13.12}$$

$$r_\beta = (u+a)t_\beta. \tag{13.13}$$

In consequence of this transformation, we have

$$V_t = \frac{V_\beta r_\alpha - u_\alpha r_\beta}{J}, \tag{13.14}$$

$$V_r = \frac{V_\alpha t_\beta - V_\beta t_\alpha}{J}. \tag{13.15}$$

To ensure that a plane between (r,t) and (α,β) has one-to-one correspondence to each other, it is essential that the Jacobian transformation $J = \partial(r,t)/\partial(\alpha,\beta) = -at_\alpha t_\beta$ is neither zero, nor undefined. Since the doubling up or overlapping of fluid particles is prohibited from physical considerations, a breakdown of thc solution will arise $J = 0$ if and only if $t_\alpha = 0$, when two adjoining characteristics convert into a shock wave. Hence, $J = 0$ will also give us the condition for the steepening of the wavefront into the shock wave, which is given by Courant and Friedrichs (1948).

Using equations (13.12–13.15) in (13.1), (13.2), and (13.4) yields

$$at_\alpha \rho_\alpha - \rho t_\beta u_\alpha + \rho u_\beta t_\alpha + \frac{m\rho u t_\alpha t_\beta}{r(\alpha,\beta)} = 0, \tag{13.16}$$

$$\rho a u_\alpha t_\beta - t_\beta p_\alpha + p_\beta t_\alpha + \frac{\mu\chi u}{K} = 0, \tag{13.17}$$

$$ap_\alpha t_\beta - \gamma p\left(t_\beta u_\alpha - t_\alpha u_\beta + \frac{maut_\alpha t_\beta}{r(\alpha,\beta)}\right) + \frac{\gamma mpu}{r} + \frac{(\gamma-1)\mu\chi}{K}u^2 = 0. \tag{13.18}$$

Equations (13.16–13.18) may be written as

$$p_\beta + \rho a u_\beta + \frac{m\rho a^2 u t_\beta}{r(\alpha,\beta)} + \frac{\mu\chi u}{K}\left(1+(\gamma-1)u\right) + \frac{\gamma mpu}{r(\alpha,\beta)} = 0 \tag{13.19}$$

Across a characteristic shock, the Rankine–Hugoniot conditions are given as

$$[u] = 0, \quad [\rho] = 0, \quad [p] = 0, \quad t = \beta \quad \text{at } \alpha = 0. \tag{13.20}$$

Further, the gas is homogeneous and at rest ahead of the wavefront, equation (13.20) is written as

$$u_\beta = 0, \quad p_\beta = 0, \quad \rho_\beta = 0, \quad t_\beta = 1 \text{ at } \alpha = 0. \tag{13.21}$$

On using equations (13.20) and (13.21) in (13.17) at the wavefront, we get

$$p_\alpha = a_0 \rho_0 u_\alpha \text{ at } \alpha = 0. \tag{13.22}$$

Equations (13.12) and (13.13) at the wavefront may be written as

$$r_\beta = a_0, \, r_\alpha = 0 \; at \; \alpha = 0. \tag{13.23}$$

To determine the amplitude of the acceleration wave $\Pi = [u_r]$ at the wavefront from equations (13.13) and (13.21), we get

$$[u_r] = \Pi = -u_\alpha / a t_\alpha \neq 0 \quad \text{at} \quad \alpha = 0. \tag{13.24}$$

Differentiating equations (13.19), (13.22), (13.12), and (13.13) with respect to α and β, we have

$$\frac{u_{\alpha\beta}}{t_\alpha} = \frac{a_0}{2\rho_0}\left[\frac{\mu\chi}{K} + \frac{m\rho_0}{\beta}\right]\pi \quad at \quad \alpha = 0, \tag{13.25}$$

$$\frac{t_{\alpha\beta}}{t_\alpha} = \frac{\gamma+1}{2}\pi \quad \text{at} \quad \alpha = 0. \tag{13.26}$$

Differentiating equation (13.24) with respect to β, and using it in equations (13.25–13.26), we get

$$\frac{d\Pi}{d\beta} + \frac{1}{2}\left(\frac{\nu\chi}{K} + \frac{m}{\beta}\right)\Pi + \frac{\gamma+1}{2}\Pi^2 = 0 \quad \text{at} \quad \alpha = 0, \tag{13.27}$$

where $\nu = \mu/\rho_0$ is the kinematic viscosity.

Introducing the non-dimensional variables ζ, τ, and Ω in the following form

$$\zeta = \Pi/\Pi^*, \; \tau = (\beta - \beta^*)/2\beta^*, \; \Omega = (\gamma+1)\Pi^* \beta^*, \tag{13.28}$$

where Π^* and β^* are parameters.

Substituting equation (13.28) in (13.27), we get the first-order nonlinear differential equation (Riccati-type equation) in the following form

$$\frac{d\zeta}{d\tau}+\left(\frac{\nu\chi}{2K}+\frac{m}{1+2\tau}\right)\zeta+\Omega\zeta^2=0, \tag{13.29}$$

Equation (13.29) is a nonlinear ODE; it can be easily transformed to a linear ODE in $1/\zeta$ by dividing with ζ^2:

$$\frac{d}{d\tau}\left(\frac{1}{\zeta}\right)-\left(\frac{\nu\chi}{2K}+\frac{m}{1+2\tau}\right)\frac{1}{\zeta}=\Omega, \tag{13.30}$$

Equation (13.30) is a first-order linear ODE whose solution is given by

$$\zeta=\frac{\phi(\tau)}{(1+\Omega\,Q(\tau))}, \tag{13.31}$$

where $Q(\tau)=\int_0^\tau \frac{e^{-\nu\chi\tau/2K}}{(1+2\tau)^{m/2}}d\tau$ and $\phi(\tau)=\frac{e^{-\nu\chi\tau/2K}}{(1+2\tau)^{m/2}}$.

In Equation (13.31), the coefficient of Darcy term has become $\delta=\chi/\mathrm{Re}$, where $\mathrm{Re}=a_0(0)l\rho_0/\mu$; we defined $l=K/L$ for convenience. We also note that the continuum assumption demands $K_n<0.01$, where $K_n=\sqrt{K}\,/\,L$ is the Knudsen number, $\sqrt{K}$ is the average molecular free path, L is the physical length scale, and l is the characteristic length scale (Nield and Bejan (1999) and Singh et al. (2010)). Ω is known as the initial amplitude of acceleration wave; $\zeta\neq 0$ denotes the value of ζ at $\alpha=0$. The function $\phi(\tau)$ is nonzero, finite, and continuous on $[0,\tau)$, and it tends to zero as $\tau\to\infty$ with $Q(\tau)<\infty$. If $\zeta(\tau)\to 0$ as, $\tau\to\infty$, is implying the decays of expansion waves. If $\Omega<0$ implies the compression waves, it follows from (13.31) that there are three conditions in Boillat and Ruggeri (1965):

1. Let $|\Omega|<\Omega_c$, where $\Omega_c=1/\phi(\infty)$ is the critical amplitude of acceleration wave. It is nonzero and continuous in $\tau\in[0,\infty)$; if $\tau\to\infty$, then $\Omega\to 0$ and $\phi(\tau)$ becomes zero; then, the decay of the wavefront is the decay of an expansion wave.
2. Let $|\Omega|=\Omega_c$. Then ζ is finite, nonzero, and continuous in the $[0,\infty]$. Again, $\zeta\to 0$ as $\tau\to\infty$ implies that the wave decays, which is illustrated in Figures 13.1–13.3, respectively.
3. Let $|\Omega|>\Omega_c$, then $\phi(\tau_c)=1/\Omega$ in the finite time $\tau_c>0$. ζ is finite, nonzero, and continuous on $[0,\infty)$ and $|\zeta|\to\infty$ as $\tau\to\tau_c$. This means that compression waves steepen into the shock wave in a finite time $\tau=\tau_c$, only when initial discontinuity increases a critical value.

13.4 RESULTS AND DISCUSSION

In this paper, we have analyzed the evolutionary behavior of acceleration waves in Darcy-type porous media. Computations of variables ζ and τ, associated with acceleration waves, have been carried out for planar ($m = 0$), cylindrical ($m = 1$), and spherical ($m = 2$) symmetric flows by taking $\nu = 0.00001496$, $\chi = 0.476$, $K_n = 0.00421$, $L = 1$, and values of $\Omega = -2$, $\Omega = -1$, and $\Omega = 1$. The steepening of compression wave is slowed down in case of porous medium as compared to the non-porous case in the planer, cylindrical, and spherical symmetric flows, which is illustrated in Figures 13.1–13.3. It may also be noted that the compression waves steepen into a shock only if the magnitude of the initial slope of the wavefront is greater than a critical value; an expansion wave is relaxed in the entire above planes, which are also shown in Figures 13.1–13.3.

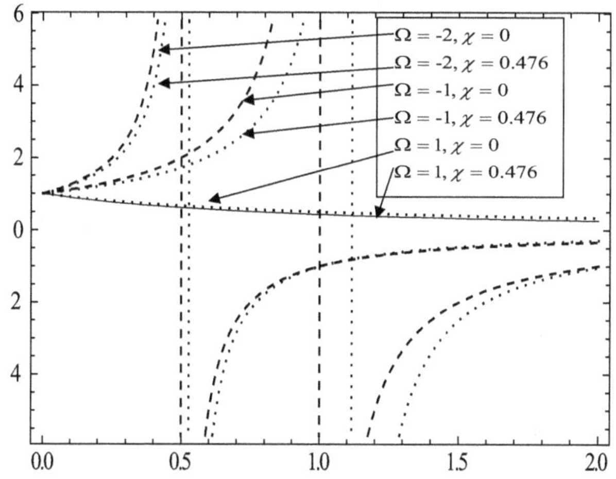

FIGURE 13.1 Evolution of the amplitude of acceleration waves influenced by Darcy-type and non-Darcy type porous media for plane ($m = 0$) flows.

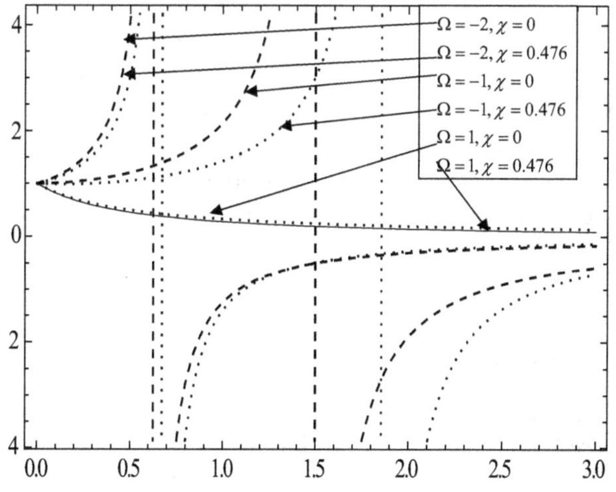

FIGURE 13.2 Evolution of the amplitude of acceleration waves influenced by Darcy-type and non-Darcy type porous media for cylindrical ($m = 1$) flows.

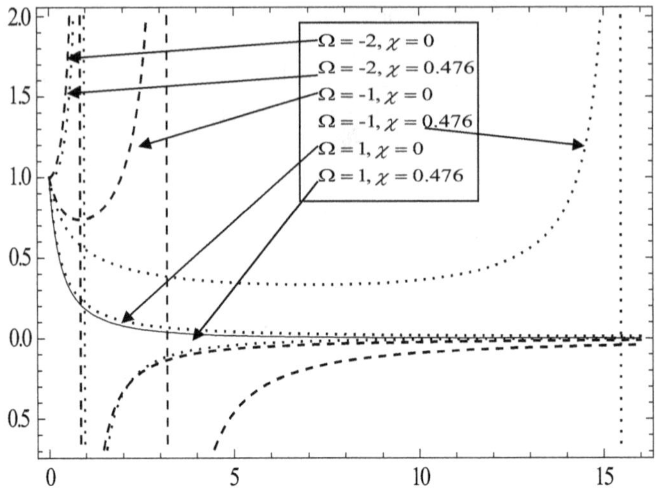

FIGURE 13.3 Evolution of the amplitude of acceleration waves influenced by Darcy-type and non-Darcy type porous media for spherical ($m = 2$) flows.

13.5 CONCLUSIONS

In the present chapter, we have used the characteristics of the governing quasilinear system as the reference coordinate system to determine an exact solution to the equations that describe unsteady planar, cylindrical, and spherical symmetric flows of Darcy-type porous media. The evolution of weak discontinuity in a state characterized by the exact solution is obtained. The analyses show that the rate of steepening of compression wave into a shock wave of Darcy-type porous media is slowed down as compared to ideal gases and expansion wave is relaxed in both an ideal gas and Darcy-type porous media.

REFERENCES

Berjamin, H. (2021). Nonlinear plane waves in saturated porous media with incompressible constituents. *Proceed of Royal Society A* 477: 20210086.

Boillat, G. and Ruggeri, T. (1965). On the evolution law of weak discontinuity for hyperbolic quasilinear systems. *Wave Motion* 1:149–152.

Courant, R., and Friedrichs, K. O. (1948). *Supersonic Flow and Shock Waves*, Wiley Interscience, New York.

Hsiao, L. (1999). Initial boundary value problem for the system of compressible adiabatic flow through porous media. *Journal of Differential Equations* 159: 280–305.

Jeffrey, A. (1976). *Quasilinear Hyperbolic System and Wave*, London.

Jordan, P.M. (2005). Growth and decay of acoustic acceleration waves in Darcy-type porous media. *Proceed of Royal Society A* 461: 2749–2766.

Jordan, P. M. (2019). Finite-amplitude acoustics under the classical theory of particle-laden flows. *Evolution Equations & Control Theory* 8: 101–116.

Jordan, P. M. (2020). Poroacoustic traveling waves under the Rubin–Rosenau–Gottlieb theory of generalized continua, *Water* 12: 807. doi: 10.3390/w12030807.

Mentrelli, A., Ruggeri, T., Sugiyama, M., Zhao, N. (2008). Interaction between a shock and acceleration waves in a perfect gas for increasing shock strength. *Wave Motion* 45: 98–517.

Morse, P. M. and Ingard, K. U. (1968). *Theoretical Acoustic*. McGraw-Hill, New York.

Nield, D. A. and Bejan, A. (1999). *Convection in Porous Media*, Springer, Berlin.

Pan, R. (2006). Darcy's law as long-time limit of adiabatic porous media flow, *Journal of Differential Equations* 220:121–146.

Pandey, M. and Sharma, V. D. (2007). Interaction of a characteristic shock with a weak discontinuity in a non-ideal gas. *Wave Motion* 44: 346–354.

Pascal, H. (1986). Pressure wave propagation in a fluid flowing through a porous medium and problem related to the interpretation of Stoneley's wave attenuation in acoustical and well logging, *International Journal of Engineering Science* 24: 1553–1570.

Ram, R. (1978). Effect of radiative heat transfer on the growth and decay of acceleration waves. *Applied Scientific Research* 34: 93–104.

Ville, A. D. (1996). On the properties of compressible gas flow in a porous media. *Transport in Porous Media* 22: 287–306.

Virgopia, N. and Ferraioli, F. (1982). Evolution of weak discontinuity waves in self- similar flows and formation of secondary shocks. The point explosion model. *Journal of. Applied Mathematical Physics* 33: 63–80.

Singh, M., Singh, L. P. and Husain, A. (2010). Propagation of nonlinear travelling waves in Darcy-type porous media. *Acta Astronautica* 67:1053–1058.

Smirnov, N. N. Safargulova, S. I. (1991). Propagation of weak disturbances in the combustion of compressible porous fuels. *Combustion, Explosion and Shock Wave*, 27: 26–34.

Whitham, G. B. (1974). *Linear and Nonlinear Waves*. Wiley, New York.

14 Numerical Solution of Fisher's Equation by using Fourth-Order Collocation Scheme Based on Modified Cubic B-Splines

Brajesh Kumar Singh and Mukesh Gupta
Babasaheb Bhimrao Ambedkar University

CONTENTS

14.1 INTRODUCTION

Fisher [1] in 1936 proposed a model for the propagation of a mutant gene in the form of nonlinear parabolic partial differential equation (PDE):

$$\frac{\partial \phi}{\partial \tau} = \nu \frac{\partial^2 \phi}{\partial x^2} + \rho \ \mathrm{F}(\phi(x,\tau)), \quad x \in \mathbf{R},\ \tau \geq 0. \tag{14.1}$$

where x and τ are spatial and temporal coordinates, respectively. ν is diffusion coefficient, ρ is reaction coefficient, and F is reaction term.

In the field of science and engineering, Fisher's equation (14.1) has a vital role in describing different phenomena such as in study of nuclear reactor theory [2], flame propagation in a medium [3], Brownian motion [4], autocatalytic chemical reactions [5], tissue engineering [6], neurophysiology [7], and combustion [8].

DOI: 10.1201/9781003222255-14

To explain the phenomena modeled in the form of PDE, its analytical/numerical solution plays a crucial role. In that way, many researchers used a variety of methods on Fisher's equation (14.1) for its analytical/numerical solution. Olmos et al. [9] used a pseudo-spectral approach for the numerical solution of Fisher's equation. Tang and Weber [10] used Galerkin finite element method, whereas least squares finite element method was used by Carey and Shen [11]. Many authors used the collocation approach [12–15] and differential quadrature approach [16–18] for the numerical solution of Fisher's equation. Some other methods are reported in [19–23].

In the literature, the collocation scheme based on modified cubic B-splines is of second-order approximation for double derivative and fourth-order approximation for single derivative, but recently, authors in [24] have developed a fourth-order collocation technique using modified cubic B-splines as base functions (mCBCT4), in which single-order derivative and double-order derivative are of order four. In this chapter, our main task is to use mCBCT4 on Fisher's equation (14.1) to find its numerical solution.

The rest of the chapter is structured in the following manner: In Section 14.2, fourth-order accuracy of spatial approximation via collocation technique using modified cubic B-splines as base functions is reported. The implementation of the technique on Fisher's equation and implementation of initial/boundary conditions is described in Section 14.3. Two different test problems of Fisher's equation are solved in Section 14.4 to validate the accuracy and effectiveness of the technique. In Section 14.5, we summarize the concluding remarks on the results.

14.2 FOURTH-ORDER SPATIAL APPROXIMATION WITH MODIFIED CUBIC B-SPLINES

The collocation technique is a widely known process for approximating the smooth functions and their derivatives in a given finite domain. Set finite domain $[a,b] \subset \mathbf{R}$ by $\Omega_{a,b}$, where $\mathbf{R}$ is the set of reals and $\Psi : \Omega_{a,b} \to \mathbf{R}$ is a smooth function. Let us consider a uniform partition of the domain $\Omega_{a,b}$ defined by $\Omega_{a,b} := \{a = x_0 < x_1 < x_2 < \ldots < x_{n-1} < x_M = b\}$, with discretization step size $\delta x := x_j - x_{j-1} = \dfrac{x_M - x_0}{M} = \dfrac{b-a}{M}$. Throughout this chapter, for time level τ, the value of ϕ at the grid point x_i is denoted by $\phi(x_i,\tau) := \phi_i(\tau) \equiv \phi_i$.

Cubic B-splines (cB-splines) of a given degree are smooth functions with minimal support. The cB-spline ${}_c\mathcal{B}_i := \mathcal{B}_i(x)$ at a grid point $x \in \Omega_{a,b}$ is defined by

$$\mathcal{B}_i = \frac{1}{\delta x^3}\begin{cases} (x - x_{i-2})^3, & x \in [x_{i-2}, x_{i-1}) \\ (x - x_{i-2})^3 - 4(x - x_{i-1})^3, & x \in [x_{i-1}, x_i) \\ 4(x - x_{i+1})^3 - (x - x_{i+2})^3, & x \in [x_i, x_{i+1}) \\ -(x - x_{i+2})^3, & x \in [x_{i+1}, x_{i+2}) \\ 0, & \text{elsewhere.} \end{cases} \tag{14.2}$$

Denote the values of cB-spline $\mathcal{B}_i$ and its first two derivatives at the grid point x_j by $\mathcal{B}_{ij} := \mathcal{B}_i(x_j)$, $\mathcal{B}'_{ij} := \mathcal{B}'_i(x_j)$, and $\mathcal{B}''_{ij} := \mathcal{B}''_i(x_j)$, respectively, and read them as in Table 14.1.

TABLE 14.1
Values of B-Splines and Their Derivative at Different Grid Points

	x_{i-2}	x_{i-1}	x_i	x_{i+1}	x_{i+2}
$\mathcal{B}_i$	0	1	4	1	0
$\delta x \cdot \mathcal{B}_i'$	0	−3	0	3	0
$\delta x^2 \cdot \mathcal{B}_i''$	0	6	−12	6	0

The set $\{\mathcal{B}_r(x) : r \in \{-1\} \cup \Delta_{M+1}\}$ forms a base for ϕ over the computational domain $\Omega_{a,b}$, where the set Δ_M is defined by $\Delta_M = \{0,1,2,\ldots,M\}$. In order to reduce the computational complexity and to handle the Dirichlet boundary conditions smoothly, the cB-splines are modified in such a way that the resulting coefficient matrix remains diagonally dominant for producing unique solutions to the unknowns. We utilize modified form of cB-spline base functions along with Taylor's approximations to produce fourth-order spatial approximations for $\phi(x,\tau)$ and its spatial derivatives $\phi'(x,\tau)$ and $\phi''(x,\tau)$; the modified form of cB-splines is read as [25]:

$$\begin{aligned}
&\widetilde{\mathcal{B}_0}(x) = \mathcal{B}_0(x) + 2\mathcal{B}_{-1}(x),\\
&\widetilde{\mathcal{B}_1}(x) = \mathcal{B}_1(x) - \mathcal{B}_{-1}(x),\\
&\widetilde{\mathcal{B}_j}(x) = \mathcal{B}_j(x), \qquad j \in \Delta_{M-2} \setminus \{0,1\}\\
&\widetilde{\mathcal{B}}_{M-1}(x) = \mathcal{B}_{M-1}(x) - \mathcal{B}_{M+1}(x),\\
&\widetilde{\mathcal{B}_M}(x) = \mathcal{B}_M(x) + 2\mathcal{B}_{M+1}(x),
\end{aligned} \tag{14.3}$$

and thus the base for ϕ over the computational domain $\Omega_{a,b}$ reduces to $\{\widetilde{\mathcal{B}_r}(x) : r \in \Delta_M\}$. In the collocation approach, the exact value of $\phi(x,\tau)$ is approximated via the polynomial $\Phi(x,\tau)$ obtained by spanning of the modified cB-splines as:

$$\Phi(x,\tau) = \sum_{i=0}^{M} \zeta_i \, \widetilde{\mathcal{B}}_i(x), \tag{14.4}$$

where $\zeta_i = \zeta_i(\tau)$'s are time-dependent unknown coefficients to be evaluated via utilization of different B-splines $\{\widetilde{\mathcal{B}_r}(x) : r \in \Delta_M\}$ for each grid x_i in collocation and the boundary/initial conditions of the differential equation.

In addition, the approximation $\Phi(x,\tau)$ satisfies the boundary conditions as well as the following interpolatory conditions:

a. $\Phi(x_j,\tau) \equiv \phi(x_j,\tau) = \zeta_{j-1} + 4\zeta_j + \zeta_{j+1}$, for all $j \in \Delta_M$, and

b. $\Phi''(x_j,\tau) = \phi''(x_j,\tau) - \frac{1}{12}\delta x^2 \phi^{(4)}(x_j,\tau)$, for $j = 0, M$

In case when $\phi(x,\tau)$ is sufficiently smooth and $\Phi(x,\tau)$ be unique cubic spline interpolant satisfying above end conditions, then for each grid $x_j,\ j\in\Delta_M$ [26], we get

$$\begin{aligned}&(*)\quad \Phi'(x_j,\tau)=\phi'(x_j,\tau)+O(\delta x^4);\\ &(**)\quad \Phi''(x_j,\tau)=\phi''(x_j,\tau)-\frac{1}{12}\delta x^2\phi^{(4)}(x_j,\tau)+O(\delta x^4).\end{aligned} \tag{14.5}$$

Using Taylor's expansion with finite difference formulae for the value of $\phi^{(4)}(x_j,\tau)$ at x_j, we get

$$\begin{aligned}\phi^{(4)}(x_0,\tau)&=\frac{2\Phi''(x_0,\tau)-5\Phi''(x_1,\tau)+4\Phi''(x_2,\tau)-\Phi''(x_3,\tau)}{\delta x^2}+O(\delta x^2),\\ \phi^{(4)}(x_j,\tau)&=\frac{\Phi''(x_{j-1},\tau)-2\Phi''(x_j,\tau)+\Phi''(x_{j+1},\tau)}{\delta x^2}+O(\delta x^2),\qquad j\in\Delta_{M-1},\{0\},\\ \phi^{(4)}(x_M,\tau)&=\frac{2\Phi''(x_M,\tau)-5\Phi''(x_{M-1},\tau)+4\Phi''(x_{M-2},\tau)-\Phi''(x_{M-3},\tau)}{\delta x^2}+O(\delta x^2).\end{aligned} \tag{14.6}$$

On utilizing (14.6) in (14.5), $\phi''x_j, j\in\Delta_M$ is approximated as follows:

$$\begin{aligned}\phi''(x_0,\tau)&=\frac{14\Phi''(x_0,\tau)-5\Phi''(x_1,\tau)+4\Phi''(x_2,\tau)-\Phi''(x_3,\tau)}{12}+O(\delta x^4),\\ \phi''(x_j,\tau)&=\frac{\Phi''(x_{j-1},\tau)+10\Phi''(x_j,\tau)+\Phi''(x_{j+1},\tau)}{12}+O(\delta x^4),\qquad j\in\Delta_{M-1},\{0\}\\ \phi''(x_M,\tau)&=\frac{14\Phi''(x_M,\tau)-5\Phi''(x_{M-1},\tau)+4\Phi''(x_{M-2},\tau)-\Phi''(x_{M-3},t)}{12}+O(\delta x^4).\end{aligned} \tag{14.7}$$

Equation (14.7) after utilizing (14.4) reduces to

$$\begin{aligned}\phi''(x_0,\tau)&=\sum_{i=0}^{M}\zeta_i\left(\frac{14\widetilde{\mathcal{B}}''_i(x_0)-5\widetilde{\mathcal{B}}''_i(x_1)+4\widetilde{\mathcal{B}}''_i(x_2)-\widetilde{\mathcal{B}}''_i(x_3)}{12}\right),\\ \phi''(x_j,\tau)&=\sum_{i=0}^{M}\zeta_i\left(\frac{\widetilde{\mathcal{B}}''_i(x_{j-1})+10\widetilde{\mathcal{B}}''_i(x_j)+\widetilde{\mathcal{B}}''_i(x_{j+1})}{12}\right),\qquad j\in\Delta_{M-1},\{0\}\\ \phi''(x_n,\tau)&=\sum_{i=0}^{M}\zeta_i\left(\frac{14\widetilde{\mathcal{B}}''_i(x_M)-5\widetilde{\mathcal{B}}''_i(x_{M-1})+4\widetilde{\mathcal{B}}''_i(x_{M-2})-\widetilde{\mathcal{B}}''_i(x_{M-3})}{12}\right).\end{aligned} \tag{14.8}$$

Utilizing the properties of modified cB-splines, equation (14.8) is simplified as

$$
\begin{aligned}
\phi''(x_0,\tau) &= \frac{-5\zeta_0 + 14\zeta_1 - 14\zeta_2 + 6\zeta_3 - \zeta_4}{2\delta x^2}, \\
\phi''(x_1,\tau) &= \frac{10\zeta_0 - 19\zeta_1 + 8\zeta_2 + \zeta_3}{2\delta x^2}, \\
\phi''(x_j,\tau) &= \frac{\zeta_{j-2} + 8\zeta_{j-1} - 18\zeta_j + 8\zeta_{j+1} + \zeta_{j+2}}{2\delta x^2}, \qquad j \in \Delta_{M-2}, \{0,1\}, \qquad (14.9) \\
\phi''(x_{M-1},\tau) &= \frac{10\zeta_M - 19\zeta_{M-1} + 8\zeta_{M-2} + \zeta_{M-3}}{2\delta x^2}, \\
\phi''(x_M,\tau) &= \frac{-5\zeta_M + 14\zeta_{M-1} - 14\zeta_{M-2} + 6\zeta_{M-3} - \zeta_{M-4}}{2\delta x^2}.
\end{aligned}
$$

We will utilize these spatial approximation formulas in the next section for numerical simulation of Fisher's equation (14.1).

14.3 IMPLEMENTATION OF mCBCT4 ON FISHER'S EQUATION

Fisher's equation (14.1) at grid point x_j with boundary conditions can be written as the following system of ODEs.

$$
\begin{aligned}
\phi_\tau(x_0,\tau) &= \dot{\psi}_1(\tau) \\
\phi_\tau(x_j,\tau) &= \nu\phi_{xx}(x_j,\tau) + \rho \mathrm{F}(\phi(x_j,\tau)), \qquad j \in \Delta_{M-1}, \{0\} \qquad (14.10) \\
\phi_\tau(x_M,\tau) &= \dot{\psi}_2(\tau).
\end{aligned}
$$

On utilizing fourth-order spatial approximations, we get

$$
\begin{aligned}
& 6\dot{\zeta}_0 = \dot{\psi}_1(\tau), \\
& \dot{\zeta}_0 + 4\dot{\zeta}_1 + \dot{\zeta}_2 = \nu \frac{10\zeta_0 - 19\zeta_1 + 8\zeta_2 + \zeta_3}{2\delta x^2} + \rho \mathrm{F}(\zeta_0 + 4\zeta_1 + \zeta_2) \\
& \dot{\zeta}_{j-1} + 4\dot{\zeta}_j + \dot{\zeta}_{j+1} = \nu \frac{\zeta_{j-2} + 8\zeta_{j-1} - 18\zeta_j + 8\zeta_{j+1} + \zeta_{j+2}}{2\delta x^2} + \rho \mathrm{F}(\zeta_{j-1} + 4\zeta_j + \zeta_{j+1}), \\
& \dot{\zeta}_{M-2} + 4\dot{\zeta}_{M-1} + \dot{\zeta}_M = \nu \frac{10\zeta_M - 19\zeta_{M-1} + 8\zeta_{M-2} + \zeta_{M-3}}{2\delta x^2} + \rho \mathrm{F}(\zeta_{M-2} + 4\zeta_{M-1} + \zeta_M), \\
& 6\dot{\zeta}_M = \dot{\psi}_2(\tau). \qquad (14.11)
\end{aligned}
$$

The above systems (14.11) can be expressed in compact matrix form as

$$\begin{bmatrix} 6 & 0 & & & & & \\ 1 & 4 & 1 & & & & \\ & 1 & 4 & 1 & & & \\ & & \ddots & \ddots & \ddots & & \\ & & & 1 & 4 & 1 & \\ & & & & 1 & 4 & 1 \\ & & & & & 0 & 6 \end{bmatrix} \begin{bmatrix} \dot{\zeta}_0 \\ \dot{\zeta}_1 \\ \vdots \\ \dot{\zeta}_j \\ \vdots \\ \dot{\zeta}_{M-1} \\ \dot{\zeta}_M \end{bmatrix} = \begin{bmatrix} \Gamma_0 \\ \Gamma_1 \\ \vdots \\ \Gamma_j \\ \vdots \\ \Gamma_{M-1} \\ \Gamma_M \end{bmatrix}, \tag{14.12}$$

where

$$\begin{aligned} \Gamma_0 &= \dot{\psi}_1(\tau), \\ \Gamma_1 &= \nu \frac{10\zeta_0 - 19\zeta_1 + 8\zeta_2 + \zeta_3}{2\delta x^2} + \rho \mathrm{F}(\zeta_0 + 4\zeta_1 + \zeta_2), \\ \Gamma_j &= \nu \frac{\zeta_{j-2} + 8\zeta_{j-1} - 18\zeta_j + 8\zeta_{j+1} + \zeta_{j+2}}{2\delta x^2} + \rho \mathrm{F}(\zeta_{j-1} + 4\zeta_j + \zeta_{j+1}), \\ \Gamma_{M-1} &= \nu \frac{10\zeta_M - 19\zeta_{M-1} + 8\zeta_{M-2} + \zeta_{M-3}}{2\delta x^2} + \rho \mathrm{F}(\zeta_{M-2} + 4\zeta_{M-1} + \zeta_M), \\ \Gamma_M &= \dot{\psi}_2(\tau). \end{aligned} \tag{14.13}$$

At each time level $\tau > 0$, first we utilize the well-known Thomas algorithm in system (14.12) of $M+1$ equations with $M+1$ unknowns $\dot{\zeta} = \left(\dot{\zeta}_0, \dot{\zeta}_1, \ldots, \dot{\zeta}_{M-1}, \dot{\zeta}_M\right)$. Thereafter, the initial values of system (14.12) of $M+1$ equations with $M+1$ unknowns together with initial solution vector (as evaluated in Section 14.3.1) are evaluated via SSP-RK43 scheme.

14.3.1 Initial Solution Vector for Systems

In order to start the calculation of (14.12), we require the initial solution vector $\tilde{\zeta}^0$, whose evaluation procedure is done via initial/boundary conditions of the considered problem. The initial solution vector $\tilde{\zeta}^0$ is computed from IC $\phi(x,0) = g(x)$ as follows:

$$\begin{aligned} &\phi(x_0,0) = \psi_1(0) \Rightarrow 6\zeta_0 = \psi_1(0), \\ &\phi(x_j,0) = g(x_j) \Rightarrow \zeta_{j-1} + 4\zeta_j + \zeta_{j+1} = \phi(x_j), \quad j \in \Delta_{M-1}, \{0\}, \\ &\phi(x_n,0) = \psi_2(0) \Rightarrow 6\zeta_M = \psi_2(0), \end{aligned}$$

which can be written in compact matrix form as

$$\begin{bmatrix} 6 & 0 & & & & & \\ 1 & 4 & 1 & & & & \\ & 1 & 4 & 1 & & & \\ & & \ddots & \ddots & \ddots & & \\ & & & 1 & 4 & 1 & \\ & & & & 1 & 4 & 1 \\ & & & & & 0 & 6 \end{bmatrix} \begin{bmatrix} \zeta_0 \\ \zeta_1 \\ \vdots \\ \zeta_j \\ \vdots \\ \zeta_{M-1} \\ \zeta_M \end{bmatrix} = \begin{bmatrix} \psi_1(0) \\ g(x_1) \\ \vdots \\ g(x_j) \\ \vdots \\ g(x_{M-1}) \\ \psi_2(0) \end{bmatrix}.$$

Solving the above system via the well-known Thomas algorithm returns initial solution vector for system (14.12).

14.4 NUMERICAL SIMULATION OF FISHER'S EQUATION USING mCBCT4

In this section, we will validate accuracy and efficiency of mCBCT4 by simulating Fisher's equation (14.1) with different initial and boundary conditions. Further, the efficacy of the technique is shown in terms of different kinds of error norms.

Example 14.1

As a first test problem, we consider Fisher's equation (14.1) with $F(\phi(x,\tau)) = \phi(1-\phi)$ in the computational region $\Omega_{-0.2,0.8}$ and the exact solution as

$$\phi(x,\tau) = \frac{1}{\left[1+\exp\left(\sqrt{\frac{\rho}{6}}x - \frac{5\rho}{6}\tau\right)\right]^2} \tag{14.14}$$

The initial and boundary conditions for the problem is directly taken from the exact solution (14.14).

First of all, we consider $\rho = 2000$ and compute the L_2 and L_∞ errors at different values of τ, which are reported in Table 14.2. Table 14.2 shows that we are getting comparable results. Next, we consider $\rho = 10000$ and the computed results are reported in Table 14.3 and Table 14.4. Comparison of L_2 and L_∞ errors at different values of τ in Table 14.3 shows that our results are better than those of CTB-DQM [17], EMCB-DQM [18], ASD [19], CN [19], [20], [21], [22], but comparable to DSC [19]. Numerical solution for $\rho = 2000, 10000$ at different values of τ is plotted in Figure 14.1(a) and Figure 14.1(b), respectively.

Example 14.2

In this test problem, we have considered Fisher's equation (1) in the computational region $\Omega_{0,1}$ with $\rho = 1$ and $F(\phi(x,\tau)) = \phi(\phi - \eth)(1-\phi)$. The initial and boundary conditions for the considered test problem is directly extracted from the exact solution (14.15).

$$\phi(x,\tau)=\frac{1}{2}\left[(1+\eth)+(1-\eth)\tanh\left(c-\frac{\sqrt{2}}{4}(1-\eth)(x-m\tau)\right)\right], \tag{14.15}$$

TABLE 14.2
Comparison of Errors of CTB-DQM [17], EMCB-DQM [18], Fourth-Order CBCM [15], and the Proposed Method with $\delta x = 0.025$ and $\rho = 2000$

	CTB-DQM		EMCB-DQM		CBCM	Present	
	$\delta\tau = 0.00001$		$\delta\tau = 0.00001$		$\delta\tau = 0.00001$	$\delta\tau = 0.000001$	
τ	L_2	L_∞	L_2	L_∞	L_∞	L_2	L_∞
0.0010	9.09E-04	5.18E-03	9.09E-04	5.18E-03	5.78E-04	1.01E-04	9.07E-05
0.0015	4.49E-04	2.45E-03	4.49E-04	2.45E-03	2.87E-04	5.02E-05	3.53E-05
0.0020	2.09E-04	1.11E-03	2.09E-04	1.11E-03	1.33E-04	3.65E-05	2.08E-05
0.0025	9.49E-05	4.93E-04	9.57E-05	4.93E-04	6.02E-05	4.08E-05	1.61E-05
0.0030	4.26E-05	2.17E-04	4.61E-05	2.17E-04	2.67E-05	4.54E-05	2.08E-05
0.0035	2.00E-05	9.50E-05	3.00E-05	9.50E-05	1.80E-05	6.51E-05	2.48E-05
0.0040	1.56E-05	4.15E-05	2.88E-05	7.23E-05	1.14E-05	1.22E-04	5.66E-05

TABLE 14.3
Comparison of Errors with CTB-DQM [17], EMCB-DQM [18], DSC [19], ASD [19], CN [19], [20], [21], [22] with $\rho = 10000$

	τ	0.0005	0.0010	0.0015	0.0020	0.0025	0.0030	0.0035
Present	L_2	1.40E-05	1.84E-05	1.54E-05	5.65E-05	1.14E-04	1.79E-04	2.50E-04
	L_∞	7.23E-06	7.40E-06	7.41E-06	1.73E-05	3.44E-05	5.41E-05	7.52E-05
CTB-DQM	L_2	1.14E-05	9.07E-05	3.41E-04	6.90E-04	1.10E-03	1.54E-03	1.99E-03
	L_∞	4.38E-05	4.11E-04	1.49E-03	3.96E-03	4.62E-03	6.53E-03	8.31E-03
EMCB-DQM	L_2	2.13E-05	2.08E-05	1.50E-04	3.74E-04	6.48E-04	9.48E-04	
	L_∞	1.05E-04	7.37E-05	6.87E-04	1.62E-03	2.76E-03	4.04E-03	
DSC	L_2	1.24E-06	6.53E-07	5.92E-07	8.35E-07	1.16E-06	1.43E-06	1.64E-06
	L_∞	6.28E-06	3.03E-06	1.98E-06	3.23E-06	4.46E-06	5.44E-06	6.22E-06
ASD	L_2	2.09E-03	6.07E-03	1.06E-02	1.53E-02	2.02E-02	2.68E-02	8.95E-02
	L_∞	1.07E-02	2.88E-02	4.93E-02	7.10E-02	9.37E-02	1.24E-01	3.42E-01
CN	L_2	1.92E-03	1.17E-02	2.65E-02	4.36E-02	6.18E-02	8.04E-02	9.91E-02
	L_∞	1.03E-02	5.55E-02	1.25E-01	2.04E-01	2.80E-01	3.60E-01	4.48E-01
[20]		1.10E-02		1.49E-01		3.44E-01		5.08E-01
[21]		2.55E-03		1.62E-02		8.65E-02		6.98E-02
[22] MMDAE	L_∞					9.25E-03		
[22] Method of lines	L_∞					9.34E-03		

where $m = \frac{1+\eth}{\sqrt{2}}$.

Table 14.5 reports the computed error norms for $\eth = 0.5$ at different values of τ. Comparison with [17] and [23] in Table 14.5 shows that we are getting better

TABLE 14.4
Comparison of the Evaluated Solutions with $\rho = 10000$ to Exact Solution and those from [13], [14], and [17]

	$\tau=0.001$					$\tau=0.003$				
x	Present	[13]	[14]	[17]	Exact	Present	[13]	[14]	[17]	Exact
–0.2	1.000000	1.00000	1.00000	1.000000	1.00000	1.000000	1.00000	1.00000	1.000000	1.00000
–0.1	0.999992	0.99999	0.99999	0.999992	0.99999	1.000000	1.00000	1.00000	1.000000	1.00000
0.1	0.972091	0.97203	0.97199	0.972092	0.97209	1.000000	1.00000	1.00000	1.000000	1.00000
0.2	0.293930	0.28644	0.29002	0.293583	0.29376	1.000000	1.00000	1.00000	1.000000	1.00000
0.3	0.000388	0.00032	0.00035	0.000379	0.00038	0.999994	0.99999	0.99999	0.999994	0.99999
0.4	0.000000	0.00000	0.00000	1.11E-07	0.00000	0.999659	0.99961	0.99964	0.999655	0.99966
0.5	0.000000	0.00000	0.00000	3.18E-11	0.00000	0.980162	0.97625	0.97839	0.979806	0.97995
0.6	0.000000	0.00000	0.00000	9.06E-15	0.00000	0.393038	0.32714	0.36191	0.386180	0.38895
0.7	0.000000	0.00000	0.00000	2.58E-18	0.00000	0.000774	0.00040	0.00057	0.000717	0.00074

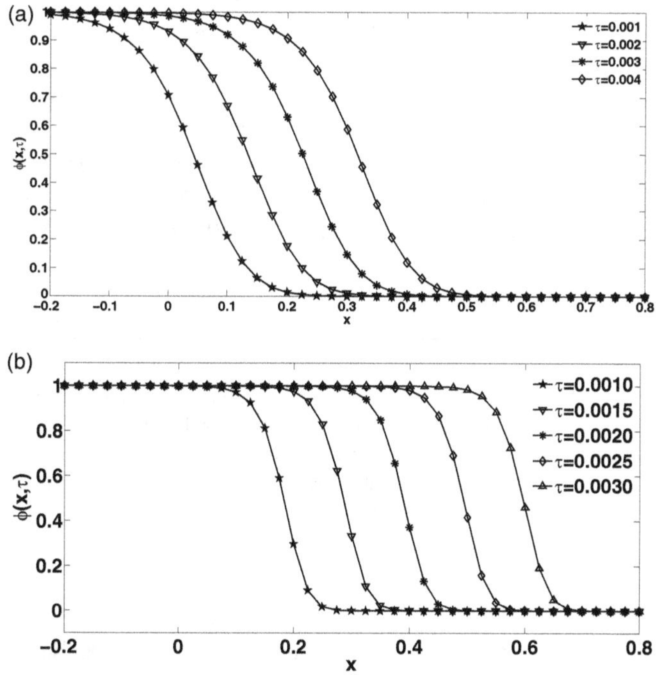

FIGURE 14.1 2D plot of numerical solution of Example 14.1 at different values of τ. (a) $\rho = 2000$; (b) $\rho = 10000$.

TABLE 14.5
Comparison of Evaluated Error Norms for Example 14.2 with $\eth = 0.5$ to those from [23] and [17]

		$\tau = 0.2$	$\tau = 0.5$	$\tau = 1.0$	$\tau = 3.0$	$\tau = 5.0$
	L_2	3.44E-08	2.86E-08	1.21E-08	4.20E-08	7.44E-08
Present	L_∞	4.48E-08	3.38E-08	1.53E-08	4.93E-08	8.65E-08
	RMS	3.42E-08	2.84E-08	1.20E-08	4.18E-08	7.40E-08
	L_2	7.90E-07	7.23E-07	6.19E-07	3.20E-07	1.58E-07
[23]	L_∞	2.41E-06	2.23E-06	1.94E-06	1.04E-06	5.23E-07
	RMS	3.05E-06	2.80E-06	2.43E-06	1.29E-06	6.43E-07
	L_2	1.04E-07	1.11E-07	1.01E-07	6.04E-08	3.20E-08
[17]	L_∞	1.41E-07	1.36E-07	1.25E-07	7.74E-08	4.16E-08
	RMS	6.14E-08	5.71E-08	4.96E-08	2.66E-08	1.34E-08

TABLE 14.6
Comparison of Evaluated Error Norms for Example 14.2 with $\eth = 1.5$ to those from [23] and [17]

		$\tau = 0.2$	$\tau = 0.5$	$\tau = 1.0$	$\tau = 3.0$	$\tau = 5.0$
	L_2	2.66E-08	1.20E-08	2.73E-08	2.12E-07	2.16E-07
Present	L_∞	3.25E-08	1.47E-08	3.72E-08	2.17E-07	2.59E-07
	RMS	2.65E-08	1.20E-08	2.72E-08	2.11E-07	2.15E-07
	L_2	7.05E-03	6.21E-03	3.34E-03	3.77E-05	3.76E-07
[23]	L_∞	2.11E-05	1.74E-05	1.10E-05	6.24E-05	6.17E-07
	RMS	5.73E-03	5.34E-03	3.06E-03	3.72E-03	3.75E-07
	L_2	9.88E-08	1.21E-07	1.35E-07	6.16E-08	1.18E-07
[17]	L_∞	1.33E-07	1.44E-07	1.53E-07	8.81E-08	1.49E-07
	RMS	8.98E-08	1.03E-07	1.28E-07	2.11E-07	1.75E-07

result for different values of τ except at $\tau = 5.0$, where the results are comparable. Table 14.6 reports the computed error norms for $\eth = 1.5$ at different values of τ. Comparison with [17] and [23] in Table 14.6 shows that we are getting comparable results. The surface plot of this example at $\eth = 0.5$ and 1.5 for $\tau \leq 5$ is depicted in Figure 14.2a and Figure 14.2b, respectively.

14.5 CONCLUSIONS

A fourth-order collocation technique based upon the modified cubic B-splines has been successfully implemented on Fisher's equation to evaluate the new numerical solutions. At first, Fisher's equation is considered at different grid points of the computational region and then the fourth-order spatial approximation of the unknown

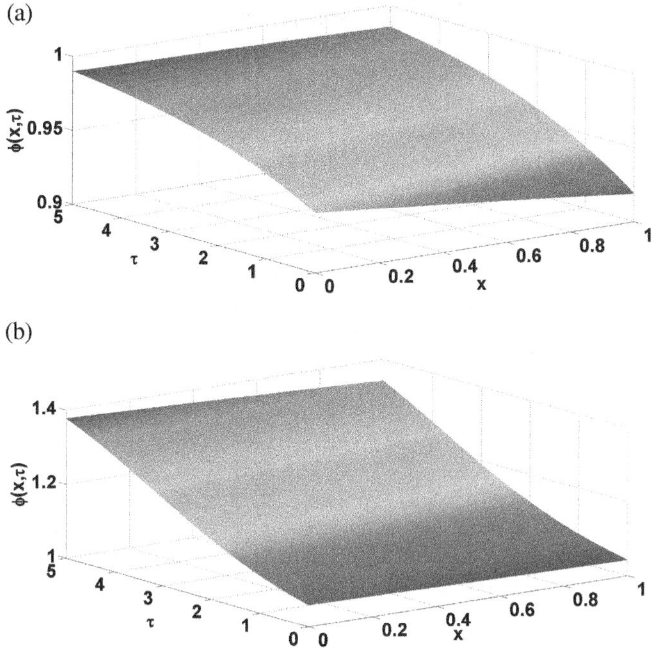

FIGURE 14.2 Surface plot of numerical solution of Example 14.2 for $\tau \leq 5$. (a) $\vartheta = 0.5$; (b) $\vartheta = 1.5$.

is implemented on it, which yields a system of first-order ODEs that we solve using SSP-RK43 method.

The numerical results, tables, and figures obtained in the numerical simulation section confirm that the technique is accurate, effective, and efficient for Fisher's equation in terms of L_2 and L_∞ error norms. Also, the numerical observation shows that the computed results from this technique are acceptable and better than some existing results.

ACKNOWLEDGMENT

The author M. Gupta is thankful to CSIR, New Delhi, India (File No. 09/961(0015)/2019-EMR-I), for their financial assistance to carry out this work.

REFERENCES

1. Fisher, R.A. (1936) The Wave of Advance of an Advantageous Gene. *Annals Eugen*, 7, 355–369.
2. Canosa, J. (1969) Diffusion in nonlinear multiplicative media. *Journal Mathematics Phys*ics, 10, 1862–1868.
3. Zeldovich, J.B., & Frank-Kamenetzk, D.A. (1938) A theory of thermal propagation of flame. *Acta Physicochimica U.R.S.S*, 9(2), 341–350.
4. Bramson, M.D. (1978) Maximal displacement of branching Brownian motion. *Communicatons Pure Application Mathematics*, 31, 531–581.

5. Fife, P.C., & McLeod, J.B. (1977) The approach of solutions of nonlinear diffusion equations to travelling front solutions. *Archieve Rational Mechanical Analysis*, 65, 335–361.
6. Maini, P. K., McElwain, D.L.S., & Leavesley, D. (2004) Travelling waves in a wound healing assay. *Applications Mathematics Letters*, 17, 575–580.
7. Tuckwell, H.C. (1988) *Introduction to Theoretical Neurobiology*. Cambridge University Press, Cambridge (UK).
8. Aggarwal, S.K. (1925) Some numerical experiments on Fisher equation. *International Communication Heat Mass Trans*fer, 12, 417–430.
9. Olmos, D., & Shizgal, B.D. (2006) A pseudospectral method of solution of Fisher's equation. *Journal Computer Applications Math*ematics, 193, 219–242.
10. Tang, S. & Weber, R.O. (1991) Numerical study of Fisher's equation by a Petrov–Galerkin finite element method. *Journal of the Australian Mathematical Society Series*, 33, 2738
11. Carey, G.F. & Shen, Y. (1995) Least-squares finite element approximation of Fisher's reaction–diffusion equation. *Numerical Methods for Partial Differential Equations*, 11, 175–186.
12. Al-Khaled, K. (2001) Numerical study of Fisher's reaction–diffusion equation by the Sinc collocation method. *Journal Computer Application Mathematics*, 137, 245–255.
13. Mittal, R.C. & Arora, G. Efficient numerical solution of Fisher's equation by using B-spline method. *International Journal Computer Mathematics*, 87(13), 3039–3051.
14. Mittal, R.C. & Jain, R.K. (2013) Numerical solutions of nonlinear Fisher's reaction-diffusion equation with modified cubic B-spline collocation method. *Mathematical Sciences*, 7(12), 1–10.
15. Rohila, R. & Mittal, R.C. (2013) Numerical study of reaction diffusion Fisher's equation by fourth order cubic B-spline collocation method. *Mathematical Sciences*, 12, 79–89.
16. Singh, B.K. & Arora, G. A numerical scheme to solve Fisher-type reaction-diffusion Equations. *MESA*, 5(2), 153–164.
17. Tamsir, M., Dhiman, N., & Srivastava, V.K. (2018) Cubic trigonometric B-spline differential quadrature method for numerical treatment of Fisher's reaction-diffusion equations. *Alex Engineering Journal*, 57, 2019–2026.
18. Shukla, H.S. & Tamsir, M. (2016) Extended modified cubic B-spline algorithm for nonlinear Fisher's reaction-diffusion equation. *Alex Engineering Journal*, 55, 2871–2879
19. Zhao, S. & Wei, G.W. (2003) Comparison of the discrete singular convolution and three other numerical schemes for solving Fisher's equation. *SIAM Journal Science Comput*er, 25, 127–147.
20. Dag, I. & Ersoy, O. (2016) The exponential cubic B-spline algorithm for Fisher equation, *Chaos, Solit. Fract.*, 86, 101–106.
21. Dag, I., Sahin, A. & Korkmaz, A. (2010) Numerical investigation of the solution of Fisher's equation via the b-spline Galerkin method. *Numerical Methods Partial Differential Equ*ation, 26(6), 1483–1503.
22. Qiu, Y. & Sloan, D.M. (1998) Numerical solution of Fisher's equation using a moving mesh method. *Journal Computer Phys*ics, 146, 726–746.
23. Verma, A., Jiwari, R. & Koksal, M.E. (2014) Analytic and numerical solutions of nonlinear diffusion equations via symmetry reductions. *Advance Differential Equations*, 2014, 229.
24. Singh, B.K. & Gupta, M. (2021) A new efficient fourth order collocation scheme for solving Burgers' equation. *Applied Mathematics and Computation*, 399, 126011.
25. Arora, G. & Singh, B.K. (2013) Numerical solution of Burgers' equation with modified cubic B-spline differential quadrature method. *Applied Mathematics and Computation*, 224, 166–177.
26. Lucas, T.R. (1975) Error bounds for interpolating cubic splines under various end condition. *SIAM Journal Numerical Ana*lysis, 11, 569–584.

15 Analytical Study of Higher-Order Time-Fractional Differential Equations with Proportional Delay for Large Time Scale

Brajesh Kumar Singh and Saloni Agrawal
Babasaheb Bhimrao Ambedkar University Lucknow

CONTENTS

15.1 INTRODUCTION

Delay differential equations (DDEs) are a section of functional differential equation used in vast area of applications in physical and biological sciences [4–9]. DDEs provide useful mathematical mechanisms to formulate distinct phenomena from real-life issues, as the unknown function in DDEs depends on the current state as well as some previous values. Moreover, the fractional model of delay differential equations attracts the researchers due to its nonlocal nature and an immense scope in various real-world phenomena of fluid mechanics, neural networks, signal processing, diffusion reaction process, and other fields of science and engineering. Recently, vigorous techniques have been developed to obtain the analytical solution to integer-order nonlinear differential equations with proportional delay; see [10–16] for more details. Moreover, it is a great task to solve nonlinear DDEs and obtain convergent

DOI: 10.1201/9781003222255-15

solution for large time scale due to their special transcendental nature. However, fractional-order DDEs have been solved by various numerical methods [18–25] and analytical techniques [17, 26–31] for the time interval $0 \le t \le 1$.

The DGJTM is a powerful technique introduced by Daftardar-Gejji and Jafari [32] for solving nonlinear problems [33–35]. In this article, we introduced a new analytical approach for the combination of DGJM and Laplace transform named as DGJTM to analyze nonlinear differential equations with proportional delay. DGJTM is a thoroughly easy method compared to other existing methods [36–39] and produces a derivative-free technique to compute nonlinear term. This method gives solutions in a series form, which converges to an exact solution using a few number of terms. Additionally, sufficient condition for convergence of DGJTM series solution to time-fractional delay differential equation is introduced to assure the uniform convergence of the new technique for large time interval [37]. In this chapter, pantograph differential equation and higher-order neutral delay differential equation have been examined by the DGJTM method for large time scale. The pantograph differential equations and neutral delay differential equations formulate a special class of differential equations, which has a wide range of applications in various applied fields of science, engineering and economics for quick hike and drop in stock exchange.

15.2 PRELIMINARIES

The Riemann–Liouville fractional integral operator and Caputo fractional differential operator [1–3] are defined as follows:

Definition: ([1–3]) i) The Riemann–Liouville fractional integral operator $\mathcal{D}_t^{-\alpha}\phi(t)$ of order $\alpha \ge 0$ for $\phi \in C_\mu, \mu \ge -1$ is defined by $\mathcal{D}_t^{0}\phi(t) := \phi(t)$ and $\mathcal{D}_t^{-\alpha}\phi(t) = \dfrac{1}{\Gamma(\alpha)}\int_0^t (t-\epsilon)^{\alpha-1}\phi(\epsilon)d\epsilon, \quad \alpha > 0, \quad t > 0.$

ii) The Caputo fractional differential operator $\mathcal{D}_t^{\alpha}\phi(t)$ of order $m-1 < \alpha \le m$ of a function $\phi \in C_\mu, \mu \ge -1$ is defined by $\mathcal{D}_t^{m}\phi(t) := \dfrac{\partial^m \phi(\epsilon)}{\partial \epsilon^m}$ and

$$\mathcal{D}_t^{\alpha}\phi(t) = \mathcal{D}_t^{-(m-\alpha)}\mathcal{D}_t^{m}\phi(t) = \frac{1}{\Gamma(m-\alpha)}\int_0^t (m-\epsilon)^{m-(\alpha+1)}\frac{\partial^m \phi(\epsilon)}{\partial \epsilon^m}d\epsilon, \quad m-1 < \alpha \le m.$$

Definition: ([40–41]) The Laplace transform for piecewise continuous function $\phi(t)$ in $(0,\infty)$ is defined as: $\mathbb{L}\{\phi(t)\} = \int_0^\infty \phi(t)e^{-st}\,dt,$

where s is a parameter. Further, the Laplace transform for Caputo fractional differential operator for any function $\phi \in C_\mu, \mu \ge -1$ is defined as:

$$\mathbb{L}\{\mathcal{D}_t^{\alpha}\phi(t)\} = s^{\alpha}\mathbb{L}\{\phi(t)\} - \sum_{k=0}^{m-1} s^{\alpha-k-1}\phi^k(0), m-1 < \alpha \le m, m \in \mathbb{N}$$

$$\mathbb{L}\{\mathcal{D}_t^{m\alpha}\phi(t)\} = s^{m\alpha}\mathbb{L}\{\phi(t)\} - \sum_{k=0}^{m-1} s^{(m-k)\alpha-1}\left(\mathcal{D}_t^{m\alpha}\phi^k\right)(0), 0 < \alpha \le 1, m \in \mathbb{N}$$

15.3 METHODOLOGY

In this section, we depict the Daftardar-Gejji and Jafari transform method (DGJTM) and its applications for nonlinear time-fractional delay differential equations. Consider the general form of nonlinear time-fractional delay differential equations as:

$$\begin{aligned} &\mathcal{D}_t^{m\alpha}\phi(t) = \mathcal{L}(\phi) + \mathcal{N}(\phi) + g, \quad 0 < \alpha \leq 1, m \in \mathbb{N} \\ &\mathcal{D}_t^{m\alpha}\phi^k(0) = C_k, \quad k = 0,1,\ldots,m-1 \end{aligned} \tag{15.1}$$

where $\mathcal{L}$ and $\mathcal{N}$ are linear and nonlinear functions, respectively, g is a nonhomogeneous function, $\mathcal{D}_t^{m\alpha}$ is the Caputo fractional derivative operator of order $m\alpha$, and $C_k(k = 0,1,\ldots,m-1)$ are the constants.

On applying Laplace transform and utilizing its properties for Caputo fractional derivative, equation (15.1) can be deformed as

$$s^{m\alpha}\mathbb{L}\{\phi(t)\} - \sum_{k=0}^{m-1} s^{(m-k)\alpha-1}\mathcal{D}_t^{m\alpha}\phi^k(0) = \mathbb{L}\{\mathcal{L}(\phi) + \mathcal{N}(\phi) + g\}, m-1 \leq \alpha \leq m \tag{15.2}$$

Now applying inverse Laplace transform on both sides of equation (15.2), we obtain

$$\phi(t) = \mathcal{G}(t) + \mathbb{L}^{-1}\left\{\frac{1}{s^{m\alpha}}\mathbb{L}\{\mathcal{L}(\phi) + \mathcal{N}(\phi)\}\right\} \tag{15.3}$$

For Daftardar-Gejji and Jafari transform method (DGJTM), linear, nonlinear and its DGJTM ith-order solution can be deformed as follows:

$$\begin{aligned} &\mathcal{L}(\phi) = \sum_{i=0}^{\infty}\mathcal{L}(\phi_i) \\ &\mathcal{N}(\phi) = \sum_{i=0}^{\infty}\left\{\mathcal{N}\left(\sum_{j=0}^{i}(\phi_j)\right) - \mathcal{N}\left(\sum_{j=0}^{i-1}(\phi_j)\right)\right\} \\ &\phi(t) = \sum_{i=0}^{\infty}\phi_i \end{aligned} \tag{15.4}$$

Substituting equation (15.4) in equation (15.3), we get

$$\sum_{i=0}^{\infty}\phi_i = \mathcal{G} + \mathbb{L}^{-1}\left\{\frac{1}{s^{m\alpha}}\mathbb{L}\left\{\sum_{i=0}^{\infty}\mathcal{L}(\phi_i) + \sum_{i=0}^{\infty}\left\{\mathcal{N}\left(\sum_{j=0}^{i}(\phi_j)\right) - \mathcal{N}\left(\sum_{j=0}^{i-1}(\phi_j)\right)\right\}\right\}\right\} \tag{15.5}$$

Here, $\mathcal{G}(t)$ represents the nonhomogeneous function with initial condition and can be evaluated as: $\mathcal{G} = \mathbb{L}^{-1}\left\{\sum_{k=0}^{m-1} \frac{1}{s^{k\alpha+1}} \mathcal{D}_t^{m\alpha}\phi^k(0) + \frac{1}{s^{m\alpha}}\mathbb{L}\{g\}\right\}$

Comparing the coefficient on both sides, the ith-order DGJTM series solution can be evaluated as:

$$\phi_0 = \mathcal{G}; \quad \phi_{i+1}(t) = \mathbb{L}^{-1}\left\{\frac{1}{s^{m\alpha}}\mathbb{L}\left\{\mathcal{L}(\phi_i) + \mathcal{N}\left(\sum_{j=0}^{i}(\phi_j)\right) - \mathcal{N}\left(\sum_{j=0}^{i-1}(\phi_j)\right)\right\}\right\}, \tag{15.6}$$

$i = 0,1,\ldots k-1$

15.4 SUFFICIENT CONDITION FOR CONVERGENCE

Let f be a piecewise continuous function defined on a rectangle R of dimension three, where $R = \{(t,\phi_1,\phi_2) \mid 0 \le t \le b, -\delta \le \phi_1 \le \delta, -\epsilon \le \phi_2 \le \epsilon\}$ and f is a bounded function, i.e., $|f(t,\phi_1,\phi_2)| \le M, \forall (t,\phi_1,\phi_2) \in R$. Moreover, if f satisfies the Lipschitz condition:

$|f(t,\phi_1,\phi_2) - f(t,\psi_1,\psi_2)| \le k_1|\phi_1 - \psi_1| + k_2|\phi_2 - \psi_2|$, then DGJTM series solution to the initial value problem (5) converges uniformly in the interval $[0,b]$.

$$\begin{aligned} &\mathcal{D}_t^{m\alpha}\phi(t) = f(t,\phi(t),\phi(\sigma t)),\ 0 < \alpha \le 1,\ m \in \mathbb{N} \\ &\mathcal{D}_t^{m\alpha}\phi^k(0) = C_k, \quad k = 0,1,\ldots,m-1, 0 < \sigma \le 1 \end{aligned} \tag{15.7}$$

Proof: For the time-fractional delay differential equation (15.7),

$$\mathcal{D}_t^{m\alpha}\phi(t) = f(t,\phi(t),\phi(\sigma t)), m-1 < \alpha \le m, m \in \mathbb{N}$$

$$\mathcal{D}_t^{m\alpha}\phi^k(0) = C_k, \quad k = 0,1,\ldots,m-1, 0 < \sigma \le 1$$

Using DGJTM, we get

$$\phi_0 = \mathcal{G};$$

$$\phi_1(t) = \mathbb{L}^{-1}\left\{\frac{1}{s^{m\alpha}}\mathbb{L}\{f(t,\phi_0(t),\phi_0(\sigma t))\}\right\}$$

$$|\phi_1(t)| = \mathbb{L}^{-1}\left\{\frac{1}{s^{m\alpha}}\mathbb{L}\{|f(t,\phi_0(t),\phi_0(\sigma t))|\}\right\}$$

$\Rightarrow |\phi_1(t)| \le M\dfrac{t^{m\alpha}}{\Gamma(m\alpha+1)}$; as f is a bounded function

$\Rightarrow |\phi_1(\sigma t)| \le M\dfrac{(\sigma t)^{m\alpha}}{\Gamma(m\alpha+1)} \le M\dfrac{t^{m\alpha}}{\Gamma(m\alpha+1)}, \forall t \in [0,b]$; since $0 < \sigma \le 1, \dfrac{b}{\sigma} \ge b$.

Further, $\phi_2(t)=\mathbb{L}^{-1}\left\{\frac{1}{s^{m\alpha}}\mathbb{L}\left\{f\left(t,\phi_1(t)+\phi_0(t),\phi_1(\sigma t)+\phi_0(\sigma t)\right)-f\left(t,\phi_0(t),\phi_0(\sigma t)\right)\right\}\right\}$

$$\Rightarrow |\phi_2(t)|\le \mathbb{L}^{-1}\left\{\frac{1}{s^{m\alpha}}\mathbb{L}\left\{k_1|\phi_1(t)|+k_2|\phi_1(\sigma t)|\right\}\right\}$$

$$\Rightarrow |\phi_2(t)|\le k_1M\frac{t^{2m\alpha}}{\Gamma(1+2m\alpha)}+k_2M\frac{\sigma^{2m\alpha}t^{2m\alpha}}{\Gamma(1+2m\alpha)}\le M(k_1+k_2)\frac{t^{2m\alpha}}{\Gamma(1+2m\alpha)},$$

and so

$$|\phi_2(\sigma t)|\le M(k_1+k_2)\frac{\sigma^{2m\alpha}t^{2m\alpha}}{\Gamma(1+2m\alpha)}\le M(k_1+k_2)\frac{t^{2m\alpha}}{\Gamma(1+2m\alpha)},\ \forall t\in[0,b]$$

Similarly,

$$|\phi_3(t)|\le M(k_1+k_2)^2\frac{t^{3m\alpha}}{\Gamma(1+3m\alpha)}$$

In general,

$$|\phi_n(t)|\le M(k_1+k_2)^{n-1}\frac{t^{nm\alpha}}{\Gamma(1+nm\alpha)},\ n=1,2,3,\ldots$$

On taking summation over n, we get

$$\left|\sum_{n=0}^{\infty}\phi_n(t)\right|\le\frac{M}{(k_1+k_2)}E_{m\alpha}\left((k_1+k_2)t^{m\alpha}\right)+\left(\mathcal{G}-\frac{M}{(k_1+k_2)}\right),\ t\in[0,b],$$

where $E_{m\alpha}(\cdot)$ represents the Mittag-Leffler function [42], which is a convergent function, and therefore, one can identify that the computed DGJTM series solution of equation (15.7) converges uniformly in the interval $[0,b]$.

15.5 NUMERICAL APPLICATION

Example 15.1

First, we consider the nonlinear pantograph delay differential equation

$$\mathcal{D}_t^{\alpha}\phi(t)=1-2\phi^2\left(\frac{t}{2}\right)$$
$$\phi(0)=0,\qquad t\in[0,1],\qquad 0<\alpha\le 1.\tag{15.8}$$

In special case, when $\alpha=1$, the exact solution to this problem is $\phi(t)=\sin t$.

By applying DGJTM as described in equations (15.5–15.6) on Example 15.1 equation (15.8), we get:

$$\phi_0(t) = \mathbb{L}^{-1}\left\{\frac{1}{s^\alpha}\mathbb{L}\{1\}\right\};$$
$$\phi_{i+1}(t) = \mathbb{L}^{-1}\left\{\frac{1}{s^\alpha}\mathbb{L}\left\{-2\left[\left(\sum_{j=0}^{i}\phi_j\left(\frac{t}{2}\right)\right)^2 - \left(\sum_{j=0}^{i-1}\phi_j\left(\frac{t}{2}\right)\right)^2\right]\right\}\right\}, \quad i = 0,1,\ldots k-1 \tag{15.9}$$

With the aid of Mathematica, on solving the above recurrence relation (15.9), the DGJTM iterative terms are obtained as:

$$\phi_0(t) = \frac{t^\alpha}{\Gamma(\alpha+1)}; \quad \phi_1(t) = -\frac{2^{1-2\alpha}\Gamma(2\alpha+1)t^{3\alpha}}{\Gamma(\alpha+1)^2\Gamma(3\alpha+1)};$$
$$\phi_2(t) = \frac{2^{3-6\alpha}\Gamma(2\alpha+1)\Gamma(4\alpha+1)t^{5\alpha}}{\Gamma(\alpha+1)^3\Gamma(3\alpha+1)\Gamma(5\alpha+1)} - \frac{2^{3-10\alpha}\Gamma(2\alpha+1)^2\Gamma(6\alpha+1)t^{7\alpha}}{\Gamma(\alpha+1)^4\Gamma(3\alpha+1)^2\Gamma(7\alpha+1)}\cdots \tag{15.10}$$

In this series, ith-order DGJTM series solutions for $i \geq 3$ at $\alpha = 1$ are obtained as:

$$\phi(t) = \sum_{i=0}^{4}\phi_i(t) = -\frac{t^{31}}{1062664199886151693758358595882188800} +$$
$$\frac{t^{29}}{2003250224799783128854752657408000} - \frac{25183t^{27}}{2083660861517231320180010778624000} +$$
$$\frac{246223t^{25}}{139133337441054441785524224000000} - \frac{334309t^{23}}{1909762064484959514525696000 0} +$$
$$\frac{140087t^{21}}{113226999870649773588480000} - \frac{12371477t^{19}}{1920815176377094373376000} +$$
$$\frac{59828773t^{17}}{2386977923860424294400 0} - \frac{2005361t^{15}}{274239191619993600 0} + \frac{64711t^{13}}{408094035148800} -$$
$$\frac{8177t^{11}}{326998425600} + \frac{t^9}{362880} - \frac{t^7}{5040} + \frac{t^5}{120} - \frac{t^3}{6} + t, \tag{15.11}$$

which is closed form of exact solution $\phi(t) = \sin(t)$ at $\alpha = 1$. It is also noteworthy that we have extended the approximate solution of Example 15.1 for large time scale $t \in (0,15)$ using DGJTM, while other iterative techniques such as ADM/LVIM, HAM and DTM [14,16] compute the analytical solution in the interval $t \in (0,1)$.

Example 15.2

Consider the second-order nonlinear neutral functional delay differential equation

$$\mathcal{D}_t^{2\alpha}\phi(t) = -\phi(t) + 5\phi^2\left(\frac{t}{2}\right)$$
$$\phi(0) = 1, \qquad \phi'(0) = -2, \qquad t \in [0,1], \qquad 0 < \alpha \le 1. \tag{15.12}$$

In special case, when $\alpha = 1$, the exact solution to this problem is $\phi(t) = \exp^{-2t}$.

By applying DGJTM as described in equations (15.5–15.6) on Example 15.2 equation (15.12), we get

$$\phi_0(t) = \mathbb{L}^{-1}\left\{\frac{1}{s}\phi(0) + \frac{1}{s^{\alpha+1}}\phi'(0)\right\};$$
$$\phi_{i+1}(t) = \mathbb{L}^{-1}\left\{\frac{1}{s^{2\alpha}}\mathbb{L}\left\{-\phi_i(t) + 5\left[\left(\sum_{j=0}^{i}\phi_j\left(\frac{t}{2}\right)\right)^2 - \left(\sum_{j=0}^{i-1}\phi_j\left(\frac{t}{2}\right)\right)^2\right]\right\}\right\}, i = 0,1,\ldots k-1 \tag{15.13}$$

With the aid of Mathematica, on solving the above recurrence equation (15.13), the DGJTM iterative terms are obtained as:

$$\phi_0(t) = 1 - \frac{2t^\alpha}{\Gamma(\alpha+1)}; \qquad \phi_1(t) = 2t^{2\alpha}\left(\frac{2}{\Gamma(2\alpha+1)} + t\left(\frac{5t}{\Gamma(2\alpha+3)} - \frac{4}{\Gamma(2\alpha+2)}\right)\right)$$

$$\phi_2(t) = \frac{1}{4}t^{4\alpha}\left(\frac{5\,2^{5-2\alpha} - 16}{\Gamma(4\alpha+1)} + \frac{5\,4^{3-2\alpha}\Gamma(4\alpha+1)t^{2\alpha}}{\Gamma(2\alpha+1)^2\Gamma(6\alpha+1)} + \right.$$
$$t\left(\frac{32\left(-5\,4^{-\alpha} - \frac{5\,4^{-\alpha}\Gamma(2\alpha+2)}{\Gamma(2\alpha+1)} + 1\right)}{\Gamma(4\alpha+2)} - \frac{5\,2^{7-4\alpha}\Gamma(4\alpha+2)t^{2\alpha}}{\Gamma(2\alpha+1)\Gamma(2\alpha+2)\Gamma(6\alpha+2)} + \right.$$
$$5t\left(\frac{5\,4^{1-\alpha} + \frac{2^{5-2\alpha}\Gamma(2\alpha+3)}{\Gamma(2\alpha+2)} - 8}{\Gamma(4\alpha+3)} + \frac{2^{4-4\alpha}\left(5\Gamma(2\alpha+2)^2 + 4\Gamma(2\alpha+1)\Gamma(2\alpha+3)\right)\Gamma(4\alpha+3)t^{2\alpha}}{\Gamma(2\alpha+1)\Gamma(2\alpha+2)^2\Gamma(2\alpha+3)\Gamma(6\alpha+3)} + \right.$$
$$\left.\left.\left.\frac{5\,2^{-4\alpha}t\left(\frac{5\Gamma(4\alpha+5)t^{2\alpha+1}}{\Gamma(6\alpha+5)} + \Gamma(2\alpha+3)\left(-\frac{4^{\alpha+1}\Gamma(2\alpha+4)}{\Gamma(4\alpha+4)} - \frac{16\Gamma(4\alpha+4)t^{2\alpha}}{\Gamma(2\alpha+2)\Gamma(6\alpha+4)}\right)\right)}{\Gamma(2\alpha+3)^2}\right)\right)\right) \tag{15.14}$$

In this series, the ith-order DGJTM series solutions for $i \geq 3$ at $\alpha = 1$ are obtained as:

$$\phi(t) = \sum_{i=0}^{3} \phi_i(t) = \frac{3125t^{22}}{21330047071393918848} - \frac{625t^{21}}{605967246346420224} + \frac{69875t^{20}}{1918896280096997376} - \frac{51875t^{19}}{59965508753031168} + \frac{1437875t^{18}}{93893362389614592} - \frac{19925t^{17}}{93148176973824} + \frac{53905t^{16}}{21917218111488} - \frac{3197t^{15}}{133177540608} + \frac{1063235t^{14}}{5193924083712} - \frac{575803t^{13}}{370994577408} + \frac{16096343t^{12}}{1569592442880} - \frac{481753t^{11}}{8174960640} + \frac{112199t^{10}}{371589120} - \frac{1247t^9}{884736} + \frac{6505t^8}{1032192} - \frac{8t^7}{315} + \frac{4t^6}{45} - \frac{4t^5}{15} + \frac{2t^4}{3} - \frac{4t^3}{3} + 2t^2 - 2t + 1,$$

which is closed form of exact solution $\phi(t) = \exp(-2t)$ at $\alpha = 1$. It is also noteworthy that we have extended the approximate solution of Example 15.2 for large time scale $t \in (0,15)$ using DGJTM, while the other iterative techniques such as ADM/LVIM, HAM and DTM [14,16] compute the analytical solution in the interval $t \in (0,1)$.

Example 15.3

Let us consider the third-order nonlinear time-fractional neutral functional delay differential equation as

$$\mathcal{D}_t^{3\alpha}\phi(t) = 2\phi^2\left(\frac{t}{2}\right) - 1;$$
$$\phi(0) = 0, \quad \phi'(0) = 1, \quad \phi''(0) = 0, \quad t \in [0,1], \quad 0 < \alpha \leq 1. \tag{15.15}$$

In special case, when $\alpha = 1$, the exact solution to this problem is $\phi(t) = \sin t$.

By applying DGJTM as described in equations (15.5–15.6) on Example 15.3 equation (15.15), we get:

$$\phi_0(t) = \mathbb{L}^{-1}\left\{\frac{1}{s}\phi(0) + \frac{1}{s^{\alpha+1}}\phi'(0) + \frac{1}{s^{2\alpha+1}}\phi''(0) - \frac{1}{s^{\alpha}}\mathbb{L}\{1\}\right\};$$
$$\phi_{i+1}(t) = \mathbb{L}^{-1}\left\{\frac{1}{s^{3\alpha}}\mathbb{L}\left\{2\left[\left(\sum_{j=0}^{i}\phi_j\left(\frac{t}{2}\right)\right)^2 - \left(\sum_{j=0}^{i-1}\phi_j\left(\frac{t}{2}\right)\right)^2\right]\right\}\right\}, i = 0,1,\ldots k-1 \tag{15.16}$$

With the aid of Mathematica, on solving the above recurrence equation (15.16), the DGJTM iterative terms are computed as

$$\phi_0(t)=\frac{t^{\alpha}}{\Gamma(\alpha+1)}-\frac{t^{3\alpha}}{\Gamma(3\alpha+1)};\qquad \phi_1(t)=\frac{2^{1-6\alpha}t^{6\alpha}\left(\frac{\Gamma(6\alpha+1)t^{3\alpha}}{\Gamma(9\alpha+1)}-\frac{8^{\alpha}t\Gamma(3\alpha+1)\Gamma(3\alpha+2)}{\Gamma(6\alpha+2)}\right)}{\Gamma(3\alpha+1)^2}+\frac{t^{3\alpha+2}}{\Gamma(3\alpha+3)};$$

$$\phi_2(t)=\frac{2\times4^{-9\alpha-1}t^{6\alpha+2}}{\Gamma(3\alpha+3)}\left(\frac{2t^{3\alpha}}{\Gamma(3\alpha+1)^2}\left(\begin{array}{l}\frac{2^{15\alpha}t\Gamma(3\alpha+4)}{\Gamma(6\alpha+4)}+\frac{2\Gamma(6\alpha+1)\Gamma(12\alpha+3)t^{6\alpha}}{\Gamma(9\alpha+1)\Gamma(15\alpha+3)}+\\ 64^{\alpha}\Gamma(3\alpha+1)\left(-\frac{64^{\alpha}\Gamma(6\alpha+3)}{\Gamma(9\alpha+3)}-\frac{\Gamma(3\alpha+2)\Gamma(9\alpha+4)t^{3\alpha+1}}{\Gamma(6\alpha+2)\Gamma(12\alpha+4)}\right)\end{array}\right)\right)+$$

$$\frac{2\times4^{-3\alpha-2}\Gamma(6\alpha+5)t^{9\alpha+4}}{\Gamma(3\alpha+3)^2\Gamma(9\alpha+5)}+\frac{2\times2^{-30\alpha}t^{9\alpha}}{\Gamma(3\alpha+1)^4}\left(\frac{4\Gamma(6\alpha+1)^2\Gamma(18\alpha+1)t^{12\alpha}}{\Gamma(9\alpha+1)^2\Gamma(21\alpha+1)}-\frac{2^{21\alpha}t^2\Gamma(3\alpha+1)^3\Gamma(3\alpha+2)\Gamma(6\alpha+3)}{\Gamma(6\alpha+2)\Gamma(9\alpha+3)}\right.$$

$$-\frac{4^{3\alpha+1}\Gamma(3\alpha+1)\Gamma(6\alpha+1)t^{6\alpha}\left(64^{\alpha}\Gamma(6\alpha+2)\Gamma(12\alpha+1)\Gamma(18\alpha+2)+\Gamma(3\alpha+2)\Gamma(15\alpha+1)\Gamma(15\alpha+2)t^{3\alpha+1}\right)}{\Gamma(6\alpha+2)\Gamma(9\alpha+1)\Gamma(15\alpha+1)\Gamma(18\alpha+2)}+$$

$$\left.4096^{\alpha}\Gamma(3\alpha+1)^2t^{3\alpha+1}\left(\frac{2^{6\alpha+1}\Gamma(3\alpha+2)\Gamma(9\alpha+2)}{\Gamma(6\alpha+2)\Gamma(12\alpha+2)}+\frac{2^{3\alpha+1}\Gamma(6\alpha+1)\Gamma(9\alpha+2)}{\Gamma(9\alpha+1)\Gamma(12\alpha+2)}+\frac{\Gamma(3\alpha+2)^2\Gamma(12\alpha+3)t^{3\alpha+1}}{\Gamma(6\alpha+2)^2\Gamma(15\alpha+3)}\right)\right)$$

In this series, the ith-order DGJTM series solutions for $i\geq 2$ at $\alpha=1$ are obtained as:

$$\phi(t)=\sum_{i=0}^{3}\phi_i(t)=t-\frac{t^3}{6}+\frac{\left(5t^4-576t^2+24192\right)t^5}{2903040}$$

$$+\frac{t^9}{214290943114569591029760000}\left(\begin{array}{l}60775t^{12}-76876800t^{10}+43644496896t^{8}-\\ 148748858818856t^{6}+3390897869291520t^{4}-\\ 5368439932924723200t^{2}+221448147233144832000\end{array}\right),$$

which is closed form of exact solution $\phi(t)=\sin(t)$ at $\alpha=1$. It is also noteworthy that we have extended the approximate solution of Example 15.3 for large time scale $t\in(0,15)$ using DGJTM, while the other iterative techniques such as ADM, LVIM, HAM and DTM [14,16] compute the analytical solution in the interval $t\in(0,1)$.

15.6 RESULT AND DISCUSSION

Example 15.4

Fig15.1A and Fig15.1B reports a comparison of the absolute error in sixth-order DGJTM solutions with exact solutions and solutions via existing methods like – ADM, HAM and DTM. Findings from Fig15.1A and Fig15.1B confirms that DGJTM provides more accurate solutions, for large time scale, converging very fast as compare to ADM, HAM and DTM. Fig15.1C depicts the behavior of sixth order DGJTM solution at different values of the fractional order (α), which confirms the strong agreement of proposed results with the exact for $\alpha=1$.

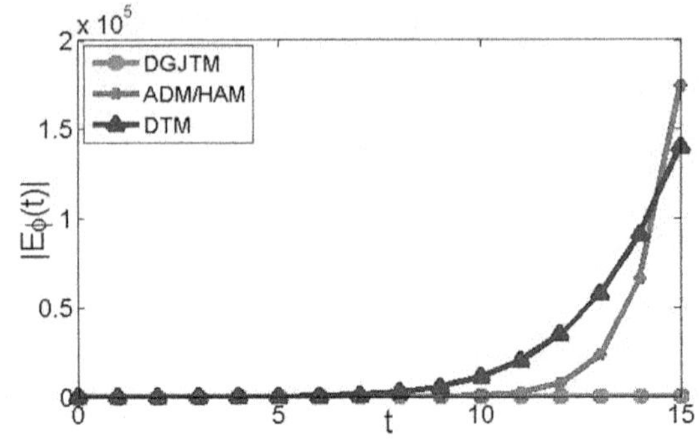

FIGURE 15.1A Comparison of absolute error in (sixth–order) DGJTM solution of Example 15.1 with ADM/HAM and DTM for $t \in (0, 15)$.

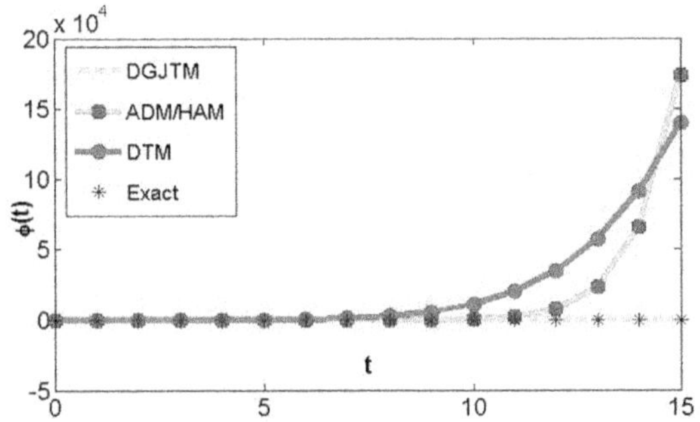

FIGURE 15.1B Comparison of (sixth–order) DGJTM solution of Example 15.1 with ADM/HAM, DTM and exact solutions for $t \in (0, 15)$.

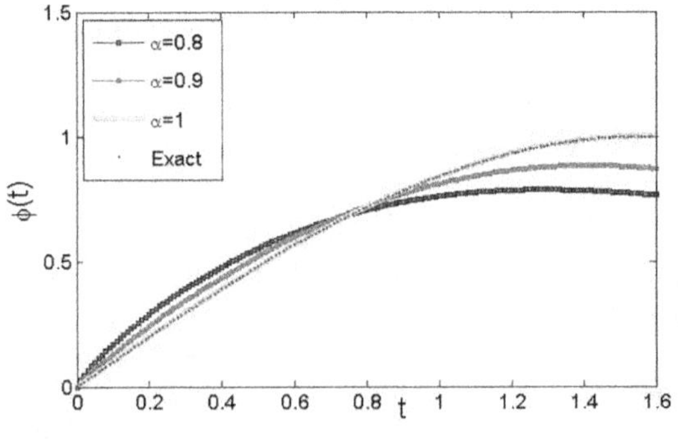

FIGURE 15.1C 2D plots of (sixth–order) DGJTM solutions of Example 15.1 at different values of α.

Example 15.5

Fig15.2A and Fig15.2B reports a comparison of the absolute error in sixth-order DGJTM solutions with exact solutions and solutions via existing methods like – ADM, HAM and DTM. Findings from Fig15.2A and Fig15.2B confirms that DGJTM provides more accurate solutions, for large time scale, converging very fast as compare to ADM, HAM and DTM. Fig15.2C depicts the behavior of sixth order DGJTM solution at different values of the fractional order ($\alpha = 1$)
which confirms the strong agreement of proposed results with the exact for $\alpha = 1$.

Example 15.6

Fig15.3A and Fig15.3B reports a comparison of the absolute error in sixth-order DGJTM solutions with exact solutions and solutions via existing methods like – ADM, HAM and DTM. Findings from Fig15.3A and Fig15.3B confirms that DGJTM provides more accurate solutions, for large time scale, converging very fast as

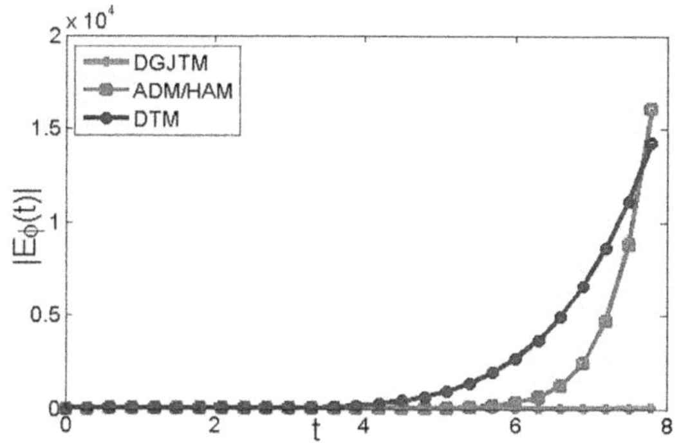

FIGURE 15.2A Comparison of absolute error in (sixth–order) DGJTM solution of Example 15.2 with ADM/HAM and DTM for $t \in (0, 15)$.

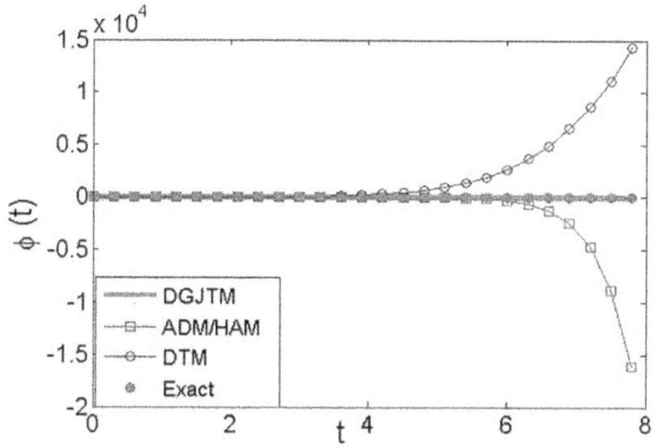

FIGURE 15.2B Comparison of (sixth–order) DGJTM solution of Example 15.2 with ADM/HAM, DTM and exact solutions for $t \in (0, 15)$.

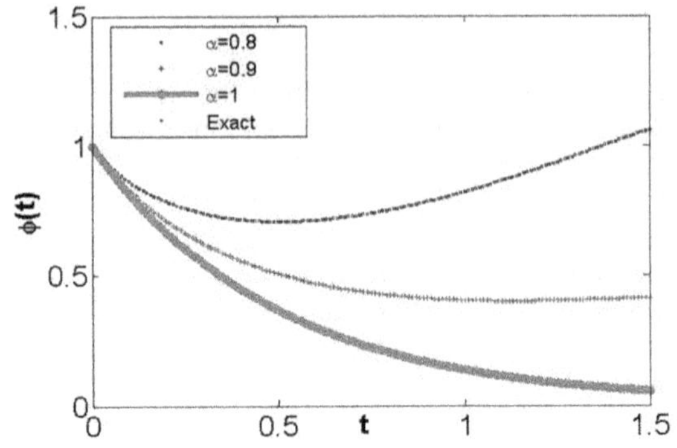

FIGURE 15.2C 2D plots of (sixth–order) DGJTM solutions of Example 15.2 at different values of α.

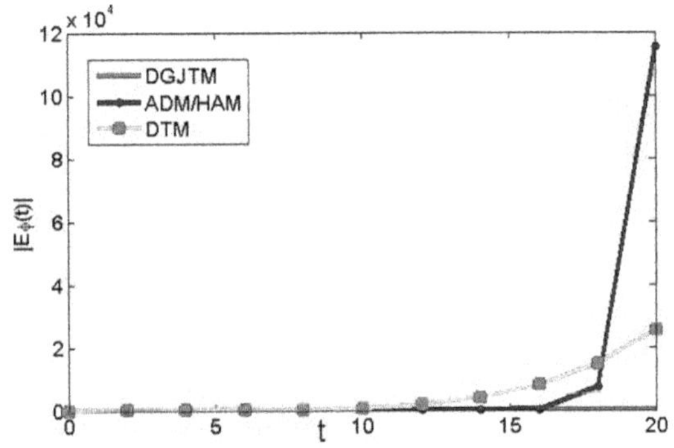

FIGURE 15.3A Comparison of absolute error in (sixth–order) DGJTM solution of Example 15.3 with ADM/HAM and DTM for $t \in (0, 15)$.

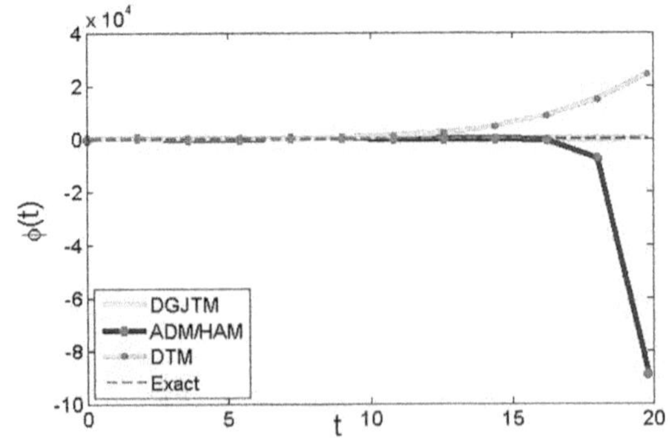

FIGURE 15.3B Comparison of (sixth–order) DGJTM solution of Example 15.3 with ADM/HAM, DTM and exact solutions for $t \in (0, 15)$.

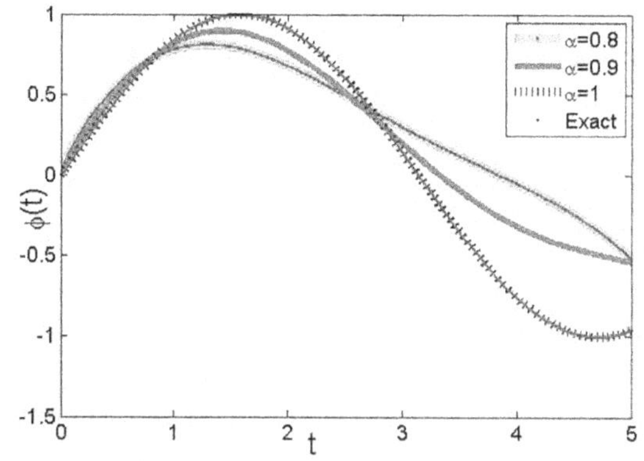

FIGURE 15.3C 2D plots of (sixth–order) DGJTM solutions of Example 15.3 at different values of α.

compare to ADM, HAM and DTM. Fig15.3C depicts the behavior of sixth order DGJTM solution at different values of the fractional order (α), which confirms the strong agreement of proposed results with the exact for $\alpha = 1$.

15.7 CONCLUSIONS

In this chapter, we have used Daftardar-Gejji and Jafari transform method (DGJTM) to compute an approximate series solution to the time-fractional pantograph differential equation and the higher-order time-fractional neutral functional differential equation with proportional delay for large time scale. It is demonstrated that the proposed series solution converges uniformly in any finite interval. In addition, the computed results are compared with the solutions obtained via other existing methods, graphically. It is examined that the DGJTM series solution provides better agreement with exact solutions for large time scale as compared to other methods such as ADM, VIM, HAM and DTM.

ACKNOWLEDGMENTS

The authors are grateful to the editor and anonymous reviewers for their comments and suggestions. S. Agrawal thanks Babasaheb Bhimrao Ambedkar University, Lucknow, India, for their financial assistance to carry out her research work.

REFERENCES

1. Podlubny, I. (1999) *Fractional Differential Equations*, Academic Press, San Diego.
2. Miller, K. S., and Ross, B. (1993) *An Introduction to the Fractional Calculus and Fractional Differential Equations*, Wiley, New York, 1993.
3. Caputo, M., and Mainardi, F. (1971) Linear models of dissipation in anelastic solids, *Rivista del Nuovo Cimento*, 1 161–98.
4. Barzinji, K., Maan, N., and Aris, N. (2014) Fuzzy delay predator–prey system: existence theorem and oscillation property of solution, *International Journal Mathematics Analysis (Ruse)*, 8, 829–847.

5. Batzel, J. J., and Tran, T. H. (2000) Stability of the human respiratory control system I. Analysis of a two-dimensional delay state-space model, *Journal Mathematics Biol*ogy, 41, 45–79.
6. Bellen, A., and Zennaro, M. (2003) *Numerical Methods for Delay Differential Equations*, Oxford University Press, New York.
7. Muhsen, L., and Maan, N. (2014) Modeling of human postural balance using neutral delay differential equation to solvable lie algebra classification, *Life Science J*ournal, 11, 1145–1152.
8. Muhsen, L., and Maan, N. (2016) Lie group analysis of second-order non-linear neutral delay differential equations, *Malaysian Journal Mathematics Sci*ence, 10, 117–129.
9. Cooke, L., Driessche, D., and Zou, X. (1999) Interaction of maturation delay and non-linear birth in population and epidemic models, *Journal of Mathematics Biology*, 39, 332–352.
10. Yang, Y., and Tohidi, E. (2019) Numerical solution of multi-Pantograph delay boundary value problems via an efficient approach with the convergence analysis, *Computer and Application Mathematics*, 38:127, https://doi.org/10.1007/s40314-019-0896–3.
11. Sakar, M. G. (2017) Numerical solution of neutral functional-differential equations with proportional delays, *International Journal Optimation Control Theory Application IJOCTA*, 7, 186–194.
12. Vanani, S. K., Aminataei, A. (2008) On the numerical solution of neutral delay differential equations using multiquadric approximation scheme, *Bulletin Korean Mathematics Society*, 45, 663–670.
13. Ordokhani, Y., Babolian, E. Rahimkhani, P., and Legendre, M. (2018) Wavelet operational matrix of fractional-order integration and its applications for solving the fractional pantograph differential equations. *Numerical Algor*ithm, 77:1283–1305.
14. Bereketolu, H., and Karakoc, F. (2009) Solutions of delay differential equations by using differential transform method, *International Journal Computer Math*ematics, 85(5), 914–923.
15. Abazari, R., and Ganji, M. (2011) Extended two-dimensional DTM and its application on nonlinear PDEs with proportional delay, *International Journal Computer Math*ematics, 88(8), 1749–62.
16. Smardal, Z., Diblikl, J., and Khan, Y. (2013) Extension of the differential transformation method to nonlinear differential and integro-differential equations with proportional delays, *Advances in Difference Equations*, 69.
17. El-Ajoua, A., Oqielata, M. N., Al-Zhourb, Z., and Momani, S. (2019) Analytical numerical solutions of the fractional multi-pantograph system: Two attractive methods and comparisons, *Results in Physics*, 14: 102500. https://doi.org/10.1016/j.rinp.2019.102500.
18. Peykrayegan, N., Ghovatmand, M., and Noori Skandari, M.H. (2020) On the convergence of Jacobi-Gauss collocation method for linear fractional delay differential equations, *Mathematics Methods Applications Sci*ence, 2020: 1–17.
19. Moghaddam, B.P., Mostaghim, Z.S., Pantelous, A.A., and Tenreiro Machadoc, J.A. (2021) An integro quadratic spline-based scheme for solving nonlinear fractional stochastic differential equations with constant time delay, *Communications Nonlinear Science Numerical Simulations* 92: 105475.
20. De Wolff, B.A.J., Scarabel, F., Verduyn Lunel, S. M. and Diekmann, O. (2021) Pseudospectral approximation of hopf bifurcation for delay differential equations, *SIAM Journal Applications Dynamics Syst*em, 20(1): 333–370. https://doi.org/10.1137/20M1347577.
21. Jackson, M., and Chen-Charpentier, B.M. (2016) Modelling plant virus propagation with delays, *Journal of Computational and Applied Mathematics.* http://dx.doi.org/10.1016/j.cam.2016.04.024.

22. Jackiewicz, Z., and Zubik-Kowal, B. (2006) Spectral collocation and waveform relaxation methods for nonlinear delay partial differential equations, *Applications Numerical Mathematics*, 56(3–4): 433–443.
23. Zubik-Kowal, B., and Mead, J. (2005) An iterated pseudospectral method for delay partial differential equations, *Applications Numerical Math*ematics, 55, 2005.
24. Moghaddam, B.P., and Mostaghim, Z.S. (2017) Modified finite difference method for solving fractional delay differential equations, *Bol. Soc. Paran. Mat.*, 35(2): 49–58, doi:10.5269/bspm. v35i2.25081.
25. Al Habees, A., Maayah, B. (2006) Solving fractional proportional delay integro differential equations of first order by reproducing kernel Hilbert space method, *Global Journal of Pure and Applied Mathematics*, 12(4): 3499–3516.
26. Singh, B.K., and Agrawal, S. (2020) A new approximation of conformable time fractional partial differential equations with proportional delay, *Applications Numericals Maths*, 157. https://doi.org/10.1016/j.apnum.2020.07.001.
27. Singh, B.K., and Agrawal, S. (2019) Study of nonlinear time fractional generalized burger equation with proportional delay via q-HAM, *In Proceedings of "International Conference on Applied Mathematics & Computational Sciences" (ICAMCS-2019)* October 17–19, 2019, https://doi.org/10.21467/proceedings.100.15.
28. Singh, B. K., and Kumar, P. (2017) Fractional variational iteration method for solving fractional partial differential equations with proportional delay, *International Journal Differtial Equations*, (5206380), 11, https://doi.org/10.1155/2017/5206380.
29. Singh, B. K., and Kumar, P., Homotopy perturbation transform method for solving fractional partial differential equations with proportional delay, *SeMA Journal*, 75 (2018) 111–125, https://doi.org/10.1007/s40324-017-0117–1.
30. Singh, B. K., and Kumar, P. (2017) Extended fractional reduced differential transform for solving fractional partial differential equations with proportional delay, *International Journal Applications Computer Math*ematics, 3: 631–649. https://doi.org/10.1007/s40819-017-0374–9.
31. Hattaf, K., and Yousfi, N. (2016) A numerical method for a delayed viral infection model with general incidence rate, *Journal of King Saud University– Science* 28, 368–374.
32. Gejji, V.D., and Jafari, H. (2006) An iterative method for solving nonlinear functional equations, *Journal Mathematics Analysis App*lications, 316(2), 753–763.
33. Gejji, V.D., Bhalekar, S. (2008) Solving fractional diffusion-wave equations using the New Iterative Method, *Fractional Calculation Applications Anal*ysis, 11, No. 2, 193–202.
34. Daftardar-Gejji, V., Sukale, Y., and Bhalekar, S. (2015) Solving fractional delay differential equations: A new approach, *Fractional Calculation Applications Analysis*, 16, No. 2, 400–418.
35. Bhalekar, S., Patade, J. (2016) Analytical solutions of nonlinear equations with proportional delays, *Applications Computer Math*ematics, 15(3), 2016.
36. Baleanu, D., Darzi, R., and Agheli, B. (2018) An optimal method for approximating the delay differential equations of noninteger order, *Advances in Difference Equations*, 284, 2018.
37. Wang, L., Wu, Y., Ren, Y., and Chen, X. (2019) Two analytical methods for fractional partial differential equations with proportional delay, *IAENG International Journal of Applied Mathematics*, (49): 2019.
38. Toan, P.T., Thieu, N., and Razzaghi, M. (2021) Taylor wavelet method for fractional delay differential equations, *Engineering with Computers*, 37: 231–240.
39. Zhao, J., Jiang, X., and Xu, Y. (2021) Generalized Adams method for solving fractional delay differential equations, *Mathematics and Computer in Simulation*, 180, 401–419.

40. Eriqat, T., El-Ajou, A., Oqielat, M. N., Al-Zhour, Z., and Momani, S. A (2020) New attractive analytic approach for solutions of linear and nonlinear neutral fractional pantograph equations, *Chaos, Solitons and Fractals*, 138: 109957.
41. Miller, K.S., Ross, B. (1993) *An Introduction to the Fractional Calculus and Fractional Differential Equations.* Wiley, New York.
42. Gorenflo, R. and Mainardi, F. (2000) On Mittag-Leffler function in fractional evaluation process. *Journal Computer Applications Mathematics* 118: 283–299

Index